Forbidden Science 6

Also by Jacques Vallee

In English:
Anatomy of a Phenomenon (Regnery, Ace, Ballantine)
Challenge to Science (Regnery, Ace, Ballantine)
Passport to Magonia (Regnery, Contemporary, Daily Grail)
The Invisible College (E. P. Dutton, Anomalist Books)
Messengers of Deception (And/or, Bantam, Daily Grail)
The Edge of Reality (with Dr. J. A. Hynek, Contemporary)
*Dimension*s (Contemporary, Ballantine, Anomalist Books)
Confrontations (Ballantine, Anomalist Books)
Revelations (Ballantine, Anomalist Books)
The Network Revolution (And/or, Penguin, Google)
Electronic Meetings (co-author, Addison-Wesley)
Computer Message Systems (McGraw-Hill)
A Cosmic Samizdat (Ballantine)
Forbidden Science, Vols. 1 to 5 (Documatica Research, Anomalist Books)
The Four Elements of Financial Alchemy (TenSpeed)
FastWalker, novel (Frog, Ltd.)
The Heart of the Internet (Hampton Roads, Google)
Stratagem, novel (Documatica Research)
Wonders in the Sky, with Chris Aubeck (Tarcher-Penguin)
Trinity: The Best-Kept Secret, with Paola Harris (StarworksUSA and Documatica Research)

In French:
Le Sub-Espace, novel (Hachette – Jules Verne Prize)
Le Satellite Sombre, novel (Denoël, Présence du Futur)
Alintel, novel (Le Mercure de France)
La Mémoire de Markov, novel (Le Mercure de France)
Les Enjeux du Millénaire, essay (Hachette Littératures)
Au Coeur d'Internet (Balland, Google)
Stratagème (L'Archipel)
Science Interdite, Vol. 1 (Marseille : O.P.Editions)
Science Interdite, Vol. 2 (Geneva : Aldine)
Trinity: Le Secret the mieux gardé, with Paola Harris (StarworksUSA and Documatica Research)

Forbidden Science 6

Scattered Castles

**The Journals of
Jacques Vallee
2010-2019**

Anomalist Books
Charlottesville, Virginia

Forbidden Science 6: Scattered Castles
Copyright © 2025 by Documatica Research, LLC
ISBN: 978-1-949501-36-0
Second Edition

Front cover image: Godruma/iStock
Back cover author photo: Jim Block
Cover design by Seale Studios

All rights reserved. No part of this book may be reproduced or transmitted in any form or by any means, electronic or mechanical, including photocopying, recording, or by any information storage and retrieval system, without the written permission of the Publisher, except where permitted by law.

The author can be contacted at:
P.O. Box 641650
San Francisco, CA. 94164

The author's website: www.jacquesvallee.com

For information about Anomalist Books, go to AnomalistBooks.com or write to:
Anomalist Books, 3445 Seminole Trail #247, Charlottesville, VA 22911

Contents

Introduction 7
Part Twenty-One: Crossing the Chasm 9

       1. Lancaster. 12 March 2010
       2. Esalen. 16 May 2010
       3. Cluny Monastery. 24 September 2010
       4. Riyadh, Saudi Arabia. 23 January 2011
       5. Mabillon. 7 May 2011

Part Twenty-Two: LoneStars 109

       6. Hummingbird. 5 January 2012
       7. L'Escoublère. 24 June 2012
       8. Mabillon. 19 February 2013
       9. Hummingbird. 15 May 2013
     10. Hummingbird. 1 January 2014

Part Twenty-Three: Field Work 285

     11. Hummingbird. 7 January 2015
     12. Hummingbird. 1 June 2015
     13. Hummingbird. 1 October 2015
     14. Hummingbird. 7 February 2016
     15. Buenos Aires. 15 September 2016

Part Twenty-Four: Solitudes 431

     16. Hummingbird. 1 January 2018
     17. Toronto. 7 May 2018
     18. Hummingbird. 8 October 2018
     19. Hummingbird. 14 January 2019
     20. Mabillon. 6 September 2019

Reflections 516
References 520
Index 532

Figures

1. The Buttes, in Antelope Valley near Palmdale, California, 2010
2. In stealth mode: Antigravity experiments in Silicon Valley, 2010
3. Jeff Kripal, Mitch Horowitz, and the Striebers, Esalen, May 2010
4. With Flamine and alchemist Michel C. in Morvan, June 2010
5. In Palo Alto with Paul Baran, Kaili Kan, Flamine, August 2010
6. Leaders of parapsychology research in Paris, March 2011
7. Prof. Salisbury and Junior Hicks: Oil Company, Utah, July 2011
8. With Douglas Trumbull and Ariel Waldman, Orlando, Oct. 2011
9. NARCAP meeting in Mountain View, California, April 2012
10. LoneStars meeting in Toronto, Canada, 2012
11. The Korolyov (Korolev, Russia) Aerospace Forum, Dec. 2013
12. Flamine at the esoteric fraternity Thomas Lake Harris, Feb.2014
13. Historic five-nation UFO conference at CNES, Paris, July 2014
14. Dr. Garry Nolan at home, inspecting UFO test samples, 2014
15. Donna Cicera and Dr. Beverly Morgan: psychic surgery, 1977
16. Manoël di Queiros Meneza abduction case, July 2015
17. Conversation with Juan Oscar Perez, Argentina, Sept. 2016
18. The research group in Venado Tuerto, Argentina. Sept. 2016
19. Visiting a satellite imagery company in Prague. April 2018
20. Demonstrating AI case retrieval with Paul Hynek, 2019

Chart 1. The Tel Aviv (Uri Geller) encounter, Spring of 1954

Introduction

> I gaze in defeat
> At the stars in the night
> The light in my life burnt away:
> There will be no tomorrow…
> Then you sigh in your sleep
> And meaning returns with the day
> (David Bowie, "How Does the Grass Grow?" 2013)

The world accelerated after 2010. In the eyes of someone who travelled between the cozy, oddly pretentious business circles of Europe and the hottest laboratories from Silicon Valley to Texas, Tokyo or Shanghai, there was the joy of expectation mixed with fear.

The resulting upheaval verified what philosopher Alfred North Whitehead wrote in *Adventures of Ideas*: "It is the business of the future to be dangerous…The major advances in civilization are processes that all but wreck the societies in which they occur."

During the years covered by this journal, as I strived (and failed) to recover from recent deaths, I continued to manage two venture portfolios, working with trusted partners, guiding startups in three specific areas: medicine, space, and software for AI and computer networking. These companies did change the societies in which they occurred, with lasting results: four even qualified as 'unicorns' (1) but much time was also spent with a discreet team which defined a new, dangerous outpost: they explored psychical research from remote viewing to the survival of death. We also researched UFOs—intensely and sometimes secretly.

Today, in 2024, the long-term value of these novel fields remains wrapped in the great puzzle of existence. We can only hope that future generations will be brave enough to review what we've uncovered, to face the 'dangerous futures' we were striving to identify, and to clarify the mysteries we could only sketch out.

Our team was small but agile and driven. It included physicists and biologists, businessmen and scholars. I served as the 'information science guy,' so I argued—and often failed—to strive for serious screening and reliable patterns. We were passionate but not blind: none of us was naïve about spooky influences on the fields we surveyed, or about the heavy bias of politics, greed, and religion even in the great halls of science. The future perils were obvious.

Our group, in Silicon Valley slang, worked in 'stealth mode.' As a result, many of our colleagues in academia and most journalists had scant ideas of progress. Information has filtered out since then, thankfully, but these journals may represent the only on-the-spot, continuously curated record of the results achieved among brave people on the swampy fringes of Academe.

It is my hope that this volume will give some idea of the day-to-day process of innovative research, stumbling through bureaucracies, shredding obsolete beliefs, and harnessing the high technology rush that lit up this decade to define the future.

I sorely missed the warm advice from Janine. After her death it only came as an echo in meditation and dreams or, more often than I've been brave enough to record, in the tiredness of despair. Samuel Johnson had observed the ravages of grief with the precision that can only come to a poet:

> He that outlives a wife whom he has long loved, sees himself disjoined from the only mind that has the same hopes and fears and interest… *The continuity of being is lacerated.*

Life, thankfully, brought a blessing, another voice that healed, another mind to guide me, when Flamine became my wife a few years later: she re-defined who I was and what I did. But the last word should be left to writer Isabelle Eberhardt, who faced danger all over the world, only to meet a heart-wrenching, unlikely death in a stupendous flash flood in the vastness of a desert:

> Only the tomb can take this richness from me…
> And who can tell? If Fate gives me time
> To re-create a few fragments, perhaps it will live on,
> In the minds of others?

Jacques F. Vallée
San Francisco, January 2024
Paris, July 2024

Part Twenty-One

CROSSING THE CHASM

Qui sait s'il est permis d'éveiller ceux qui dorment ?
(Who can say if we're allowed to wake up the sleepers?)

Maurice Maeterlinck, *Aglavaine et Sélysette*

1

Lancaster, California. The Oxford Inn. Friday 12 March 2010.

The sun is setting over the Angeles Crest Mountains whose summits still bear traces of snow. A local restaurant has given me a cup of coffee, and I'm getting oriented: the massive expanse of Edwards AF Base is to the north, Lockheed Skunk Works just a few blocks away.

Two months have passed since the terrible night when the generous and vibrant woman I loved stopped breathing: she slipped through my arms. I have traveled like a lost soul between California and Normandy, where she wanted her ashes to rest. Bitter loneliness now, and those conversations I hold with her, in an empty void.

It took me two dreadful months to put some order in my thoughts and to regain the will and the ability to write. I've lived in my busy study with the amateurish stained-glass, feeling like a lonely monk.

Janine wouldn't agree with this capitulation. She had urged me to resume my research, no matter what else I might do in the coming years. So, I flew down from San Francisco to L.A. this afternoon, rented a car, and drove alone to this town in the Mojave, anticipating a meeting with a Mr. Saël, who presents himself as "a quiet science writer based in Italy." He comes here for a few days to see his family.

This man wrote to me in November 2009, at a time when I had little inclination to follow up on research. His letter was elegant and filled with information. I learned about an extraordinary series of events that his family had witnessed in the Antelope Valley of California.

Mr. Saël wrote: "In 1979 in Lake Los Angeles, my brother was living at the foothill of the Three Sisters Buttes. Late one night, he and two other men were all sitting in the back of a truck with two of their sisters watching the night sky. Suddenly, four UFOs appeared."

Anticipating my hasty guess that the objects might be devices from Edwards Air Force Base, he added: "They were clearly ships, not just indistinct lights in the distance. The four of them began shooting beams of light at each other. When a ship would be hit by a beam, it would vanish. The witnesses thought they witnessed a battle in which crafts were being destroyed. But *the vanished ship would reappear,* and the same pattern would repeat. Somewhere in the distance, they saw a small truck drive up a hill. One ship shot a beam of light at the truck. The witnesses heard an explosion."

I was shocked to read this, because reports of blatant hostility by UFOs against witnesses are rare in North America and Europe. While there are instances of paralysis, temporary blindness, or burns, this kind of attack sounded like the deadly stories I had heard in Brazil.

The letter went on: "As soon as it was light enough, they drove to where they'd seen the truck in flames. On the side of the truck was the name of a company based in Palmdale."

New vehicles had arrived: more trucks and a NASA van with a satellite dish on top. A man in uniform approached the witnesses and said, "Follow me." He picked up a five-gallon metal gasoline can and led them to the burned truck, placed the gasoline container next to it, and said: "You see what happened, don't you? The owner of this truck set fire to it, to collect insurance, right?" Then they were threatened and instructed not to report anything about the incident.

Lancaster is a desert town, a loose assemblage of shops, recently constructed houses, and low, undistinguished office buildings along Highway 14 that goes from L.A. to Mojave. This is the modern brain of American aerospace, if its heart is in Alabama, at Huntsville. The local news is all about EADS-Northrop-Grumman's loss of a big contract for Air Force tankers, and about Lockheed's plan to build the Joint Strike Fighter. Locally, the slow economy is sustained by military development.

Once out of the city, driving along Palmdale Boulevard, the landscape is mostly gray dirt, Joshua trees gesticulating in the strong wind, and distant mountains highlighted with swathes of snow. Spring adds color: pink and white trees in full bloom.

On the way here from rainy San Francisco, I reviewed what had passed during those months when I wasn't able to write. In spite of the agonizing wait for decisions from NASA about the formation of the new venture fund we proposed to them (Red Planet Capital), new contacts opened up in the aerospace world.

In 2009, my partner Dr. Peter Banks had approached decision makers at the National Reconnaissance Office (NRO), resulting in a meeting for the two of us with Martin ('Marty') Faga at Mitre (he's a former director of both Mitre and NRO), followed by a meeting at the Aerospace Corp. in

Chantilly with Vincent Boles, general manager of the advanced technology division of NRO (2), and Dr. Philip Fawcett. Former Aerospace CEO Bill Ballhaus participated by phone.

After that first session, where Peter and I presented Red Planet Capital (3) and our recent work, two executives took me to their secure classified intelligence facility (SCIF) for a detailed presentation of the NRO requirements that left me impressed with the scope of their thinking—wider and more perceptive than anything I'd ever heard.

The facility where I was led wasn't just a secure room but an entire section of that floor, with secretaries' workstations, separate offices, and classified video links. The main thing I learned, apart from some details of technology, was about the office's open philosophy: while venture capitalists strive to get their hands on intellectual property that is disruptive of current markets, the man said with a wink, NRO is looking beyond that level for technologies that are not only unknown to others, but *so advanced others cannot even imagine they can exist.*

One past example (no longer classified but misunderstood) is the ability to run interferometry between two satellites in formation, to detect minor gravity changes inside the Earth. This technique, based on 'station keeping,' could detect underground facilities anywhere.

Such visits, in the rarefied atmosphere of the top agencies, put the DIA's interest in the Utah ranch known as Skinwalker Ranch into new light for me.

During the last weekend in November 2009, I'd exchanged news and ideas with colleagues at BAASS, a classified research group founded by Robert Bigelow (4). I'd told them about *Wonders in the Sky* and suggested hypotheses about the new picture of the phenomenon that emerged from the computer analyses.* They now have over 50 employees, with an increase in scientific staff, and a tenth database in Capella containing some 350 incidents from the Ranch, which has become a target of intense

---

*   Nearly fifteen years have passed since I wrote those lines. Much information has been revealed and published about the AAWSAP-BAASS project (although not everything by any means), and I can now describe more freely my role in the project. In agreement with Robert Bigelow, I headed up the computer information activity, starting with the construction of the CAPELLA 'data warehouse' of a dozen interconnected worldwide databases. The project, coordinated with Dr. Colm Kelleher and specialized consultants, was designed in three phases.

interest by the Sponsor's staff: six of their personnel have now had paranormal experiences there.

The problem is that our contacts in DC, including our research leader, Jim Lacatski, don't have full visibility with the agencies that are thought to have privileged information about UFOs, like AFOSI (5). They made it clear they couldn't deal with us unless we gained the status of an SAP (special access program), and even then, it wasn't obvious they'd feel like sharing anything of value.

I also sought contact with Dr. Herzenberg's team at Stanford Hospital to help a witness studied by BAASS who still suffers extreme effects from an encounter with a strange light. (6)

Other groups were also busy. Early in January I had lunch with Bernie Haisch and his gifted artist wife. They gave me sad news about a bankrupt web company created by Joe Firmage that lost the money of its investors. Joe himself has let go of his house in California, but he still believes that his invention will provide free energy. (7)

The next day, when I called Bob Bigelow to apologize for my long silence, I found out BAASS was lost in the government transition. In a situation reminiscent of Red Planet Capital, the follow-on money had not come through, mysteriously tied up in bureaucracy. MUFON (8) has been in turmoil too, with few cases of interest: their director, James Carrion, was removed for daring to question the Roswell crash.

The larger world keeps spinning in dangerous ways. The European community is challenged by the financial bankruptcy of Greece, made worse by the awareness that Spain, Portugal, Italy, and Ireland are similarly exposed.

When I flew to France for my work, I took a day off to visit Chartres and see the famous labyrinth again, long of interest to me; its ancient spiritual journey was revived in recent years by the books of American writer Lauren Artress, Episcopalian priest and psychologist. At Chartres I met a psychically gifted woman named Flamine de Bonvoisin who knew Lauren and was involved in Veriditas, her spiritual group. Familiar with San Francisco, she held academic degrees in psychology. We spoke in the shadow of the cathedral; she told me of the group's programs in California, where they'd built the fine labyrinths at Grace Cathedral and in Petaluma.

Her artist mother, a gifted musician, had died of glioblastoma like

Janine, and the pain was still there.

Returning to California a few days ago, I was aware that the most difficult phase of the grieving process was just beginning. I could make no rash moves: The global crisis temporarily stumped innovation, although the decision by the Obama administration to delay the manned space program had freed up the imagination of private firms that envisioned aggressive commercial developments in LEO (Low Earth Orbit). What seemed to the media as a defeat for American prestige simply fueled the energy of Elon Musk and others. (9) Back home in Silicon Valley, reacting against depression, I scheduled meetings with colleagues and former mentors who'd known me for years. They were immediately helpful.

Steven Millard invited me to lunch (at the posh Burlingame Country Club) with Paul Baran and Federico Faggin. Ian Sobieski, head of the Band of Angels, thought I should advise startups; he introduced me to several incubators. Gary Tauscher, a Yahoo! manager, almost convinced me to restart InfoMedia, since there was nothing equivalent on the web, thirty years later, for sophisticated projects. Paul Saffo and I discussed research foundations and listed references. Peter Beren thought I should help syndicate self-publishers into a new firm. Rob Swigart loved the notion of starting a for-profit version of the Institute for the Future, with a fresh model for research. And Paul and Béatrice Gomory, over dinner at the Yacht Club, advised calling professor Elisabeth Paté-Cornell, and promised support.

Two months after Janine's passing, the pain hasn't left. I don't want this sorrow to become one fact among other facts (what my friends call 'turning the page,' as they urge me to do, as if that was possible) because that pain is the only thing that still links me to her.

Lancaster. Saturday 13 March 2010.

Mr. Saël has the look of an Italian intellectual, with his high forehead and the long scarf he wears with elegance. The first unusual incident in his life occurred when he was three or four, living near the Tuolumne River. He recalls that he slipped and fell into shallow water, swallowed, and lost consciousness. When he came to, he found himself dry and at peace, lying on his back on shore. Everyone had left. Having lost sight of him, the family thought he'd gone to the house. He never knew who saved him.

As a child, his brother recalls a typical abduction experience, lying on a table with mechanical 'Grays' around him. The family has a history of striking encounters going back to 1934 when Saël's father met a stranger dressed in workingman's clothes, while on a train going to Bakersfield. He told the elder Mr. Saël (born 1919) that humanity had to change or millions would soon die. That was the time of collectivization in the USSR. Millions did die, and of course the Nazi came to power in Germany and slaughtered more millions.

In 1970 Saël and his siblings were in Modesto when they saw a figure of a man with a long beard. All four had different reactions. His sister, who was about two, has no recollection of it. Philip has nightmares; for him the experience was negative, involving a pursuit of some kind. Saël recalls a Germanic accent, something about the 'Secret of the Way,' which was meaningless to him. The third brother only recalls that the man stared into his eyes. He got the message that this world was not the real world: *There is another, unseen world.*

By 1975, the family had moved to a farmhouse in the desert near Palmdale Blvd (we drove to the site today), where Saël had a vision preceded by an ecstatic feeling. He attributes his gifts to this contact, especially his facility with physics and music. The Green Man is majestic, he tells me, reminding him of the Jinn of the Arabs.

Lancaster. Sunday 14 March 2010.

Mr. Saël came over for breakfast with the news that he'd found that the date he'd given me for the sighting in Palmdale was wrong: the incident happened in 1975, the same year when he saw the Green Man in the mist. Resetting the incident in time brings up an interesting coincidence, because the only similar case I ever heard of was in the same area. I'd made a note of it and published it in the second volume of *Forbidden Science* under the date of 23 February 1976.

The story goes like this: When I'd met 'Jim Irish,' the Hollywood expert hired by NSA to track UFOs, I'd asked why he quit:

"One day, in the desert near Barstow, there'd been sightings," he said. "The team we sent never came back. We found two bodies, reduced to ashes, near a burned-out car: two guys from our group, one of them a close friend of mine. The report said they burned to death because the car had a gasoline leak and they lit up a cigarette! But in this business, you just don't smoke on the job..."

Barstow is just to the east of Antelope Valley; the dates place the two incidents in the same period. Saël was struck by the confirmation.

Hummingbird. Sunday 21 March 2010.

BAASS is working hard to acquire SAP status, indispensable to be invited into whatever advanced study might exist. The thrust of the work in Vegas hasn't changed. In Washington, Jim Lacatski bravely defends the project. In Nevada, Bob remains reluctant to delegate much spending authority.

The staff is now up to 65 people, many showing healthy skepticism. One rare bit of classified data the project ever saw was a comment from the Russian ambassador to Brazil, overheard to say he took the subject "very seriously indeed."

No kidding! I could have told them the same thing, free of charge.

At the ranch in Utah, twelve independent witnesses have now reported creatures like wolves, walking on their hind legs.

Hummingbird. Monday 22 March 2010.

John Alexander called me tonight to say he was working on a book, which led him to call Colonel Coleman, now in his 80s and living in Florida (10). Coleman had some 200 unreleased cases in his files and was thinking of publishing them. John was especially interested in "precognitive sentient cases, like those at the ranch," and events in history that seemed to anticipate future technology. In a striking case I'd published, a witness saw a hexagonal flying object with rivets in the 1920s, in Oregon. John wanted to quote it.

While I was speaking with the Colonel, Bob Bigelow tried to reach me. When I returned the call, he said he was eager to get me involved again, specifically in a series of remote viewings designed to advance the probability of a genuine contact: There's a manager who is double-blind to the targets, and he'll calibrate the remote viewers, but so far, *they have been warned away by the phenomenon; they got clear messages to back off.* We need a multi-level process, other attempts, he said.

I thought the coincidence with John Alexander's call was curious. He'd just spoken of precognitive sentient cases at the ranch. Bob went on: "There've been indications in previous attempts, and some by-products.

Fig. 1: *The Buttes, in Antelope Valley, near Palmdale, 2010*

Fig. 2: *In Stealth mode: antigravity experiments in a private Silicon Valley lab, 2010*

We've backed off with the cameras."

We agreed I'd contact him after my trip, but I may be too caught up with other business. He spoke of experiments with abnormal plant growth, and new reports of the two-legged wolves. Well, wolves must eat something, yet these creatures never seem to kill chickens or cats. There's a theory in Magonia lore that they're non-physical entities incarnated into corpses, hence the putrid smell. (11)

Hummingbird. Tuesday 23 March 2010.

Lunch today (at Boardwalk in Los Altos: sloppy Joe and fries with a Coke) with my physicist friend and his two sons. They were eager to show me their latest experimental target, an aluminum disk they jokingly called 'The 'Dropa,' incorporating their latest geometry for anti-gravity. They've finally detected fusion, thanks—as often in science—to an experimental error, a target left under tension; the temperature rose and the emissions were multiplied 1000 times. They reviewed their calculations and now believe they can fix their previous failures.

Washington. Monday 29 March 2010.

The National Academy of Sciences, where I met Peter Banks between two rain showers, occupies a fine building next to the State Department at 22nd and Constitution. Its dark paneling, painted ceilings, and intimate library create a dignified atmosphere. We reviewed our presentation scheduled for tomorrow with the NRO science director and we spoke of MEDEA, very advanced climate research with newly formed teams within Intelligence.

Only a few months ago, I was here with Janine. I took a picture of her on the shore of the Potomac. We'd walked to the Capitol the night of the historic healthcare reform vote. Love involves not just the heart but every cell in the body; the loss of love resonates in the same way, in the agony of absence. Coming to Washington to show you the city remains one of the good things I was able to do, but the memory stolen from imminent death tears at me with relentless cruelty.

Arlington. Tuesday 30 March 2010. The Hilton Garden Inn.

For the first time, last night, I had a dream about Janine; it woke me up from a deep sleep. I was waiting for her in an apartment in Paris, on a high floor. I'd prepared a suitcase, left open on our bed; I was eager to see her, back from her travels. She climbed the stairs easily, wearing a summer dress. I greeted her with a gentle embrace, but our meeting was matter of fact, nothing special. She was taking the suitcase for another trip; it all seemed very natural.

I had trouble getting back to sleep. I'd enjoyed a pleasant dinner at the Cosmos Club with Peter and Mary, and I was thinking about our early-morning trip to Chantilly and the presentation to the NRO. I finally took a pill and slept uneasily for a few hours.

In the morning, the rain had stopped but there was an intense wind as we drove off to the massive campus in Chantilly, all light blue metal and curved corridors. There was a last-minute question about my clearance, which they had neglected to check.

My personal data were logged on a federal system called *Scattered Castles*, a wonderful code name for the newest database of classified access.

After that, everything went smoothly. Jim Arnold, a heavyset man who'd obviously reviewed our backgrounds, listened to our briefing, asked clear questions, and took notes.

The topic of the detection of anomalies came up again: how can one discover connections, cluster facilities, recognize unusual behavior? *What groups are unique, within unique groups?* How can wide data streams be processed for 'information extraction' amidst human replications?

New contacts at the top echelon of NASA may help us pave the way to put Red Planet Capital back on track. Both Lori Garver, the associate administrator, and Bill Ballhaus, newly-appointed vice-chairman of the technology committee, are reportedly convinced of its value, amply demonstrated by what we have already generated, and by the new red book of 180 technology projects I left on Jim Arnold's table.

Hummingbird. Thursday 8 April 2010.

An interesting experience, last night, at the Golden Gate Spiritualist Church, good guardians of an Anglo-American psychical tradition that emphasizes mediumship. First, we heard a series of readings that could have applied to almost anything. But then Reverend James Ehrhart, an older pastor, took the pulpit. He pointed at me, and said a woman was coming through, *a motherly figure, but not my mother*. He was puzzled by what she said: There were "rods" I could use, spiritual rods as in dowsing; I should take advantage of these tools, I had the power and "there was still plenty of time."

I know that motherly figure: an older UFO witness I interviewed in the 1970s, who showed me how to practice dowsing with those exact same 'spiritual rods,' in the same terms. I hadn't published the case.

Hummingbird. Saturday 10 April 2010.

Returning from the foothills of the Sierra, I need to review what I learned today. It's a three-hour drive through the lush Central Valley, but the landscape becomes attractive as you reach Calaveras County with its rocky hills, old mines, and busy wineries.

I had a flat tire in Copperopolis but managed to fix it and climb up to Arnold in time to pick up Robert. His house at 4,000 feet, neatly lodged among the pine trees and redwoods, was still surrounded with hardened piles of snow. A fresh storm was expected in the evening.

Robert is a tall, thin man with balding forehead and a gray goatee, a loner—he describes himself as a hermit. I had a list of questions about the details of the main event in 1975.

"It happened in the summer, in June or July, about 10 p.m.," Robert told me once we'd ordered brunch. "Steve had come over to sell some equipment the next morning. So, Larry and I were sitting in the back of the truck, Steve and his girlfriend sat on the front seat, and all of us were just watching the night sky. That's when it happened."

At my request, Robert made a drawing of what they saw: a large lens-shaped object: "The first disk must have been 60 to 80 feet wide and about 15 feet thick. There were no lights and no sound; it was just a silvery metal like aluminum all over. It didn't glow. It was shaped like two saucers stuck together."

I pulled out my cellphone and showed him the site I'd photographed. Robert laughed: "You know, I don't even have a picture of the place." So I asked him to point at the house and he confirmed: "We were facing North, and this came from the West (left of the photo), it floated in at 70- or

80-feet altitude and stayed there for about half a minute. Then another one came in, about 50 feet higher to the left, and 30 seconds later a third one came to the right, and it was identical. They formed a triangle, with the first one at the bottom."

So, there were only three disks, not four as Saël had recalled.

The next phase was an incredible display: "The bottom disk shot a beam of light, up to the one above to the right, which in turn shot to the left, exchange was repeated about ten times, as if they were charging up some form of energy." At that point the bottom disk shot a beam directly down to the ground, creating a big glow behind the ridge, at a point the witnesses couldn't see. When I probed him about the nature and color of the beams, Robert simply said it was whitish, clear, and fluorescent, "and the left one shot below to the first object."

"What was your reaction?" I asked.

"It was strange, but it didn't impress me that much. I've recalled all that, when Saël reminded me of it, but I'd pushed it out of my mind, including what happened the next morning. Larry seemed to go into shock and to this day he doesn't remember any of it. At the time he just had a blank stare on his face. And I don't remember what we did after that, when or how I went to sleep."

"How did the sighting end?"

"These things just started to move, and they vanished."

"Did you walk up to the place where the big light had been seen?"

He laughed: "Oh no, we weren't about to go up there in the dark! We did that the next morning, to check it out. We drove up from the other side. We knew the desert, we knew those hills very well, so we carefully checked for traces.

"There was this little white truck, like a small Ford, with no footprints to or from it in the sand. There was a big clean hole in the top of it. On the passenger seat we saw a yellow worker's helmet filled with some mushy material and hair, that we took to be brain matter, but there was no body, no trace of burns inside or on the seats. We couldn't make sense out of it, so we drove back out and went to the first house we found, asking to use the phone. A man came to the door. He was almost 7 feet tall and looked like death warmed over, with gray skin. He said he didn't have a phone and closed the door. So, we drove over to the sheriff's office in Lancaster, and they laughed at us."

"Didn't you encounter the same tall man there?" I asked, recalling that

Saël had mentioned that detail.

"No, not at that time; we saw him later, when we had breakfast at a local restaurant. We all went back to the site, walking up through the tall grass, and we saw vehicles and people in white jumpsuits. One of them was carrying an Army gas can, pouring gasoline over the truck. When we approached, he told us, very matter of fact: 'See, some people stole this truck and took it here into the desert, and they set it on fire.' *But he was the one setting it on fire!* His tone indicated, 'take it or leave it, it's the official story.' So we just walked off."

"Did anybody threaten you, or tell you not to say anything?"

Robert answered, "No, nobody told us to stay quiet; there just was no argument; we'd heard the official line and that was enough."

Saël, who was at his parents' house, only heard about it later that day, and evidently distorted parts of the story: There were only three disks, there was no cadaver in the truck, and the tall man didn't show up at the police station. As we went over details, Robert clarified something: "Out there, we were seeing lights in the sky, and interceptors going after them, and losing them."

We also discussed his childhood experience. When Robert was six or seven years old (in 1965 or 66) he used to find himself lying on a steel table, with three beings around him. He saw his body lying there, and his recollection is that they were handling one of his eyes and pointing it at him: that's why he had the impression of being outside his body. His latest sighting was three years ago, in the Arizona desert, when Saël rented a property to work on a book he was translating.

Hummingbird. Sunday 11 April 2010.

A pleasant break: Béatrice has invited me to hear the Bridge Chamber Virtuosi, three Chinese string musicians from local orchestras who played Shubert and Erno von Dohnanyi but also Chinese and Tibetan pieces adapted for the cello, the viola, and the violin.

A Pacific storm had moved over with high winds and rain shaking the houses, and even this towering building. The road up to Robert's house would be impassable today because of the fresh snow. The obvious question now is, "who should I tell about the case?" and the answer is, "no one for now, nobody will care."

I haven't given up on the efforts in Vegas, but I cannot invest long-term energy in projects with obvious flaws. Once all of us had security clearances, the rich exchange of ideas among Hal, Eric, and me, potentially a good driver of valuable research, became stale and lost value: we couldn't share what we knew.

Reinforcing my sad impression, controversies have erupted with the abrupt departure of James Carrion from MUFON. He reportedly suspects the UFO phenomenon to be entirely based on human deception.

Hummingbird. Thursday 15 April 2010.

The last three months of grief and anxiety have been hard. The mirror sends back a sagging face with a stern look, the face I would draw if I was asked to imagine poor Don Quijote riding home after fighting the windmills. Fortunately, lively discussions with mentors bring clearer vision: François Rongère, at PG&E, told me about new systems in the Utilities; Bob Kyle sees me as a partner for a group like Paladin; Elisabeth Paté-Cornell supports my work with BAASS, mentions computers in medicine, but cautions me: no big decision for a year. "You'll be driving in fog," she said, speaking from sad experience.

Michael Geilhufe, on the way to Kenya, offered his resources. Ken Alwyn cautioned me about the need to keep flexible schedules, to enjoy my remaining life. Federico Faggin spoke of consciousness research, advising me to focus on what my heart wanted to do rather than listening to my brain. Bob Johansen told me about a foundation started by Bob Noyce's widow. Most recently, I spoke to Lisa Lockyer and Jaffer Hussain from NASA Ames supporting a restart of Red Planet. Tim Jenks, busy with the imminent IPO of NeoPhotonics, carefully listed my talents and limitations. Mark McKee was practical and direct. He said how much our association meant to him. As we finished lunch at L'Olivier two older men walked in, none other than Arthur Rock and Warren Hillman, legends of California finance.

They sat in a corner. We waived at them discreetly as we left.

Hummingbird. Sunday 18 April 2010.

After forty years, San Francisco still takes my breath away. Yesterday, with my cousin Francis Passavant (son of my uncle Charles), we toured the City from Washington Square and Chinatown, to the Presidio and Fort

Point and on to Sausalito. I was sad and weary, but Catherine joined us for lunch, bringing sunshine. I was scheduled to board Air France to Paris, but a huge volcano has erupted in Iceland, spewing abrasive dust over Western Europe; all air traffic grounded.

Hummingbird. Thursday 22 April 2010.

Under pressure from the airlines, authorities have reopened European airports but it's too late for me. Having rescheduled the trip for June, I called Bob Bigelow to report about my visit to the Utah ranch to assess the situation there: I've spoken to the previous manager, Terry Sherman (alias 'Tom Gorman').

I've been reading Flammarion again, his books on survival of bodily death, where he extensively quotes Myers and testimonials from his own readers. After a 50-year period of research, he came to the conclusion that the soul is immortal or, more accurately, exists in a dimensionless state with no time. This is close to what I feel, yet it doesn't explain the phenomena, or predict them. He goes on to say that he believes in reincarnation, but here he loses me, like the others.

Hummingbird. Thursday 6 May 2010.

In front of my office in Palo Alto, all the flowers Janine planted are blooming, blending their carpet of red and purple with the perfume of the white rosebushes. Spring spreads its glorious tapestry over the Bay Area, deepening my distress.

Jeff Kripal's book *Authors of the Impossible*, with a generous section analyzing my work, just arrived. Given Jeff's notoriety and the publisher's excellence (U. of Chicago, no less) it will shake my quiet life.

Hummingbird. Thursday 13 May 2010.

Hal, on the phone: he's been very busy with the series of scientific papers he's editing. He shares my concern about the classified requirements restricting us, hampering the collaboration between the two of us and Eric. I agreed we should restart this on our own, that it had been a poor use of my prior work not to send me to Brazil instead of the retired policemen with limited background about the situation. We struggle getting official records about the Navy's 'Tic Tac' case of the Nimitz aircraft carrier, most of it unavailable.

Hummingbird. Saturday 15 May 2010.

A nice letter has arrived from Joscelyn Godwin, professor of music at Colgate University and expert on western esotericism, commenting on my recent posts explaining crop circles on the *Boing Boing* blog. Eric and I discussed Sagan's position on the phone: he's alleged to have told Hynek that he believed in UFOs but couldn't talk openly. (12)

Eric Davis knew Sagan at JPL and LPL during his time with Voyager: "He was the popular science celebrity, so he rarely chatted with us lowly grad students. He only came by to gather the latest planetary imagery and atmospheric data to use for his TV shows or popular books. He was an affable guy, pleasant and entertaining when he talked to people around Mission Control. He was very much pro-SETI and pro-ET life, but the topic of UFOs never came up. He expected that the Pioneers or Voyagers would end up being intercepted by spacefaring ETs, and he hoped they'd find his gold plaques (Pioneer) or the golden records (Voyager) and use their data to find Earth and visit. But he also told my supervisor that he thought it would be tens of thousands of years before that might happen.

"I don't think he ever bothered to study the good book that he and Thornton Page co-edited: *UFOS: A Scientific Debate*. I know he talked to Jacques about UFOs sometime during the 1960s. He was more open-minded back then, but he took a turn as his popularity grew. My supervisor said UFOs were nonsense: Sagan was a peacenik. Very much anti-secrecy, anti-war and nuclear weapons. I don't believe he got 'read-in;' he wouldn't have accepted the secrecy, and he wouldn't have passed the background tests..."

2

Esalen. Sunday 16 May 2010.

This is the third time I've stayed at Esalen. They gave me the lovely Passage room, on the edge of the cliff, apart from the main house. There's a carefully tended garden between my bedroom and the precipice, with chairs to contemplate the waves.

Jeff Kripal leads the sessions with his interest in extraordinary abilities—those that emerge in the X-Men of comic books and in modern phenomena. He looks for a theoretical middle term between the claims of witnesses on TV and the stories of popular literature.

Esalen co-founder Mike Murphy's introduction defined mystical realism as a way to "encode actual observations into the literature."

I feel privileged (and a bit amused) when I read Jeff's insightful analysis of my work in a form usually reserved for long-dead authors. He holds a mirror to me; it provides feedback, critique, and guidance any thinker would treasure. How can I rise to that challenge? Of the four authors studied by Jeff in *Authors of the Impossible,* two are dead (Charles Fort and Frederic Myers) and Bertrand Méheust has partly retired. That leaves me exposed, when I would prefer to work in the shadows.

Jeff has accomplished something my friends had often urged me to do, merging the two sides of thought, the analytical and the speculative, which I keep apart: the stained-glass windows in one corner, computers and strategic business in another.

Esalen. Monday 17 May 2010.

As soon as he arrived, there was drama around Whitley Strieber. The staff was a bit disorganized: everybody had been retained at the gate, and he'd been unfairly lampooned in the media. His early claims of gray Aliens have made him a target, so he was rather tense until the sheer beauty of the site put things in perspective. His wife, Anne, runs a fine website built around the many letters he's received, a treasure. We spoke privately with him and Jeff once we were cleared.

Soon, Whitley became his usual eloquent and avuncular self again. He mentioned that General Arthur E. Exon, who once told Whitley about Roswell (hence *Majestic*), ended badly, senile and suspect, in a compound for high-level spooks "who knew too much and lost their marbles."

Esalen. Tuesday 18 May 2010.

This morning, I was asked to present the challenge of field research. I'd selected three cases, two of which could be explained (hence, rejected), but not without very extensive analytical work on site. My point was that the stories were equally interesting as folklore context, but that a scientist shouldn't be fooled by a superficial study.

There was a question about extraordinary claims on the web like those of SERPO (13). Yet smart people smell the rat: the networks are abused to plant fake ideas, confuse research, and initiate new memes.

Later presentations by artists and writers agreed that the cinema, even at its most controversial, always reinforces the *status quo*, while *religion serves to protect us from the Sacred.* "The Sacred can kill you or scare you," somebody said. "Often, it does both."

Whitley has told a few of us how Laurance Rockefeller once invited the Clintons to his Montana ranch and surprised them with a briefing on UFOs, received in silence. The next morning, Hillary begged Bill not to react. As they went horseback riding, Bill Clinton told Laurance, "I won't get into that UFO business."

I noted websites to see, films to watch (*Pan's Labyrinth*), and the temple at Konark. Did Ingo Swann publish a book about psychic sexuality?

Whitley now believes the Dead are involved, and Mitch Horowitz noted that the area of Whitley's cabin was in the 'burned-over' district of New York State where many paranormal events have emerged.

Hummingbird. Sunday 23 May 2010.

In the streets of San Francisco, I often see old couples shuffling along, shopping or simply strolling for their evening walk. They drag their canes and their rheumatisms. I used to take pity on such 'poor old folks.' Now I envy them, at 80 or 90, still together, enjoying each other's closeness, taking precious pleasure in caring for their sick partner. I am moved by that spectacle in ways I'd never expected.

Today I went rowing again among the blue herons, then came back to write. Ron Brinkley leaves for several months in the mountains above Yosemite, so we said goodbye over coffee at Quetzal on Polk Street and we discussed Jeff's book. Ron's ability to dig up important philosophical nuggets is remarkable.

Over lunch today, Steve Millard said Paul Baran suffered from liver cancer and subsequently had a heart attack from which he's now recovered, so he's thinking... of starting yet another company, when anyone else would retire peacefully. Amusingly, Steve also asked me if I was ready to date women again; his wife, Linda, has a long list of distinguished girlfriends, hunting for new husbands...

Hummingbird. Wednesday 26 May 2010.

This morning, as I was just stepping out, motorcycle cops briefly blocked our street. President Obama drove by our building on the way to the Marina Green, his helicopter waiting. As the long black limo passed, flags fluttering on the hood, I only caught sight of a silhouette.

My life is simple, suspended between a great void and vague professional hopes. Dave Caplan invites me to join his investment team. Lunch is often with one of my trusted colleagues, like Bob Johansen, who presented me with C.G. Jung's *Red Book*, a marvel.

In the afternoon I record updates from companies, I work on the computer, pay bills... The news is on at 5 pm, then music (Schubert's lieders, my current favorites) and catch up with my daughter. I eat late, alone, a simple, comfortable life I can endure for a while.

Hummingbird. Friday 28 May 2010.

New cases of large 'dogs' at the Utah ranch, standing on their hind legs; also 'coyotes' running upright next to a car, and other peculiar animals. The Ute shaman doesn't find it unusual.

The BAASS project comes to a bureaucratic end on 23 September and may not be renewed. At a minimum, there'll be a gap until after the November elections, when it might be revived as an SAP, but it takes a special line in the black budget to do this, and we may not be well enough connected to obtain it. Bob Bigelow begins reducing headcount as Capella goes on under Doug Kurth's able direction: my databases of pilot cases, Blue Book, and NIDS are now integrated but approval for in-depth analyses/statistics and AI analysis was declined. We've lost our calibration.

Austin. Sunday 29 May 2010.

Eric Davis was at the airport to pick me up, so we dived into the issues as soon as I landed. Everything I learned was consistent with my earlier assessment, although more specific: "There's nothing really new about the phenomenon, either from MUFON or BAASS itself," Eric said as we drove to the hotel. Nor is there any breakthrough about hardware. The newer plan is to approach the Assistant Secretary of Defense for procurement to avoid going 'dark' in October when the initial two-year period runs out. As the project lays off people, we lose the benefit of their unique training, and many old records have been discarded in Washington since the Clinton days, when a misguided effort to digitize everything went awry.

Austin. Monday 30 May 2010.

Hal Puthoff met me at breakfast, and we spent the whole day together: I had much data to tell him based on my recent field trips, and someone at an aerospace firm who knows the truth about the hardware says that most of it was returned to Wright-Field. No one works on it anymore, supposedly because none of the experiments make any sense.

There's a new theory of remote viewer Pat Price being killed by the KGB. A defector confessed, but not everyone believes him; multiple parties may have gotten very upset with Pat, as Russell Targ argues.

The Ranch is the subject of a massive report, complete with electromagnetic surveys and plant growth analyses with the clever use of special systems using plants as bio detectors. Also, large dogs walking on two legs, the 'cynocephalids.' I gave him data from other locales.

Efforts to find the full report on the Tehran UFO case within the system have been futile. It may have suffered the same fate as the other research reports destroyed under the guise of bureaucratic efficiency. Then someone called Hal in the middle of our conversation, upset again at BAASS' research on the phenomenon: people's names end up attached to some reports they don't approve, and they're upset.

Hal tells me they could never scale up the super-electron tests. The facility outside Austin, owned by Bill Church, was sold long ago.

I was tired when I came home. My suitcase, unpacked after the Austin trip, is ready to be filled again with books for my friends and toys for Maxim. After watching the French news (the Israeli blockade of Gaza, the gushing oil disaster in the Gulf), I left the TV on as it played *Quai des Brumes* with a young Jean Gabin who reminded me so much of Alain, Janine's brother, who died two years ago.

Mabillon. Wednesday 9 June 2010.

Sweet Paris Spring, conducive to evening walks through the gardens of *Les Halles* near the CNES building. Over lunch today (at Atelier Maître Albert) Yves Messarovitch introduced me to a financier with international experience, looking for partners to create a 'green institute.' There's plenty of new economic data, he said, most of it scary. France still spends beyond her means, Greece plunges towards bankruptcy. Fortunately, the soccer World Cup begins in two days...

Valognes. Thursday 10 June 2010.

When I was a teenager, my father introduced me to the novels of Barbey d'Aurevilly, who wrote so lovingly about this town. Now I read Alexis de Tocqueville (1805-1859), once the local representative at the French Assembly who went on to analyze American democracy.

I'll meet Annick here and we'll go to Yvetot-Bocage, to remember Janine at the foot of the old church. I am taking flowers to her, roses from the garden in Palo Alto, which she hadn't seen blossom. Annick brought some roses from her own garden, the same variety and color, a nice coincidence. A stubborn Norman rain fell, blurred by stubborn Norman haze. We went through Bricquebec and back to Caen. Saint Lazare station was drafty, hostile, and dirty.

Messages from Peter Banks and others awaited on my computer, and one from Jean-Claude Venturini stating that Dr. Claude Poher's paper (14) describing new experiments for the *European Journal of Physics* had been withdrawn "under pressure from the scientific community." Not much is changing in Europe; people are so afraid of the future they cling to antiquated knowledge, censoring innovators.

Annick gave me the book by Raphaëlle Bacqué, *Le Dernier Mort de Mitterrand,* about Monsieur de Grossouvre. It doesn't question his suicide.

I'll never know why he wanted to see me one more time; he was a man with universal interests, and curious friends with guns.

Annick told me Janine "admired" me and was happy with me, yet I feel unworthy of that admiration. I wasn't strong enough, or wise enough, to give her all the support she deserved.

Mabillon. Friday 11 June 2010.

Over lunch with Dominique Weinstein, we went over his pilot case statistics, expanded to 656 verified reports. The main surprise is an inversion of expected percentages: maneuvers (55%) are higher than flybys (39%) though one expects the reverse pattern. 'Maneuvers' best represent the intelligence of the phenomenon.

Michael Vaillant sends me his own statistical study of the Larry Hatch database. He considers several correlations, including the solar cycle, and concludes that my control system hypothesis (based on autocorrelation statistics years ago) remains valid over long periods.

Nevers. Monday 14 June 2010. Hotel de Diane.

Flamine accompanies me on this trip with a healthy sense of cutting ties to the obsessively demanding web and the spreading financial crisis crushing everyone. Our excursion began with lunch at Le Train Bleu, the restaurant at Gare de Lyon with its lovingly ridiculous angelots, naked goddesses, and glorious old paintings of the Colonies of Imperial France. Nevers has a great cathedral and a nice ducal palace, but its greatest beauty is the Loire itself, gliding in majesty.

Château-Chinon. Tuesday 15 June 2010.

We arrived in the middle of the town market, which blocked the heart of this town atop a steep hill. I was eager to explore the superb landscapes of the Morvan I recalled from a trip made at age seven with my aunts, from the artificial Settons Lake to the spectacular dam at Pannecière. I even found the magical spot where I used to play along the riverbed of the Yonne River, very shallow at that spot.

I needed to rewind these images from childhood. Doing it in Flamine's gentle company, in such mild filtered sun, was good for my soul. We'll spend two nights here before moving on to Autun to visit a local researcher, Michel "C", who's sent me three delightful, illustrated manuscripts of symbolic correlations.

On the silent hillside of Château-Chinon, the view from our window faces east, with nothing but forests and pastures used by white Charolais cattle. Today, we hiked along wide dirt paths to the springs of the Yonne River, on the slopes of Mont Préneley where stood the ancient city of Bibracq, conquered by Caesar around 50 AD; the region recalls the proud Gauls. Near Arleuf, we sat on the stone steps of a Gallo-Roman theater with space for 500 spectators.

Autun. Wednesday 16 June 2010. Hotel des Ursulines.

It rains over this town and its massive ramparts built on top of older Roman walls. In the cathedral, grimacing demons chase penitents beyond the top of pilasters. In one of the most interesting sculptures, an angel wakes up the Three Magi to show them the star they must follow. We have a fine view of the pastures, beyond the fortifications.

Autun. Thursday 17 June 2010.

This side trip to meet Michel 'C' is a voyage in the mystery landscape along a small river, the Arroux. The fields are fat and generous; everything grows happily here, wild flowers everywhere. Michel met us in the tiny town center and we followed him among trees and farms. After a career making precision plane models for museums, he retired to study alchemy. It rained softly while we discussed the various modes, which he said were falsely 'revealed' on various websites. The French practitioners continue to hold their knowledge secret.

To prove it, Michel asked his wife to bring us material for a Philosopher's Stone they'd received from a friend. While he doesn't have a laboratory, he keeps in touch with operating alchemists. The fragments resembled hard plastic. They were intermediate products, he said, while Flamine and I marveled at the turn of that conversation.

Nevers. Saturday 19 September 2010.

Yesterday we stopped in Avallon. Now we're back in Nevers after an ultimate excursion to Vézelay in the mild rain. We admired the lovely, sculpted devils warning good Christians against cardinal sins, but we were more impressed by Clamecy with its ancient houses and intact medieval streets at the confluence of the Yonne and the Beuvron, and a peaceful canal in the background.

For centuries, the region has specialized in the floating of wood from the forests of Morvan towards Paris. The villages are clean, well-organized, and the food excellent. Flamine found a brochure about the giant phenomena of 5 November 1990, still unexplained at the CNES.

Mabillon. Sunday 20 June 2010.

We came back from Nevers at noon, and a few hours later I was having a drink with Jean-François Boëdec near the Beaubourg museum. He looked frail, his sharp nose cutting the cold wind like a knife. He quickly came to the point: Admiral Pinon, a serious student of UFOs, wrote to President Sarkozy a couple of years ago but his letter never reached the Elysée palace, so he asked Boëdec to try again through his contacts at the Assembly. This time, Sarkozy's office responded (on 9 June 2009) with polite interest, directing two ministries to respond, but Boëdec tried in vain to convey the information to Pinon. He eventually learned of his death in his car, of an apparent syncope.

Reading my books, Jean-François was surprised that I was aware of the 1974 sighting of a UFO over the Bay of Douarnenez, which involved two gendarmes (Kalinski and Le Stunff) and military witnesses. It had never been officially transmitted to Poher.

He has also investigated another remarkable 1974 case at Landivisiau, near a naval base, strongly similar to the 1976 Tehran case: In both instances, the object was a cylinder (2 to 3 meters thick, 6 to 8 meters long) that turned into a ball of light and flew away.

Mabillon. Monday 21 June 2010.

Peter Banks has sent me urgent news from Washington. Jim Arnold, chief technologist, our contact at NRO, has drowned while on vacation in South

Fig. 3: *JV, Dr. Jeff Kripal, publisher Mitch Horowitz, Whitley and Anne Strieber. Esalen seminar on mystical realism, 17 May 2010.*

Fig. 4: *With Flamine and alchemist Michel "C". In front of Flamine: test fragments for a Philosopher's Stone. Morvan, 17 June 2010.*

Carolina. It was another setback for our project, a strange death for a rugged fellow like him.

Two reports have arrived based on the Capella statistics, which Doug Kurth wanted me to see, from the Weinstein catalogue. The differential behavior of the objects (military vs civilian planes) is high.

The digital warehouse now contains eleven databases with a total of 248,000 cases, which probably overlap in places.

Mabillon. Thursday 24 June 2010.

Professor Pierre-Eric Mounier-Kuhn agreed to meet for lunch yesterday at Balzar. A historian of computer research in France, he'd invited a friend of his from the Prime minister's staff, Xavier Delamarre, who was aware of my work.

The Genopole meetings, as usual, were very interesting to me because of the caliber of the participants and the companies presented.

Pascale Altier was there, representing the Pasteur Institute. She spoke to me for a long time about grieving, and my plans.

Hummingbird. Monday 28 June 2010.

Another report from BAASS has been sent out about the investigators' trip to Maxwell Air Force Base where old UFO records are stored, but as Hal rightly observes, "the most revealing records didn't end up in the archive." Many documents were shipped to NARA for storage in the 1990s. The Chief investigator went there with Doug Kurth. They extracted 12 unclassified or declassified documents and got six more items declassified, all relevant but low on new data: nothing on Roswell, crash retrievals, Port Moresby, the 1952 Washington overflights, or nuclear site penetrations.

Much of the recent discussion on the net had to do with rumors of internal fraud within the intelligence community. The whole 'classified' process, which looked so impressive to me from the outside, can turn very sleazy—and scary, when you see it up-close.

Middle of the night. The plaintive wail of the Alcatraz foghorn adds a sense of depth and adventure to my ruminations.

Hummingbird. Monday 19 July 2010.

Flamine sends me touching details of her personal itinerary: her father's family was from Belgium, descended from a beloved illustrator (Maurice Bonvoisin, known as 'Mars') and from Normandy through his mother, daughter of sculptor Raoul Verlet. Her paternal grandfather, a fervent Catholic, provided subsidies to poor families and remained an admirer of Maréchal Pétain as the victor of Verdun.

On her mother's side, her grandfather was a discreet and pious man from the south of France; his wife, who came from Anjou, was an attorney who spent time restoring a small castle in Mayenne, full of charm and majesty. He was as mystical as she was practical.

We have much in common. My mother's grandfather came from Belgium, my father from Normandy. We share romantic feelings about ruined castles, although mine are in dreams only, or mystical tales.

Hummingbird. Tuesday 20 July 2010.

Flamine and I spoke again about her family. Both of her parents were artistically gifted. Her mother was a professional singer with Radio France, often playing piano at home with musician friends. Her father, who managed a small real estate company, was a novelist once active in the theater. They divorced when she was 17, with two younger sisters.

The events of May 1968 brought Flamine visions of freedom and exploration. The world of ideas she embraced was that of Kazantzakis and Zorba the Greek, Henry Miller, Lawrence Durrel, Boris Vian… and later the study of Freud, Jung, and Lacan (15).

The Human Potential movement was flourishing at the time, both in France and in California, using techniques from Esalen; Flamine dreamed of leaving for San Francisco. Driven by her complex family history, she decided to become a therapist, successfully earning advanced French and American degrees in psychology. She began her professional career in Paris. Her beloved mother died in 1992, while her father, who had remarried, suffered a cardiac accident two years ago that confined him to a wheelchair, fortunately sparing his mind.

Later the same day.

Hal called this morning as I was about to introduce a lunchtime lecture on remote viewing by Ed May at the Institute for the Future. He was disappointed the project had never seen any useful classified data.

"We've stopped learning," he said matter-of-factly. "Like you, I miss the days when we could talk as a group. Maybe we should go to Bob to set up special meetings."

I had to respond: "To be frank with you, I'm not terribly impressed with the resources of the project. Once the current staff leaves, the information will be stale and the computer files useless. We need cross-calibration, that's what I miss the most in such a slippery area. Each one of us knows something the others don't; we need to keep one another sane, well-informed..."

"You're right," he said, "but we also need a place to talk about human effects, physics theory, important things that aren't classified."

Hummingbird. Tuesday 27 July 2010.

The renovation is complete here: the research books have found their place again, and the blue bedroom is quiet, ready for long evenings of music, reading, and contemplation. On Sunday I arranged for Luisa Teish to meet Jeff Kripal and his wife, Julie, over dinner in Emeryville.

Hummingbird. Sunday 8 August 2010.

The routine of my solitary life has almost become comfortable, between frugal microwave dinners and afternoon coffee breaks.

In an hour or so I will drive down to the airport to pick up Flamine. The apartment is ready, and San Francisco is swathed in great patches of white-gray fog that muffles the sounds of summer. Last night, recalling foggy nights like this, I enjoyed Bela Lugosi in *The Devil Bat*.

Hummingbird. Saturday 14 August 2010.

Busy week: on Monday we had lunch with Peter Beren and Ed May (with his wife Diane) in the picturesque wine cellar of Hotel Mac in Point Richmond. The next day, lunch in Palo Alto with Paul Baran, Steve Millard,

Fig. 5: *In Palo Alto: Flamine, Paul Baran, Kaili Kan, 9 Aug. 2010*

Fig.6: *Leading parapsychology researchers in Paris, March 2011. Standing, left to right: Ronlyn Schwartz, Sophie Varvoglis, Jean-Pierre Rospars, Vallée, Bertrand Méheust, Pascale Catala, Ahmed Dhoharsi. Sitting: Mario Varvoglis, Stephan Schwartz, physicist Peter Bancel, Flamine.*

and Chinese expert Kaili Kan. (Fig. 5)

I've been gathering more contacts, tracking our portfolio, and keeping alive my links with NASA, Peter Banks at the Academy, then Chris Aubeck and our publisher, catching a few more errors before *Wonders* is printed. Having Flamine here is a blessing, a warm hand in the labyrinth at Grace Cathedral: candles, music, and the sound of bells.

San Francisco is blessed by the magic light that follows the fog when it's swept away by the warming noon sun and the ocean breeze. We shop for flowers and lamps to enhance the apartment.

I speak to Olivier: I catch him walking down Eighth Avenue in New York, looking for a place where he could plug in his computer. We marvel that Maxim just turned fourteen years old...

Hummingbird. Tuesday 24 August 2010.

At Tomales Point, I spent last week constructing a plan for next year: a massive re-ordering of research files to centralize them. I also expanded contacts with venture-backed companies.

We had dinner last night with George and Mike Kuchar, about to turn 68. George looked drawn and walked with a noticeable bend in his neck, but his latest 'video weather diary' from Oklahoma was an inspired look at an American ghost town battered by thunderstorms.

Richard Niemtzow called: At 67 he's 'retired' but the Air Force has brought him back as a consultant on alternative medicine. He successfully battles cancer and remains active.

Leslie Kean was on The Colbert Report last night, promoting a new book with first-hand reports by generals and pilots. Yet the only French general she cited was Letty, who told me he'd never witnessed any unknown object, and was drawn in to the COMETA Report by Gilbert Payan. It's a move he now regrets, as he once told me over lunch. We do admire Leslie's book with its foreword by Podesta.

More recently, Jim Oberg wrote a penetrating critique of pilot sightings, explaining some of Weinstein's cases, and citing Hynek, who once noted that pilots were not necessarily the best observers.

Hummingbird. Saturday 28 August 2010.

Flamine has flown back to Paris. Leaving her at the airport I suddenly felt, washing over me, all the sorrow from which she'd sheltered me.

Once again, I also find myself skeptical about any sequel to BAASS. The only connection now is indirect, by email with Hal and Eric. A recent series of trips by Doug Kurth to data depositories have resulted in a statement that most UFO-related data "had been destroyed years ago." Allegedly, no document was found, *even at the classified level*, related to the incidents we'd flagged because of their relevance at the time. I may have the only records.

Eric observes that nothing more can be done until we get our SAP 'hunting license,' which he expects to come through efforts to "turn up the heat on CK, RTB, JL and the Tasmanian Devil." He's talking about Colm, Bob, and Jim Lacatski, but who's the Tasmanian Devil?

Someone told us that the absence of relevant records at NARA should have been obvious: "SAP Byeman records are automatically deemed not releasable under the 1946 Act or FOIA law. Most SAPs are executed in industry; *they don't exist inside the Pentagon*."

Hummingbird. Sunday 29 August 2010.

Kit Green called this morning with comments about *Forbidden Science 3*, suggesting minor clarifications, notably on the discussions with Richard Niemtzow. He reminded me that Richard held an MD degree from France. Kit helped him get the license and suggested he join the Air Force to perform his residency, which gave him a commission. When Richard became enthused about the alien autopsy stories from Leonard Springfield in the 1980s, Kit found him naïve and even got upset at him. He may have changed his mind now.

Overall, Kit was impressed with the book, confirmed his role and the many details and facts it brought back to memory. He also confirmed the project was out of money in Vegas. Bob keeps it going with his own funds and still hopes for a restart.

Now Peter Banks is calling from Washington to talk about a potential venture fund. Peter is getting advice through his previous role at ERIM, Veridian, and of course General Dynamics.

Today I started to reorganize my Blue Files and hoped to find a home for them. Flamine is unimpressed with research centers or institutes, pointing to their questionable record. She said frugal personal efforts, where you don't owe anything to anyone, often give the best results.

Also today, I received Frank Salisbury's new book about the *Utah UFO Display,* (16) with a healthy review of our Skinwalker ranch stories. I'm glad Salisbury has jumped back into the research. We've begun discussing a joint research trip to Utah.

Hummingbird. Saturday 4 September 2010.

Lunch on Thursday with the physics team that looks for a gravitation breakthrough, back from their latest series of experiments. Over sloppy joes, fries, and Cokes at a sports bar in Los Altos, they told me how their early tests had failed. They now do generate fusion: their 300keV protons penetrate 2 millimeters through lithium targets (normal penetration should not exceed 9 microns) and their 8.6 MeV alpha particles penetrate 4 millimeters, versus a standard distance of 150 to 180 microns. Although these figures still fall short of fully validating the theory, they can now build a better and less expensive device. The physics is over my poor head.

Hummingbird. Wednesday 8 September 2010.

It catches me at unexpected moments: on the way to a coffee shop, I park my car next to a rosebush in full bloom and I feel very sad, because these flowers are gorgeous, and you cannot see them.

I describe them to you in silence, through my tears. I do need rest, and I need to decide about a future course of work.

This afternoon I will fly to Paris, a dozen projects in my head. The abrupt termination of BAASS, after only an initial two-year run, remains a source of frustration, yet Bigelow's passion is genuine. We should do what I suggested in the early days: state our pre-established hypotheses and researchable issues, then make a list of *what needs to be explained.* Last, define a *frugal* approach to the data.

We also need a post-mortem of NIDS. We should list the instances of greatest concern and what we learned about them, one by one, with no attribution of guilt. Also, unexplored hints, which are many.

Paranormal research is like good software: if you can't do it with ten people you certainly can't do it with a staff of a hundred.

On the train to Normandy. Saturday 11 September 2010.

The trip to Valognes takes three hours. If I wasn't so tired, I'd enjoy Janine's country, the lush beauty of Normandy, places she loved.

Yesterday in Paris, on my second visit to the Dassault domain, we met the 85-year-old senator on the way, then sat in his son's office, under fine paintings of polo horses. Praises for rare wines, rare books. He made money with Genset. Invitation for us to come back with a project: 150 million dollars, or euros? Indifferent. Strange feeling: I thought I'd given that up.

Later, over dinner of crêpes with Flamine (at Beaubourg), I reflected that Dassault would only mean a return to things I did before, with people whose current interests are different from mine.

This train moves on, from station to station, as if I was watching Janine's life in reverse: from Paris to Caen, then Bayeux, Valognes, and finally Yvetot. It's a courageous little girl I meet at the end: a strong country girl, mischievous—and so clever!

Valognes: the town of *Monsieur de Tocqueville*, far from Paris where Sarkozy's power system swims in scandals. He's made his first big mistake by not firing his treasurer, a dead albatross.

Mabillon. Tuesday 14 September 2010.

My daughter, between her tears, calls to tell me about a dream in which Janine joined her to clean some glass fixtures. She could only stay for a short time, she said, yet she returned and said she'd soon stay for good. In my own dreams she passes by, warm and busy.

My meetings here follow an intense schedule, with a keen awareness that I don't have many years left, as I told Pascale Altier over lunch at La Coupole yesterday. She runs the business incubator for the Pasteur Institute, a place I expected to see blossom under the government stimulus for innovation. Instead, she told a tale of bureaucratic uncertainty, and her delight, during a recent vacation, about the wide-open spaces of the American West.

Claudine Brelet has just published her book about Jacques Bergier. (17)

Mabillon. Saturday 18 September 2010.

French political life is shaken by scandals every day, escalating to Brussels because of Sarkozy's expulsion of gypsies from France, of questionable legality. None of this translates into real protests: the French linger in cafés and slowly reconnect to their offices after their long summer vacation.

Government buildings are open this weekend. I will spend the day with Max and take him to the magnificent Paris Observatory.

Mabillon. Monday 20 September 2010.

Yesterday I was moved to meet Flamine's 90-year-old father, weak but of clear mind, at his street-level apartment in the Alesia section of Paris. Later, calling Gérard Deforge to wish him well (he recently fell and broke a rib), I learned about a recent sighting in Créteil, where a woman watched *and photographed* a fiery object passing alongside her balcony at slow speed, very close to the building. Such facts keep punctuating our lives. We must continue to catch them before they get carried away by the foolish tempests of the imagination.

The BAASS project never did produce the *WOW! Factor* Bigelow had promised. Like Hal and Eric, he expected to unveil the famous ET hardware. I still believe the super-secret project is real, but it may be engaged in very different activities from what my friends imagine. It's also illegal, probably. Why isn't that a big problem, gentlemen?

3

Cluny Monastery. 24 September 2010.

L'Auberge du Potin Gourmand, built in the 18th century, served as a pottery manufacture until 1970 when it was converted into an inn with a fine restaurant and nine rooms arrayed in a disorderly architectural arrangement with twisted wooden staircases leading to ivy-covered towers. Flamine and I have a room at the very top. It features sculpted furniture in dark wood and a most original ceiling made up of wine barrel ribs nailed together to form an arch over our bed.

In the same spirit, the tub in the bathroom is a half-barrel and the toilet is a *chaise percée* thankfully hoisted above a modern fixture. The heavy wooden door at the foot of the steps has a clever hidden panel for the lock: *La Chambre de Clotilde* comes out of a fairy tale.

On Wednesday, Flamine was with me among Parisian ufologists where I interviewed the witness of that recent close encounter in Créteil.

My accomplices Gérard Deforge and Jean-Claude Venturini were there. Also George Metz and Gilles Lorant: they have a rich supply of stories from their field investigations. Cluny, with its serene atmosphere, is taking us far from all that.

I turn 71 years old today, a mere speck in time compared to the massive monastery nearby, which celebrates its 1,100th birthday. Experts in medieval architecture and monastic history have converged here, welcomed by quaint shops and restaurants serving Burgundy food and good wines from Mâcon, Saône, and the Rhône region.

Feeling completely out of time, I called Graham in California from the stone ledge of the cloisters to firm up our decision to re-invest in Triformix, poised to enter the world market of high-precision lenses for optical devices. Data transmission is accelerating a thousand-fold in the computer industry worldwide, I forced myself to realize as Flamine and I walked on in the quiet evening to sit in the ancient park.

A few hours ago, we were admiring a collection of 12th century illuminated manuscripts, including a Bible that took two years for two scribes to complete. Back then, the monks of Cluny were already accelerating the transmission of knowledge…

Cluny. Saturday 25 September 2010.

French archaeology buffs and professionals arrived today. They wander across the vaulted rooms, appreciative of the high naves. We're mere students among 130 knowledgeable people, like our friends Philippe and Kathryn Favre, dedicated to learning every detail of the site and the current digs.

The relationship of the spiritual life of Cluny to its political power in the 11th and 12th century is striking. The role of the *scriptorium* in disseminating knowledge touches me in special ways: as if I had once worked there? I would do it again!

Mabillon. Monday 27 September 2010.

This trip was filled with intense experiences. I will return to California with memories of renewed friendships (Claudine Brelet, Philippe Favre, Gérard Deforge and his friends), new meetings (George Metz, Laurent Dassault) and resolved issues of bureaucracy. But it is our retreat in the lovely, ivy-covered attic of Potin Gourmand that gave me the perspective I was trying to achieve.

We even found a tapestry of the Holy Grail to complete this apartment before I had to leave.

Palo Alto. Wednesday 6 October 2010.

Autumn has assumed her sumptuous yellow and purple dress: Silicon Valley is sweeter than ever. My car gets blanketed with leaves. There are small oranges, new fruits on the guava tree, and buds on the rosebushes we had planted in the front yard. In Vegas, my friends still hope for a decision that keeps getting delayed. The research is in urgent need of reorganization, Federal funding at an end.

Books on ufology have become popular again, in spite of the shrinking number of reports. Following Mark Pilkington's *Mirage Men* with a long interview of Dr. Green, a book by Nick Redfern called *Final Events* exposes a rogue group of fundamentalist Christians within the Intelligence community, 'The Collins Elite.' It tried to recruit Kit. I was given a list of rumored members... I cannot verify.

The Collins Elite supposedly preaches the tale that the 1947 UFO wave was triggered by the magical rites of Jack Parsons, influenced by Aleister Crowley with help from L. Ron Hubbard. Those evil rites with stoned strumpets supposedly opened a door through which demons flew in, disguised as spaceships? "Give me a break!" my daughter says, laughing. *This makes no sense at all:* neither Crowley nor Parsons were Satanists, contrary to the obsessions of the fundamentalists. Neither are the scientologists. In my extensive contacts with Hymenaeus Alpha and Israel Regardie, both of whom knew Crowley well, aliens and saucers played no role.

Now John Schuessler tells us he's the target of death threats from a Christian cult in Houston. "It was the only way they could get rid of the demons they befriended," he said. The FBI raided their place when they planted a car bomb against the mayor. Who's the evil guy now?

Hummingbird. Saturday 9 October 2010.

While I was mulling over the Lancaster events and wondering how I could reconnect with the man I called Jim Irish (in *Forbidden Science 2*), I found his daughter in Burbank: "I do remember him telling us kids about a project he did with you," she said right away, adding: "He once wired a house for sound to record what was going on. I remember the individual to be recorded was a 'Brian Scott.' He was going into trances and speaking in a horrible hoarse voice, who scared me to death: I was maybe 14 at the time…"

Hummingbird. Friday 15 October 2010.

There are long and frequent exchanges of email among our group now, arguing about the abuse of the Internet by groups spreading disinformation. One such group turned out to be a decoy, setup as a social experiment sponsored by a large organization. It is widely suspected, with evidence, that our own Intelligence agencies use it to track public opinion. "It's largely Freudian, Jungian, and traditional work; the CIA is suspected of research into social attitudes since the Fifties," someone speculates on the Internet. "The linkage to the conservative agenda suggests who the current sponsor could be, moving into web tracking, futures research, and right-wing agendas."

Hummingbird. Sunday 17 October 2010.

"Jacques is back!" exclaims my friend Maurice Gunderson when I tell him we have an opportunity to develop a new venture fund, and why he should join me. Maurice is an aviation expert who agrees we'd be a strong team. Despite the encouraging progress, however, I'm not so sure I am 'back.' I often wake up in the middle of the night with worries about my work and anguish about my children, thrown into a world of such turmoil. Then I think about the beautiful intellectual effort that feeds Silicon Valley, and my old confidence returns.

I wrote a piece called "I, Product" for David Pescovitz' popular *Boing*

Boing blog. Peter Beren thinks I should turn it into a book. It argues that folks are deluded when they fancy themselves as 'users' of Google: *They are simply the stuff that Google sells to the ad industry.* Why else should Google be free?

Hummingbird. Saturday 23 October 2010.

Last night, the Foundation for Mind-Being research, a group of middle-aged, well-meaning parapsychology buffs that includes Arthur Hastings, Russell Targ, and me, met in Palo Alto to hear a panel of three women who had experienced near-death experiences: Nadia McCaffrey, Yolaine Stout, and Ellie Schamber. They mentioned the usual tunnel of light, the greeting by a shining being, the feeling of universal love, but what seemed to follow was *a melting into an ocean of undefined bliss*. (They acknowledged that it was possible to have an experience identical to theirs in a normal state.)

Their hypothesis was that 'souls' got reborn from that ocean, and that we are all one, which has always made sense to me. In a song that brings my tears, Jolie Holland sings (in *Escondida*, quoted with permission: a song called "Goodbye, California"):

> When I'm dead and gone
> My immortal home will hold me in its bosom
> Safe and cold, no more desires
> Will light their fires or disturb my immaculate calm...

Yet the evidence for apparitions and other phenomena after death often denies this. I'm still struggling with the line of separation between brain effects (perceptions such as remote viewing of past events, or faraway people) and those effects related to physiological death. Also, of course, the unknown nature of time. Those who've gone through an NDE speak of the great peace, the awareness that death doesn't matter.

It's been raining all day. I pored over maps of Utah, assembling warm clothes for the long-planned trip with Frank Salisbury, and taking short naps that leave me defeated, disoriented, and moody.

Roosevelt, Utah. Tuesday 26 October 2010.

The sky was clear in Salt Lake City where I met Frank, but the wind got chilly over Daniel's Canyon, and we had a dusting of snow East of Heber. Along the way, Frank told me about his career in plant biology, his studies at CalTech, and his work as an American expert on the Mir space station team. He moved to Utah in 1966.

In Roosevelt, we were joined by Junior Hicks, who offered the hospitality of an unoccupied house. There's an increasing proportion of 'orbs' and strange animals in the area: they resemble wolves walking on two legs, he said, like the mythical Skinwalkers.

Roosevelt. Wednesday 27 October 2010.

Our first visit to the site of a recent Skinwalker sighting was on Cedarview Road, which branches off the highway to Neola. It involved a being who crossed the road in the glow of a car headlights: no effect on the engine, the creature larger than a wolf, gray in color, its eyes shining red in the headlights. (It's a natural effect of retina reflection, Frank told me: the eyes were not red.) The second sighting was about a week later, on Neola highway, and the third one on Cedarview again.

From Neola we went on to visit a man whose sighting dates back to 1972. He's a tall Westerner with a commanding presence, a John Wayne character, but he has no use for the movie star swagger. His sighting took place years ago, yet he remembers it vividly. He was on horseback as a guide with a group of hunters when he saw the object "kicking up a lot of dust and lighting up the whole landscape." They watched it for a full five minutes, but they could find no trace after it flew off.

Naturally, we asked him about Skinwalkers. "The Indians are very serious about it," he answered, "but they won't talk. One of the creatures was seen recently by Indian police. It was like a man with a wolf's head, who jumped over roofs, laughing like a hyena. They chased it in their car but it ran off through the countryside; they lost it going over sixty miles an hour." None of that gets on the record.

Later the same day.

We made three stops today, the first one in Vernal with a dynamic businessman who described for us a UFO that chased him near Yellow Hill, which turned green under the illumination. As he fled, the object exerted a drag on the truck. This isn't the first time such interference happened. In Nevada, Dr. Hynek and I once investigated a similar case near Ely, where the truck left the ground, breaking its axle, as told in our book, *The Edge of Reality*.

Our last visit of the day was to Ron Cutch, a Ute Indian who serves as supervisor and guard at the Tribe building near Bottle Hollow. He saw "a spinning basketball with rainbow colors, moving at the height of a telephone pole." Half an hour later, two military helicopters circled the lake. After they left, the object shot out of the water and flew off.

Mr. Cutch recalls a visit by the security from Bigelow's ranch in black SUVs, all in black with dark glasses, hardly standard in rural Utah! They planted cameras on his property.

Roosevelt. Thursday 28 October 2010.

Oranna Felter is a vibrant Indian woman of mixed blood of Cherokee, Cheyenne, and Ute. A gifted writer knowledgeable in her tradition, she's witnessed some extraordinary things. Just before Memorial Day two years ago (2008), she went to the Fort Duchesne cemetery to change the flowers and flags on her family graves. As she and her daughter got out of their truck, they saw a big dog with red eyes (in plain daylight, this time) sitting on a pile of rocks. When it got up and walked on four legs the creature was three feet high, but as they went away it raised itself on two legs, jumped the fence, and disappeared in the sagebrush.

Roosevelt. Friday 29 October 2010.

This morning, we headed for the Fort Duchesne cemetery, and we located the piles of rocks. This was a hard trip for Junior Hicks as he read out the names, a good man sorry for his former students buried there: drugs, alcohol, and car accidents take a heavy toll. From Fort Duchesne we drove south to the ranch owned by Elly and Alex Barney, who'd seen an object over the Deseret power line, back in the sixties. Last year the whole family, including children and grandchildren, saw a bright, pinkish, round object of large size, 200 to 300 feet above the same power lines.

Later, Frank located Philip Garcia. At about 16, he heard a hum and saw a small, hairy man staring at him. It had a round face, big eyes, and "a face full of hair, like a cat." It did a flip out the window—the closest description to an elf I've ever heard.

About 1997, up the hill above Skinwalker ranch, his dog found half of a baby calf, neatly cut, dropped from the sky, legs broken. Garcia called Officer Pete Pickup, who took pictures. "After that, you could see those strange things flying around every night," he said.

Hummingbird. Sunday 31 October 2010.

After these remarkable interviews, Frank and I had a leisurely, spectacular drive over the mountains. I reflected that one of the greatest pleasures of the journey had been the opportunity to work with him and with Junior, both 'retired' yet so valiantly engaged in their passion. Frank spoke to me of modern biology, of circadian rhythms, of the Gauquelin theories of astrology (he knew professor Brown at Northwestern, whom I had coaxed into writing a foreword to *Cosmic Clocks*), and of the sorrows of old age: Frank's second wife suffers from Alzheimer and will have to be placed into a home. I drove Frank to his house and went on to return the truck and flew home.

Hummingbird. Thursday 4 November 2010.

Over lunch with the physics 'twins' today. They spoke of filing new patents. With dry humor, they value rejections as encouraging validation. Their theory dovetails with all the tests of general relativity, but that isn't the point. They have a new proton gun from Germany and plan new experiments.

At Stanford, the Herzenbergs invited me to their lab. (18)

Hummingbird. Sunday 14 November 2010.

Dinner last night with David Pescovitz, Mitch Horowitz, Jody Radzik, and Jay Kinney, ex-editor of *Gnosis* Magazine, discussing the sad state of research. My mood brightened when I got a long, beautiful letter from Dan Tolkowsky. He begins with a remark about logic: "Logic is a human artifact, devised by humans to help them understand what's going on around them, and assist in navigating their way. However, any attempt to use it 'as is' to understand pre-filtered matters and phenomena is misguided and fundamentally flawed, hence the need for a new method of engagement. We must question the notion of God, as commonly presented."

Hummingbird. Monday 15 November 2010.

Over lunch with Helen Stewart, operations manager at NASA-Ames, I casually mentioned my trip to Utah and was surprised to learn she'd lived in Fort Duchesne and once had a house in Vernal, facing the mountains. She's very familiar with the Skinwalker stories. Most recently, the wife of a tribal police investigator saw a seven-foot creature from 20 yards away (Colm Kelleher records it as the morning of 11 September 2009). It leapt off her tall barn, cleared a fence, and landed 25 yards away. This was half a mile away from the Bigelow ranch.

That same night, Helen saw a four-legged humanoid animal running through her garden. It was very dark brown, had skin (*not hair*), and covered 40 yards in four graceful zigzag leaps.

Hummingbird. Friday 19 November 2010.

This week was exhausting: always my Palo Alto office in the morning, long visits to several companies, then a meeting with Tim Jenks at Neo-Photonics to review the pending IPO of his company (following the re-introduction of General Motors to trading yesterday), and the restructuring of the finances of AlterG.

This evening, I confess I lacked the energy to work on any of that, so I watched a pleasantly silly spoof on horror movies called *Tremors*: huge carnivorous worms live underground in Nevada and seize unsuspecting cowboys and big-breasted women geologists to pull them inside their holes and eat them: a very satisfying Western story.

Hummingbird. Sunday 21 November 2010.

Two more weeks and I'll land in Paris. Flamine writes she dreams of the time when we'll be able to stay together.

Once again, the trip to Utah has shaken my model of the phenomenon.

Skinwalkers cannot be related to spacecraft and orbs—*or can they?* I had carefully separated these categories in my mind (although my case classification does embrace them all) but I must revisit the issue, starting as usual from the shelves of my library. In *The Werewolf,* Montague Summers writes: "In North America we meet once more the werewolf, as also the *werebuffalo.*" (19)

In the classic *Man into Wolf,* Robert Eisler mentions a case recorded by Alberta Hannum in *Spin a Silver Coin* (20): "There was a Navajo man near the trading post at Klagaton whom the other Navajos feared with a fear almost worse than death. He was called *The Werewolf.* He was believed to turn himself into a wolf at night and raid their flocks..."

But the best reference comes again from Helen Stewart. To my surprise, she recommends John Perkins' recent book *Shapeshifting*.

Eric Davis tells me Bob now believes he'd hired too many investigators. Evidently, all of us had accepted MUFON's inflated claims about 'hundreds of cases a month.'

Hummingbird. Saturday 27 November 2010.

We celebrated Thanksgiving at my daughter's warm house in the hills, by a fire that swept away the chills of an early winter. Elsewhere, North Korea is rattling sabers and Europe sinks in denial of its financial woes. To be frank, California is doing even more poorly than those Europeans we mock. Unemployment is at its worst here, and those programmers who still work are on reduced salaries.

Douglas Trumbull came to see me on Thursday. He'd shipped ahead his micro-scanner, so we spent the morning inspecting the film of the Créteil case I had hand-carried back from France. The frames had been exposed all right, but they only showed distant headlights.

This reminded me of Mike Kuchar's observation of a huge flying mass over the Mission district, where I could not find a single corroborating witness after canvassing the area for two days.

What does it mean, when someone clearly observes an object long enough to grab a camera and take two pictures, yet doesn't record anything interesting on the film, except for distant streetlights?

My own life goes on in routine fashion: breakfast at 7 am watching the bleak European news, in bed at 10:30 pm, with coffee breaks during the day and the tasty cornbread cakes Catherine made for me.

Other news is sad: Steve Millard's wife, Linda, is ill; George Kuchar has an advanced form of prostate cancer; and Smokey Wallace, one of the colorful cohort of systems programmers with whom I worked at SRI, has died. Characteristically, he asked that his friends remember him "by playing bagpipe music and drinking lots of beer."

Now I see an obituary for Barbara Bartholic, dear friend and heroic investigator. She died on November 10, of a stroke. Although I hadn't seen her for some 25 years, her enthusiastic outlook remained a bright memory. In November last year, Bob and Barbara's car was violently hit from behind by a teenage carjacker. Typical free spirits, they never used seat belts. They were thrown off, hospitalized. Bob died at Christmas. Barbara had evidently survived him for nearly a year.

Hummingbird. Sunday 28 November 2010.

Last night Jeff Kripal, Bob Johansen, and I spent the evening discussing 'The Impossible.' When I brought my two friends to each other's attention, Bob recognized Jeff's original quest and went to see him in Houston. Now Jeff is on his way to Esalen again, stopping for dinner with us at Bob's home in San Mateo, with Robin and kids.

Hummingbird. Monday 29 November 2010.

Budd Hopkins, interviewed last night by George Knapp, possibly for the last time, still has enough venom in him to rant against all those who question his lurid interpretation of abductions as ugly-little-aliens-who-keep-probing-our-women. He didn't mention names, but it was clear he kept big grudges against Whitley and me.

Under the kind and paternalistic image of the old UFO guru, there's an irate spirit that will lash out at the slightest critique. It doesn't help that Hollywood is hyping a new series on Alien Hybrids… but none of that is making much impact in Paris, where I'm heading next.

Mabillon. Wednesday 8 December 2010.

After a couple of days of mild rain, the snow started falling again this afternoon. Tourists file into Notre-Dame to warm up before another foray in the ice and the mush. The mood is uncertain and subdued, Europe suspended to fates beyond its control amidst a beautiful white Christmas-time. Dominique Strauss-Kahn, ruler of the IMF in Washington, looms as the potential leader of France.

I share a quiet transition among my friends. Gabriel Mergui gets up early every day to sing the Kaddish at the synagogue, following his mother's death. I have my own private incantations but happy moments too, like an impromptu lunch with Maxim on Ile-Saint-Louis, followed by a furious battle of spinning tops, his latest game.

Mabillon. Thursday 9 December 2010.

Suddenly intensified, the snowstorm closed the airports and paralyzed Paris. Around the Genopole building the snow makes a scene of pristine beauty, unusual for this suburb with industrial ambitions. Amusingly, only two of us 'investment experts' made it to the meeting from the States, plus one brave French scientist who'd spent the night in his car on the lonely parking lot. None of the other French members ever made it.

I've had another friendly (but inconclusive) meeting with Laurent Dassault, followed by dinner with Flamine at the 'Oval Office' where the waiters greet lovers with indulgent smiles.

Mabillon. Monday 13 December 2010.

On board a riverboat with the Mergui family and their guests, Maxim, Flamine, and I spent a relaxed afternoon celebrating a bar-mitzvah. Max and I spent time with old and new friends, like Isabelle Mergui's father who was the very first president of CNES.

The sleek white ship sailed from its berth for a short excursion upstream. A cruise towards Notre-Dame had been planned, but the river was already too high for the boat to pass under some bridges.

In the evening, we went to La Défense where we met Gérard Deforge and Pauline, our witness from Créteil. I returned her now-famous camera, her uncut negative, and the verdict Douglas Trumbull and I had reached: How could she have seen the egg-shaped object she described so clearly, *when the exposed frames showed nothing more than background suburbs*, and car headlights in the distance? Was the image transmitted directly to her brain? Pauline took it in stride and the group speculated on the nature of what she saw: ordinary physics would state there was nothing there.

Mabillon. Tuesday 14 December 2010.

Jean-François Boëdec tells me about his ongoing hunt for the *rats bleus* (blue rats), as the French Air Force calls its UFOs. Their detection network (Système Grave) extends to 1,000 km altitude. Reportedly, it finds anomalies where "objects emerge from the sea."

They agree that the phenomenon is not extraterrestrial as commonly believed, and that it changes its behavior as a function of its witnesses, as I argued. They've recorded a cumulative duration of two hours of UFO observation in one year, over any given *département* of France.

Now, a surprise invitation to a conference in Saudi Arabia next month.

Mabillon. Friday 17 December 2010.

Christmas shopping, bright lights, and brave attempts at holiday cheer in department stores with police cars nearby, ready to pounce in case of a suicide bombing. I was hoping to relax but business marches on with a financing at Materna (my latest investment in medicine) and preparations for the trip to Riyadh.

Cold rain fell yesterday when Flamine and I had dinner at Jean-François and Christine Deluol's apartment. A protégé of Chirico, Deluol devotes his time to painting, after a career managing shelters for children with special needs. They warmly recalled my mother. Several of Jean-François' works are scenes from Elsewhere: beings with oval faces, no eyes or features, yet burning evidence of Contact.

Deluol insisted on the decay of society and the hopelessness of the young that politicians have chosen to ignore, a big mistake.

Mabillon. Sunday 19 December 2010.

From our table by the window at *Les Editeurs* we could see the snow falling over Odéon last night. I'd brought together Claudine Brelet and Bertrand Méheust to discuss publishing Jeff Kripal's book in French.

Claudine has long worked in Mali where she earned a Dogon 'twin,' while Bertrand has spent time in Gabon, so their information is fresh and reliable. We had a wide-ranging discussion of African dynamics, the needs of the world as water becomes a strategic asset, and the deepening hypocrisy of politicians among ever-increasing dangers.

A new phase has been entered after the shock of Wikileaks, with the revelation of cyberwarfare in Iran where nuclear installations have been sabotaged by a sophisticated virtual attack using the Stuxnet worm. The operation shut down the Nataz nuclear facility.

Flamine and I walked home in the falling snow and today we woke up to a pretty Christmas scene of white roofs over Rue de Buci and bundled-up passersby. At last, I have time to think of new projects. Most dear to my heart, I find a shared understanding of the confusion of life when I speak with my son.

I attempt to rebuild a life on the treasures I preserve, first among them the love of children, second a rich tapestry of terrestrial life, relentless research against suffering. That doesn't make me a mystic...

In correspondence, Colm Kelleher mentioned a reference to terror that may be relevant in what our witnesses often report, in Wulf Haubersak's article in *Nature* (21). The work was done at Caltech.

Tonight: dinner with Gabriel Mergui, Pierre Tambourin, and his wife, Catherine, to seek new medical funding for French research.

Mabillon. Wednesday 22 December 2010.

Total eclipse of the moon yesterday, on the Winter Solstice. Again, an earthquake coincided with it, this one a 7.4 event in Japan.

I had lunch with Alain Dupas, fresh from a meeting with Director Dordain at ESA, who considers our project. I learned that Dassault had designed and built the rocket for the Israeli space effort, including the engines. It can deploy drones and space devices.

DARPA now follows up on their invitation to the 100-year Starship workshop, even asking how I want to "coordinate my personal security detail with the conference guards..."

Mabillon. Thursday 23 December 2010.

The Seine River is close to overflowing, carrying brownish-green water, tree branches and other debris, with waves a couple of feet deep and impressive, angry eddies as it rushes under the bridges. Gabriel Mergui and I work on the Genopole "Capital innovation fund."

Douvres-la-Délivrande. Monday 27 December 2010.

A friendly French diplomat, Mr. Mourier, noted that money has never been a major focus of mine. Under most circumstances, that could be a compliment, but he probably meant it in a spirit of friendly surprise.

I almost fell into the trap of apologizing for lacking the obligatory billions in the bank. Next week, I return to Silicon Valley with novel projects and a sad heart. Flamine writes it's natural to be caught between "the love I know and the love I do not yet know."

Douvres-la-Délivrande. Tuesday 28 December 2010.

The rain has returned to wash out the large white fields of Normandy. It feels safe here, playing with Maxim and his buddy, Rurik. I read Father Gabriele Amorth's *An Exorcist Tells His Story*, reflecting on his most baffling cases, an obvious parallel with abductions.

A storm has passed over Yvetot. In the cemetery, the flowers are blown away, the plants frozen dead. Yesterday, I recalled that Janine had taught me everything worth learning, but that wasn't quite true; she offered me that chance and I wasn't equal to it. I disappointed her, I am sure, when I failed to see how much more she wanted to give me.

There's still a little time for me to go over my mistakes and apply that knowledge to "the love I do not yet know…" Valognes was busy and dull. On the way back, overcome with awesome puzzlement, I caught Berlioz' *Fantastique* on the radio, directed by Seiji Ozawa.

Mabillon. Friday 31 December 2010.

Perhaps it is because of the low sky of Paris, but I feel small and pointless. My thoughts keep returning to the UFO conundrum, and then I feel even smaller because I now doubt that science as we know it today can comprehend the full problem. Based on Utah and Marley Woods, if the weird entities represent the real phenomenon, then we're indeed their toy, as Charles Fort said.

Hummingbird. Tuesday 4 January 2011.

A partial eclipse of the sun should be visible from Europe this morning. In sunny San Francisco, where I arrive home, letters from readers have piled up among the bills, Christmas cards, and business notices. Angelika Cawdor writes beautifully to thank me for *Wonders*.

I cleared up accumulated business issues today and multiple details about four new projects.

I also spoke to Fred Adler in New York, wishing him well for the New Year. He suffers from his hip again (those furious tennis matches...) but talks about selling his luxurious apartment and visiting Silicon Valley. About the economy, however, his tone has lost the buoyancy. He's clearly worried about the slow recovery and the twisting financial policies of Barak Obama.

Bob Bigelow, whom I called next with good wishes, was equally cautious, skeptical of the Administration and leery of NASA.

Las Vegas has the nation's worst employment situation and is greatly over-built.

Cavallo Point, Sausalito. Tuesday 11 January 2011.

This particular date was deliberately chosen by the DARPA managers: 11-1-11, a symbol to mark the start of an ambitious project, starships for the next hundred years. DARPA has allocated a million dollars to the effort, NASA a miserly $100,000. They invited thirty participants including Creon Levitt and Larry Lemke from Ames (also, director Pete Worden and our friend Lisa Lockyer), and David Neyland and Paul Eremenko from DARPA. The external experts include Dr. Craig Venter of genome project fame.

Also there: Jill Tarter of SETI and Barbara Marx Hubbard, whom I hadn't seen in 20 years. The debates, moderated by Peter Diamandis, range from propulsion and communication issues to the spiritual meaning of Contact. The site, once a battery emplacement and military reserve in the curve of the Marin headlands, overlooks the Golden Gate and a peaceful meadow beyond which the Bay and the shoreline of San Francisco spread their grand magic.

A lonely widower like me cuts a pitiful figure in such a landscape. As the first anniversary approaches, I cannot think of anything else.

Burbank. Friday 14 January 2011. Airport Marriott.

Why do the diplomatic services of so many countries have to behave in such a harsh, antagonistic way when someone applies for permission to visit? Today I tried to collect an ordinary visa from the Saudi consulate in LA. It fell into institutional complexity, with last-minute demands for "urgent" documents, never mentioned before.

Now I watch the sunset over the runways of Bob Hope airport. I have time to study my blue file for *Slender Panel*, the story of Brian Scott and his electromagnetic visitors. I review it in anticipation of my meeting with Joane, the daughter of 'Bruce,' the Hollywood sound expert who studied the case at my request in 1976.

Joane is bubbly and direct, a skilled manager for a studio where she oversees equipment and matériel. She shares a home with Bruce's second wife Pat, who married him in 1975, in time for the strange events at Brian Scott's house. The recordings, which she'll give me, confirm the witness' possession. Bruce saw the balls of light at the house. He had his own workspace, with special equipment.

"My father did work for *them*, a three-letter agency, as a consultant," she told me as we sat down to dinner at the Marriott. "I believe the agents operated out of a place in Petaluma called the Two-Rock Ranch, where they had an underground facility. A phone call would come in, they would give him coordinates. He'd drive there and take pictures, infrared exposures that he'd develop at home. The next morning, those were picked up. He never saw anybody."

She showed me his card with the Two-Rock Ranch as the address. "Do you recall any specific incidents?" I asked.

"Many. I heard about planes crashing in the desert; that was a big issue in those days. Also, a helicopter crashed on Magic Mountain. The rumor was that the aircraft interfered on a route the UFOs were using. Supposedly there was a geological formation, a big bowl, inside which they would dive and enter the Earth. My father was sent there repeatedly to take pictures as they went in and out of the ground."

"Did you ever hear of his colleagues getting killed?"

"Oh, yes," Joane replied. "I never knew the details, but I heard these men went down too far into that depression and got blown away."

She also remembers the Mount Baldy incident, which her father had described to me, when he was caught in a storm and big trees blocked his road, falling out of nowhere so he was forced to stop...feet away from a precipice he hadn't seen, where a bridge was washed out.

Burbank. Saturday 15 January 2011.

The weather over Southern California has been perfect. Following new fashion, LA women now walk around in those tall boots Texas cowboys call shit-kickers. More sedate, Joan and her mother wear simple sandals and live on an elegant, unpretentious side street.

Southern California, on such a day, feels like an easy-going paradise. It brings me back to my first visit in the seventies, tracking down stories of magic, the sense of imminence tempered by quiet pleasure. That feeling is just as artificial as the too-green lawns and too-clean sidewalks. I know how quickly frustration can replace inspiration, so close to Hollywood, yet I want to enjoy the illusion.

Pat and Joane took me to brunch at Marie Callender's and gave me the original recordings. Pat remembers meeting Brian Scott at her husband's studio. We'll never know the final truth, so bizarre it scared Allen Hynek and prompted him to recommend an exorcism.

Hummingbird. Monday 17 January 2011.

Our building was fully engulfed in fog this morning. I was glad to be separated from the world, my thoughts lifted above the city. Tonight, I had dinner at our daughter's house, a special treat mixed with nostalgia. A full year has passed since we lost you.

The thought of jumping into an airplane to Arabia in a few days comes as an opportune distraction. Friends confirm Riyadh is mostly urban sprawl with a clump of skyscrapers, "but there's a richer history and interesting sights in the desert, if one knows where to go."

A kind reader sent me a letter recalling his visit to a bookstore with his father when he was fourteen. They bought *Forbidden Science*. He writes: "I am jealous that you had wisdom at the age of 18 that I do not have now. You knew to look for unbounded potential within oneself, but also to be skeptical of its existence...My sense from it is that from the beginning you placed great focus on those things in life that matter: discovery and love."

Do I still have that wisdom today? That focus? Do I still have the strength? We'll find out in the work yet to come.

4

Riyadh, Saudi Arabia. Sunday 23 January 2011.

Suddenly, it seems that the debate about Contact and its implications for human thought and science has been reframed at a higher level, for serious audiences. The style, too, has radically changed.

Just weeks after our Sausalito workshop, another announcement has stunned both skeptics and UFO buffs as the kingdom of Saudi Arabia placed a plenary session on the subject (and its blatant implications for innovation) on the first day's schedule at the Global Competitiveness Forum, a prime business and financial showcase.

I landed before dawn after an excruciating series of delays in Frankfurt. The streets of Riyadh were empty and gave me no feeling for the city. Now, as delegates mill around the Four Seasons conference center, parking lots overflow and big media trucks maneuver alongside package scanners. The atmosphere is tense following an uprising in Tunisia that just ousted Ben Ali; the Arab world is nervous about Egypt's Mubarak and the future of dynasties.

Michio Kaku was at the Lufthansa gate in Frankfurt when I arrived from California; we exchanged a few words, but he preferred keeping to himself, so I sank into an exhausted wait as the airline announced delay after delay. Six hours later, they loaded us into a replacement plane with engine problems, further evidence of a failing European infrastructure, major airlines cutting costs across the board.

Here in Riyadh, the first impression is of a traditional world that mutates under its own forms of power: oil and money. The Forum takes place in a science-fiction skyscraper with futuristic shape, yet a partition divides the audience, shielding the women away. Veiled in their black *abayas*, they stay by the side, but there's a remarkable elegance to their dress, an undeniable style and grace I did not expect, and the heavily made-up eyes are smiling and free. The men are superb in their proud *djellabas* and colorful headdresses.

Our ambitious panel in the first plenary session of the afternoon went quite well. Michio and I were joined by Nick Pope, whom I'd only met briefly once, and by the always energetic and talkative Stanton Friedman, consistent and well-informed. Also on the panel was Professor El Naggar, who teaches earth sciences in Leicestershire. He gave us a Koranic interpretation of space phenomena.

Nick Pope is an ambitious man under his British reserve. He's hired a media promoter, plans to marry his American girlfriend, and hopes to settle in San Jose. Stan Friedman spoke with his usual verve and welcome humor, more aware of technology and media than his critics give him credit for. On Roswell, he remains unshaken.

These men dream of an 'imminent' disclosure of alien life. Something holds me back from joining the bandwagon. We're being seduced by images that do not conform either to exobiological data, nor to the reality our own witnesses describe, seduced and misled by a phenomenon so masterful, yet so close that I sometimes feel as if I could just extend my hand and feel its scaly skin.

Last speaker on the panel, I tried to provide a synthesis of what others had said, showed two slides of French pilot sightings to nail down the argument, and ended on a California-style, hopeful note: Let's do the physical research without jumping to premature conclusions. "We're not ready to invest in UFO research, but the basic research needs to go on." That became an often-quoted phrase.

The audience seemed attentive and curious, a mix of Western businesspeople in dark suits and Arabs in white djellabas. The session was a surprise, why this sudden burst of government interest, and in a strict Islamic country, to boot? The fact that astronomers are now discovering dozens of planets, many within habitable zones by human standards, has changed the tone of the debate.

I struck an interesting conversation with Nawalf, one of the young Arab men who serve as our guides. When he heard about the paranormal topic of our panel, he told me about an experience he'd witnessed, involving a Jinn. A member of his family had inadvertently poured hot water onto the ground in the courtyard without first saying the obligatory prayer to Allah. An invisible Jinn who lived at the spot was scalded and took possession of the man in revenge. The family had to call a special cleric versed in Koranic secrets to convince the Jinn to stop screaming and thrashing in hysterical motions, and to leave the man's body. The fellow became his normal self again.

Riyadh. Monday, Safar 20, 1432 A.H. (24 January 2011)

Presentations on the panel by Jean Chrétien of Canada and Tony Blair of England highlight current economic worries: inflation cannot be far away, they warn, impacted by high food prices, rising global demand, and the growth of middle classes. Food is a major problem; in India, 30% of the food produced goes bad before reaching the market.

Jean Marion, friend and former colleague of Philippe Favre, runs the local branch of Crédit Agricole. He picked me up and took me to his well-guarded house where we enjoyed a most pleasant lunch served by two silent attendants. The servants, in Riyadh, are all immigrants. They form about one third of the population of 27 million.

I told Jean and his wife that I was expected for dinner by local friends; what was the protocol? "You don't have to worry," Jean laughed immediately, "there won't be any women."

In the evening, I met Nawaf, who took me through the old quarter. We saw the palace of the King of the Nedj, drank green tea on the square as night fell, then he proudly showed me the city at night from the skybridge before driving off into the desert beyond city limits to meet his father, his uncle, and half a dozen other men under a vast tent. They get together once a week, he said, to eat grilled goat meat with rice and vegetables, and talk into the night, of politics and technology, of things to come.

Between Riyadh and Frankfurt. Wednesday 26 January 2011.

The Lufthansa flight, before dawn. At some point everything becomes irrelevant: the tiredness, the dislocation, the blurred sense of the future, even the sorrow. After enough delays, wanderings through cavernous airport corridors, warnings about engine quirks, and rescheduling of a flight and another, nothing matters much anymore. One becomes just another package bouncing through the system, hit from all sides by meaningless blows to your mind and your suitcases.

Frankfurt airport. Later the same day.

I checked email during yet another two-hour maintenance delay on top of our two-hour layover. A new dialogue is going on about what may have happened at Lawrence Livermore Labs during the fifties. Eric quotes

Larry Lemke, detailing his father's story; it's as I remember him telling it at Joe Firmage's house in 2000. Eric, Hal, Al Holt, and Bernie Haisch were also there.

> My dad had to have an 'R', not a 'Q' clearance, in order to be able to handle the material. That's the only way you could get into the 'Blue Room.' He didn't say whether it was metallic or not, only that it was extremely strong, and resistant to machining. He specifically referred to the grinding wheel of a surface grinder *stalling out*.
>
> He was told it was a piece of "that thing that crashed in New Mexico" by whomever ordered the testing done. His group worked in support of the Principal Investigators of the lab, including Lawrence himself (whom my dad referred to as 'Ernie'), Don Cooksey, Louis Alvarez, and Ed Teller. His direct supervisor didn't have a *need to know*; he was given a cover story that the material came from some captured Soviet spacecraft.

Larry's father also spoke about the material:

> The property that all the Roswell witnesses were describing is 'super-elasticity,' not 'thermal shape memory.' It turns out that Nitinol has yet a third property called 'pseudo-elasticity' that can mimic super-elasticity, up to a point. Non-material scientists cannot even understand what all these terms mean, much less keep the differences between them straight, I guess. The point is that the properties the witnesses report derive from the organizational structure of whatever the material was made of, at the atomic level, not simply whether it had Titanium in it or not.

One panel member who visited Lawrence Livermore in 1973 confirms there's a 'Blue Room,' and another for the same program/SAP at Wright-Patterson. In both cases the clearance is an R clearance, a material handling clearance, not a Defense Department clearance.

Hummingbird. Thursday 27 January 2011.

This morning, I woke up late, dressed in an old shirt, told my assistant not to expect me in Palo Alto, and prepared to spend a quiet day at home to recover from the tiring trip. I had begun cleaning up accumulated computer messages when Doug Trumbull called. He was in town between flights and was eager to talk.

Doug was focused and intensely interesting. He told me about his project of developing a totally virtual studio; his close association with an investor in Denver who backs the idea. After *Tree of Life*, he has many plans to deploy his 'Digital Showscan' in the 3D world.

In the Uintah Basin, the pattern of weird observations continues, and around Dulce, NIDS finds numerous reports of Bigfoot sightings, cattle mutilations, orbs, and weird creatures overlaid on classic discs and flying triangles. Around the Northern Tier and particularly Malmstrom AFB, many unreported cases also involving cattle mutilations and weird creatures.

At Marley Woods, they found the same 'weird stuff' that Doug and I have heard and seen with Ted Phillips. In the San Luis Valley of New Mexico, Chris O'Brien reports on cattle mutilations, creatures, and poltergeist, along with the classic UFOs. In Colm Kelleher's words, *"We're attempting to make sense of an unbelievably distorted dataset."*

Hummingbird. Friday 28 January 2011.

The whole world has been watching rapid developments in Tunisia with the ouster of Ben Ali, catching the Sarkozy government off-guard. Now the chaos has spread to Egypt, heart of the Arab world. Government offices have been looted or burned; the army is deployed in the streets. Videos from the scene show riots out of control.

In Palo Alto, the weather is cold, and the pace of business is high. There's an anticipation of blazing IPOs in the air, starting with social network companies like LinkedIn and Facebook, already worth billions. More modestly, our own NeoPhotonics may go public next week, which would enable us to complete our third Fund with success and some bragging rights, even back in sleepy Europe.

Hummingbird. Monday 31 January 2011.

At the Roxie theater on Saturday a small group gathered to celebrate George Kuchar's underground film career, watch his recent travel diaries, and screen Jennifer Kroot's *It Came from Kuchar!* Afterwards I invited George, Jennifer, and a friend of theirs to a dinner of pancakes and cider across the street. San Francisco was busy and moody.

From Austin, Eric Davis writes: "There's a lot going on in the background...The program contract did not get cancelled last month after all, because it was originally supposed to be a 5-year program, so they decided to keep it open, but the client in DC doesn't want to 'host' it anymore because they don't believe it has relevance to their war-related missions. So, the program is searching for a home, and the money for the next three years won't be available until it's found."

This is reminiscent of SRI days, when the remote viewing program kept bouncing between agencies and often went months without paying salaries. Lady Intelligence is a harsh mistress.

Hummingbird. Wednesday 2 February 2011.

Hal tells me, evasively, "the problem lies...with the government contracting process." The team wants me involved again in Capella but there's nothing going on, and Bob may shut everything down.

There must have been a huge misunderstanding, as in the days when the classified SRI project was compromised by a higher-up (member of the National Research Council, no less) who didn't believe in parapsychology. He just traded us against some political secrets.

The American economy continues to recover. NeoPhotonics just went public this morning, a development that comes at the right time to provide credibility for other projects.

Hummingbird. Thursday 3 February 2011.

Our small group (Colm, Hal and Eric, and Jim Lacatski, now joined by 'Jonathan Axelrod,' whom I've never met) has become intrigued about Dugway Proving Grounds, where unknown lights interpreted as 'red flares' were recently seen, and someone has claimed there was a group of aliens at Dugway. Was it delusion? Or just baiting?

Hummingbird. Saturday 5 February 2011.

'Jonathan Axelrod' (22) is one of the military officers who've spent time at the Ranch, along with several officers sent over recently. A colleague and I discussed it this afternoon as we walked along Polk Street in search of a coffee shop where we could talk quietly.

It turns out the research has led to some information about the actual hardware, but (as we elaborated in an unclassified 'core story') the contractor has made no significant progress. Results of their analyses seem to vary with every test they do. We even speculated the material might reformulate itself, so I recalled my conversation with Dr. Craig Venter at the DARPA workshop, as we shared drinks on the side, that we should send DNA to other planets, not obsolete rockets.

Later the same day.

The funding may stop but events do continue at the Ranch. I'd heard about three of the DIA personnel, all battle-scarred officers, observing a light on top of the ridge, each seeing a different geometric pattern, although they carried identical night vision devices.

What I didn't know was that they were stopped by an invisible barrier that felt like a hard gelatinous membrane, at the same time experiencing acute terror, like the description I was given by one of the surgeons in the Soissons event, or the case of Juan Perez in Argentina. All three cases displayed the same effects.

In total, six of the security people onsite approached Bob, asking to be reassigned because they were more scared than they'd been in Iraq or Afghanistan. Bob suggested it would be a good idea to do MRIs, and several guards who hadn't reported any phenomena were later added in double-blind fashion. Results were considered consistent with a new syndrome. At this moment there are sixteen cases. As a precaution, three senators, including Reid, who'd asked to visit the ranch were turned down *because their safety couldn't be guaranteed.*

Many big and small secrets linger in the background. Today, over dinner at a private table, a friend asked my opinion of a 'new physics' company created some years ago with Joe Firmage as CEO. Hal and others were listed as directors. I was never consulted, but I'd known of its plans. Egos got in the way.

Egos and greed, all around. (23)

I emerge from this with new questions about the phenomena. It may be time for a visit to Texas where spring is close and where, as the song goes, "the sage will bloom, smelling like perfume…"

Hummingbird. Wednesday 9 February 2011.

Peter Sturrock met me for lunch today (at Zibibbo on Kipling). At ninety, he's as busy as ever, with a proposal to NSF for a study of the correlation between the flux of solar neutrinos and the alteration of the rate of radioactive decay, in a study with physicists in Arizona. He's also working on the true identity of the man who wrote Shakespeare's plays, which leads him into interesting areas of cryptography. He makes a strong case that the actual author was a nobleman named Edward de Vere, Earl of Oxford.

Hummingbird. Thursday 10 February 2011.

Richard Niemtzow called out of the blue yesterday. He asked if I was still in touch with Jean-Jacques Velasco. Actually, I haven't heard from Jean-Jacques since he retired.

Richard also asked if GEIPAN was really a high-power group. I told him truthfully that their only serious project was to help document new observations at Hessdalen in Norway.

In Cairo today, a huge crowd exults: "Egypt is free!" Mubarak, their dictator for the last 30 years, has relinquished power; the army is in charge of transition, and the Arab world dreams of revolution.

Flamine has just joined me in meetings with David Pescovitz (27) and with Paul Gomory at *Ovation*. Paul had seats at a concert where we heard Hummel's trumpet concerto with Gabriele Cassone. It was a sweet evening, a cool cloudless night.

Hummingbird. Saturday 19 February 2011.

Every day now, another country erupts into bloody riots: after Tunisia and Egypt, Algeria, Bahrain and Libya have massive demonstrations. The movement catches diplomats by surprise, with no time to revise alliances. The French elite is exposed in cozy relationship with tyrants.

The rains and the wind have returned to San Francisco. Flamine has joined me to meet with Rob Swigart at IFTF (24) for a seminar about the DARPA project and the Riyadh conference.

Hummingbird. Sunday 20 February 2011.

Lunch with Colm at Mochica's on Harrison. The city was calm in anticipation of Presidents' Day, and the sun was bright. As we enjoyed lunch, Colm and I went over what remained of the project, and prospects to restart. We're convinced there are hardware 'widgets' around, but it would take months to have a new organization up and running with the right (SAP) clearances. As for Capella, the staff that transcribed and translated for the data warehouse is gone as well.

When I asked about biological material, which exists 'in a different place,' it could only be available 'at a later stage.' Yet, current events are all about strange creatures, which curiously started appearing when NIDS' presence wound down in 2003. The important phenomena began when Terry Douglas bought fifty head of prize cattle (genetically pure, show-level), as if the phenomenon was attracted to rural settings with lots of special biological material.

I said that research on 'widgets' could only give limited results, unless the other levels in our model were also actively pursued. To understand a piece of machinery you need to analyze its composition, but you also need to know about the system that produces it, the conditions under which it operates, and the civilization that needs it.

Colm asked if I thought the control system was fallible, if I had hard examples of it. The answer is yes, based on the deep data I have. The time isn't right, however, as with the biologicals. It's still early.

Hummingbird. Monday 21 February 2011.

A Dassault secretary called me this morning: we have an appointment on March 16 to present our Runway venture project to Laurent Dassault and Olivier Costa de Beauregard, head of investments for the firm. The latter is a nephew of the celebrated relativist for whom I once organized a physics lecture at Northwestern in the mid-sixties...Maurice Gunderson will join Graham and me at Dassault to present the Runway venture fund.

Oke Shannon's notes from the May 1985 meeting of the Advanced Theoretical Physics conference, organized at the BDM facility in McLean by John Alexander, reveal a depressing set of preconceived theories playing against an array of poorly researched data points. My name appears in the notes, alongside Sturrock's; we're pigeon-holed under 'databases!' Was Shannon influenced by the Collin's Elite group? There's also a tantalizing reference to a major engineering project under Admiral Bobby Inman, unfinished.

Now to pack for Paris. A year ago today, I stood at Yvetot-Bocage to place Janine's ashes into her family's grave. We burned our messages, sending the smoke to the sky. For a year I have survived with this sorrow. Yes, I need more time. The whole world does.

Mabillon. Friday 11 March 2011.

Gabriel Mergui and I took the Thalys train to Brussels today and met with Jacques Darcy in the richly appointed buildings of the elite European Commission, where he heads up a research and innovation division with a large budget for projects that, to my American eyes, seem so long-term as to be irrelevant. The train rushed me back to Paris in time to meet Flamine at the Institut Métapsychique International for a lecture on clairvoyance by Bertrand Méheust. Tomorrow we'll meet with French parapsychologists. (Figure 6).

Mabillon. Sunday 13 March 2011.

Paris wakes up in a world shocked by the 'Arab Spring,' a new bloodbath in Libya, and a radioactive scare in Japan after the tremors that have shaken the island and the six reactors at Fukushima.

The massive nuclear crisis in Japan reminds me of all the good work we did at InfoMedia with the agencies in charge of plant safety in six countries. In those days the Japanese were assuring us "their ancient culture would never tolerate a nuclear accident…" Laugh or cry?

I went to bed late, tired after a full day of lectures on remote viewing organized by Alexis Champion and the team of IRIS. Russell Targ was there along with Paul Smith, Dominique Surel (a woman researcher from Colorado), and Stephan Schwartz. A French remote viewer, Alexis Tournier, spoke eloquently about his own technique.

Mabillon. Wednesday 16 March 2011.

This morning, I led Graham and Maurice Gunderson to the Dassault headquarters on the Champs-Elysées and we met for an hour with Olivier Costa de Beauregard and Josée Suilzer, a financial manager. We explained the plan for Runway Capital as a new strategic investment vehicle in avionics and aerospace. Afterwards, at the Siparex office, Graham and I held our final meeting with the investors in our current fund. We explained the creation of a Trust to manage our three remaining (and profitable) investments.

Logis de Moullins. Sunday 20 March 2011.

Six centuries ago, this ravishingly sculpted manor was the residence of Archbishop Michel Bureau, Abbé de Couture, and his entourage. It is a joy to be here with Flamine and Max, learning the message of the old stones. Maxim runs everywhere, delighted at the sight of the geese and the sheep. Philippe Favre and Kathryn have made us comfortable in the beauty of their medieval lair, an opportunity to rest before flying back to the States.

Hummingbird. Saturday 26 March 2011.

Steady rains fall over San Francisco, in gray blurs that wash out the landscape. I stayed home, slept a bit, put some order into the invoices that piled up during my trip, and watched the Japanese tragedy.

The recent death of Jean Deléage, who brought me into venture capital in 1981, leaves me surprised and sad.

Hummingbird. Sunday 27 March 2011.

Steve Millard sends me a sad message announcing yet another death, that of Paul Baran, my old mentor of Arpanet days. It happened yesterday, after his long fight with lung cancer.

Now I watch the stupefying tsunami in Northeast Japan, lifting cars and trucks, then houses, and hurtling them through towns, sweeping everything away in one big pool of mud, buildings, dead people.

The world reels from the scary prospects of radioactivity releases.

Hummingbird. Sunday 3 April 2011.

Spring, with blooming flowers and hundreds of trees, spreads perfume (and pollen, alas!) over the park. I keep pushing my lazy legs: yesterday, a ride on a rented bike, forcing my body out of its complacency, my sadness of recent years no longer an excuse.

The hiatus in Vegas will linger until some wheels start turning again in Washington.

My friends are certain they've located the right place and they're eager to start 'the next phase,' but the people in question show no intention to cooperate until full security is in place.

While all this goes on in Washington, drama builds up on the Utah ranch where even the hardened guards have resigned, telling Bob that the terror of wild 'goblins' day and night was too much. It seems the old homestead, already prominent in the days of NIDS, is their favorite haunting place. Magonia has landed.

Hummingbird. Saturday 8 April 2011.

Last night I had dinner at Sens with Diane Hegarty, whom I had not seen in many months. She was as vivacious and smart as ever, one of the most trusted friends we ever had.

She told me of her work with the San Francisco Handwriting society, which she now chairs, and of the abilities she discovers in herself, of keen observation of people and documents, and even works of art she's asked to analyze.

On the drive back she recalled a stunning remark Janine had made to her. This was only a few weeks before she died, but she'd enjoyed the food and the conversation. She confided, very uncharacteristically: "You know, *I'm wise…*"

Janine's intelligence shined brighter even as her illness got worse, and now, this remark when she came into an understanding of ever greater scope, as a more subtle form of consciousness emerged.

Hummingbird. Saturday 15 April 2011.

I am alone and often frustrated but my work moves on several fronts, including a legal trust to close our third high-tech investment fund.

Dan Tolkowsky writes from Tel Aviv, after meeting with a former pilot under his command, Ya'acov Avrahani. His long letter states:

> Ya'acov was born in 1929. In 1954, he was serving in the Israeli Air Force. The incident took place in April or May 1954...oddly enough, I lived in my parents' flat on Shadal Street at the time. Ya was walking in this street towards Rothschild Boulevard and turned right on the corner of Shadal and Rothschild. Approaching the corner of Rothschild and Bezalel Yaffe, he was startled to see a large white sphere above the horizon ahead of him, shining and then dropping below the horizon, while shedding luminous white 'sparks' on the way, then reappearing and approaching him.

The big sphere became a small disk, about 90 cm in diameter, moving in the opposite direction as it neared the corner of Bezalel Yaffe:

> Ya'acov saw a small boy emerging from a residential building on the corner in question, obviously seeing the disc, scared, stopping in his tracks. YA, too, was frightened. He saw the 'light' zigzagging towards him. Suddenly everything stopped—Ya'acov in the boulevard approaching Bezalel Yaffe corner, the light having stopped at the corner, the boy moving backwards towards the entrance and staircase of the building, frightened, whereupon the light changed direction, apparently pursuing the child.

The culmination of the event happened fast:

> To Ya'acov it looked as though the light had hit the boy. However, the boy disappeared in the entrance of the building, and the light 'exploded' on the façade of the building, leaving what looked like a cloud of soot on the wall. Ya'acov stood speechless, near the entrance of the building, shaken, passers-by staring at him.

Avrahani returned the next day to investigate; he asked about the boy and was told his name was Uri Geller and he lived nearby. He made contact with him, and they met in Israel about 2008. When YA spoke to Dan a couple of months ago, the latter contacted Geller in London and heard the story with minor differences, notably stating that the light was 4 meters in diameter. The first thing he noticed was that all the neighborhood cats started meowing. A sphere appeared and "shot a beam" at his forehead, forcing him to lie down, following which he ran home. Soon after, he was eating his soup when his spoon started bending…Uri may be a showman, but nobody in Israel would tell a lie to General Tolkowsky.

Austin. Monday 18 April 2011.

Hal picked me up at the airport and we went to a late dinner of tacos at a TexMex restaurant where he told me about the events affecting 'Axelrod,' whose family is now plagued by a wolf-like entity in his backyard, and a series of blue and red lights harassing his children. So perhaps we were lucky that our Spring Hill experiments failed to produce such emerging phenomena in the nineties, only displaying the two bright, unexplained apparitions we recorded.

My own research continues, leading again to Anubis, the jackal-headed god associated with mummification and the afterlife in Egyptian mythology. In connection with his funerary role, he's known as *He who is upon the mountain* (protecting the diseased and their tombs). During our recent trip, I brought the cemetery connection to Frank Salisbury's attention. The wolfman in Fort Duchesne was sitting on the pile of rocks surveying the cemetery when the two women noticed it.

Next, Hal and I traded hypotheses about the hardware and the biological materials. I'd brought over pictures of the landing gear imprint from Soissons, so I gave him the dimensions of the device.

On the way to the hotel, Hal said, "You know, I've been looking for this for 50 years and suddenly, it's happened! I haven't lost the passion for the research, but the emotion of the chase has drained out. My interest is of a different nature now. We're going to enter another world, an incredible world I never suspected, with its own set of rules. Be careful what you wish for…" I can only agree. I feel the same way.

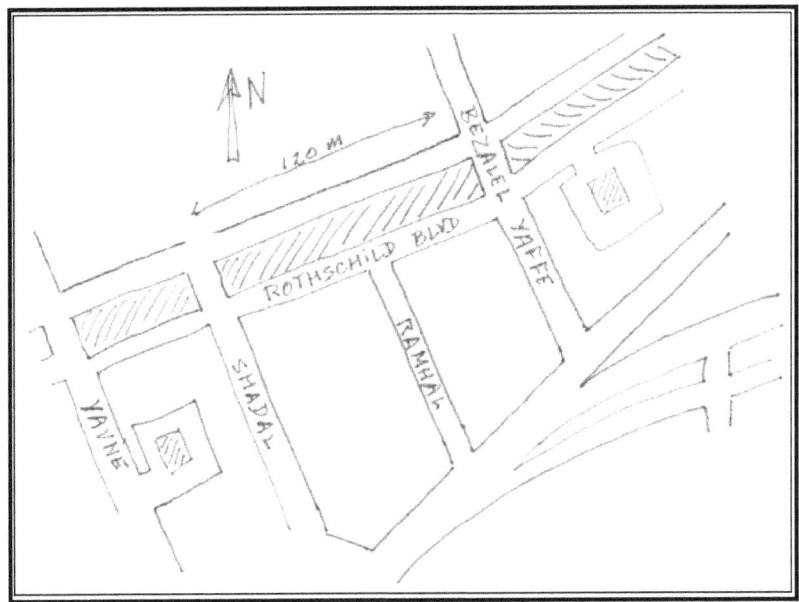

Chart 1: *The Tel Aviv event with young Uri Geller in the spring of 1954. Map of the area where Ya'acov Avrahahi saw the light.*

On the plane back to San Francisco. Tuesday 19 April 2011.

Summer will soon arrive in Texas. It is 95 degrees and the humidity is already high. At EarthTech, the parking lot is blocked by bright orange cones to keep the mass of the cars from affecting the measurements the team is taking, designed to determine the existence of dark matter, or lack of it. If recent theories about MOND (Modified Newtonian Dynamics) are correct, astrophysics doesn't need dark matter at all; there are new explanations for the speed of the stars' escape from the outer arms of galaxies. This led us to discuss the new 'information quantum' theories of Jan Walleczek.

We stopped for coffee, then resumed our discussion. Hal ran through the slides of the briefing he'd given about BAASS to his new contacts at AAWSAP (or "AATIP" for Advanced Aerial Threat Information Project). It even mentioned the Pavlov report that I knew from Russia.

The attention of our potential colleagues focused hard on two potential threats: first, the phenomenon itself; and second, old adversaries who may be more aware of it than we are. A new home agency has become known, where we might restart the research, but the bureaucracy has "a few more iterations to complete..."

We wrapped up our discussions, coming back to the fact that a world of secrecy existed, of which we had no idea even during the psychic project. It could confirm what Larry says about Livermore...Run this way, even a complex project can only be discovered by a rare accident. But is the scheme effective in science?

Hummingbird. Wednesday 20 April 2011.

An English correspondent reports on a series of supposed Alien contacts at Windsor, in a parkland area owned by the British Royal family. It involved a cab driver who helped an older lady with errands around town, and a blonde woman in her company. She's supposed to have come from a spaceship, and there's even a metallic sample associated with this, soon to be shipped to the US for analysis.

I remember Gordon Creighton telling me that the Royal family had a history of involvement with UFOs. Lord Mountbatten, a senior UK admiral and Allied WW2 Commander, pushed hard to get various secret UFO intelligence projects set up. There are rumors that he once met an alien, but such stories often serve as blind alleys.

Hummingbird. Thursday 21 April 2011.

Lunch (at the Elks' Club in Los Altos) with the happy band of the Gravity Company. The manager has aged quite a bit, and so does our lead physicist, but they remain optimistic. The math suggests a craft could cover one light-year in 174 seconds. Most physicists, including relativists, state that this is impossible. Einstein looked for that solution all his life and couldn't find it, so he had to go to a tensor formulation that never satisfied him. "In fact, lots of physicists were looking, but they never reconsidered the structure itself."

Later the next day.

In a more somber mood, Hal told me he was starting to notice the death of friends younger than him; I have the same experience, alas.

I've started to pursue a 'theory of everything–else' involving coincidences, entanglements (of people, events, and things) and the meta-levels of reality.

I told Hal that, if the hardware is a product of a metasystem, any classic analysis may be meaningless: "What would be the signatures? If you were a superconducting gas cloud the size of a solar system, and you wanted to look like a minivan, who could stop you? The next day you might want to look like a Jeep, or a Mack truck. All of which would be meaningless in terms of your actual physical nature."

Hal is a classical physicist who doesn't need metasystems, while an information scientist considers them intuitively.

I speculate that somebody or something is busy recompiling our reality. This idea had come to me in *SubSpace,* and even more in *Dark Satellite* in 1961, where spoons got bent and blue spheres expanded to engulf witnesses who found themselves inside strange craft designed to alter the topology of the universe…The inside was bigger than the outside…But nobody reads those books anymore; or notices those coincidences, born in ancient layers of time, the Sixties...

Hummingbird. Saturday 23 April 2011.

Dan Tolkowsky has conveyed my questions to Ya'acov, but the man did not experience any unusual effects after the incident, and it didn't change his life.

The light was "flexible," with an apparent diameter of 90 cm, but it expanded and contracted, uniform, with no sound.

Dan adds: "The smear covered part of the building. When he returned the following day there was no sign of it. Could it have been an optical illusion, due to the glare of the disc?"

This is the Easter weekend. The Middle East is in acute turmoil, with a new bloodbath in Syria where the goons of the government shoot at funeral crowds with real bullets.

There is trouble in Yemen and a full-scale civil war in Libya, where the US has pulled its ships and its planes but now re-enters the action with drones and missiles.

Hummingbird. Friday 29 April 2011.

This lingering distress builds up. I must not be in such bad shape, however, because I scrambled all the way up Russian Hill to see George Kuchar's latest class movie.

I've cleaned up the archives at my Palo Alto office and thrown away some 200 old UFO videotapes, boxes of letters, and old magazines.

Next week I leave for Normandy on a writing and spiritual retreat. I'll bring fresh flowers to Yvetot and take long walks in the countryside. To Annick I'll give her sister's jewelry, and my encouragements.

5

Mabillon. Saturday 7 May 2011.

Summer comes early this year. Crowds mill around the shops of Paris. In the Latin Quarter, cafés overflow with people in slacks and open collars, or T-shirts and naked thighs. I treated myself to rhubarb pie and coffee at the Danton in the middle of the noise and bustle. The pollution stings the eyes and scratches the throat.

Catching up with the work begun in California, an old report brings me back to a remark Aimé Michel once made after a paranormal conference: he'd found the speakers "all terribly marginal, repeating the same tired things…" (7 December 1977). He could write the same lines today. Ufologists are still pondering Bentwaters and Roswell, having gained no fresh perspective on the historical and multicultural nature of the phenomenon, or the basic work it demands.

This afternoon I dutifully filed the foreign travel form demanded by our security officer in Vegas. This raised the question of what, if anything, I'd learned as a contributor to Bigelow's ambitious quest. Equipped with high-level clearance and privy to conversations behind the scenes, we've obtained much new data about significant phenomena. The Ranch experiments are an ambitious trial, but the major value was in the improvement and extension of observations, and in the compilation of basic datasets, interrupted too early.

To go further in a scientific way, we would have to recompile the patterns we've extracted from the data. That would take a long-term effort and a grasp of datamining since the available record has now blossomed into complex files. We would also need to continue the improvement of data capture across the spectrum.

John Alexander thinks he's proven there was no hidden UFO study. Hal, Eric, and I strongly believe there is one. Is that all? A single study, protecting captured hardware? Assuming we get cleared to step inside a hangar full of crushed alien stuff, will that advance our knowledge, in the absence of all the other correlations the problem demands?

Again, the wrong questions, at the wrong level. Even that hangar may not hold anything more than tarnished slides of alien tissue and another set of problems with no answer. Hal's work on physics explanations is brilliant (although untestable, so far), while studies of abductees' brains show insults, truly unexplained.

Everything works as if the phenomenon projected its images directly into the observer's skull, which could explain why Douglas Trumbull and I found no physical object on Pauline's authentic photographs from Créteil, but that only augments the complexity of the future problem. What else does it do in the brain?

Mabillon. Sunday 8 May 2011.

Meetings are planned for the end phase of our Euro-America funds. The mottled sky filters a heavy sun but the crowds remain lively.

Flamine comes over in an elegant gray dress and red shawl. We discuss her notes from the latest meeting at the IMI institute. Then we have a simple dinner in the kitchen, aware that we could simply forget everything, stay here and be happy among the attractions of Paris. But I will call Bill Calvert tonight and plan our return to Brazil.

Douvres-la-Délivrande. Wednesday 11 May 2011.

The Norman countryside, warm and sleepy like me, looks forward to some quiet days before the bustle of summer. Annick has created a little paradise here: the bedroom is pink and red with a tiny bookshelf, a small TV, and a large flat bedside table where I keep my laptop.

California seems far away. We speak of our families, the memories of laughter with Alain. Then, the days of sorrow.

The news remains filled with terror plots, blasts in the Middle East, dictators mauling down protesters with tanks, and idiots 'heroically' capturing innocent hostages in faraway deserts.

Douvres-la-Délivrande. Sunday 15 May 2011.

Waking up at dawn, fighting my laziness, I managed to complete a numbing task. The publisher Editions Aldane, in Geneva, have sent me a complicated draft contract for a French translation. Then, as a bit of rain comes to Normandy, nice message on my phone from David Pescovitz. (24)

Suddenly, at 8 am, Annick burst out of her room with stunning news: Dominique Strauss-Kahn arrested by New York police! The charge: attempted rape of a woman in a downtown hotel. All comments spin, a whirlwind of speculation, the press picked up by a tornado, its trusted spokesmen unsure of anything they say.

Annick, solid Norman woman who detests the Socialists, can hardly contain her jubilation. As if scandals were a specialty of the Left…

Douvres-la-Délivrande. Monday 16 May 2011.

The world tries to adjust to the upheaval of DSK's sudden erasure from the world scene. He's stunned to find himself in a Manhattan jail like a murderer. While New Yorkers appreciate the punishment of privilege, Parisian pundits laugh at America's absurd sex obsessions, and French candidates in next year's election rejoice.

I don't stay close to French politics, so I'm not shaken by the news. I treasure this simple village and the well-tended garden where I throw a ball to a dog. I will keep writing. There's a change of leadership at CNES: Yvan Blanc, who was supposed to re-invent GEIPAN, has asked an engineer named Xavier Passot to replace him.

Douvres-la-Délivrande. Thursday 19 May 2011.

When the sun shines so early and joyful over this modest paradise of flowers and bright bushes, Fedora likes to sit on the front steps with me to catch the morning warmth. This quiet period is good for both of us. For the first time in a long while I slept through the night without pain or worry. Future concerns can be handled at their own pace, but I need to call Dr. Salisbury, and plan our next investigation trip through the Uintah Basin in July.

Douvres-la-Délivrande. Tuesday 24 May 2011.

As I edit the third volume of *Forbidden Science*, I find it maddening to see so little progress in paranormal research over those ten years, 1980 to 1989, especially when balanced against the vibrant technical and social breakthroughs, the rise of entrepreneurs, and the building of revolutionary technical architectures across the world.

Tonight, we had dinner in Bayeux for Annick's 76th birthday, with two of her friends. The city boasts a splendid cathedral and tons of charm, little shops along the river with its medieval water wheels still turning, antique machinery glistening in the evening light.

Mabillon. Sunday 29 May 2011.

On the train back from Normandy, as I sat next to a pleasant-looking retired French couple, my cellphone rang and I spoke for a few minutes to Bill Calvert, calling from Illinois. Later, as we disembarked, I overheard the man telling his wife, "The *strong tall fellow* next to me, he was American, you know." I never thought of myself as a "strong fellow," but perhaps the years have been kind.

Over lunch with Olivier and my grandson, carrying his green and blue parrot on his shoulder, we admired the beauty of this sky, the trembling trees, the pulse of Paris.

Going over my Spring Hill experiments and the conversations with researchers twenty years ago, it is obvious that we'd already come close to the reality of the phenomenon and the ugly political games played around it. Then, Flamine joined me for a frugal dinner on the balcony, watching the sun set on rue de Buci below.

Mabillon. Wednesday 1 June 2011.

The market on rue Mouffetard, at the edge of the Latin Quarter, straddles centuries with its narrow course among shops of abundance, overflowing produce on both sides of the street, along with newer displays selling bright toys and crafty doodads, designer objects and jewelry for refined tourists. It's the side alleys that interest me: the shadows and cool stone walls, dark and subtle, leading to courtyards no bigger than someone's living room; greenery; a table, four chairs. It's easy to imagine the friendly evening conversations.

I remember my mother shopping here, well into her nineties, arguing that the salmon slices were not thin enough (they must be presented between sheets of special paper, so they never stick together, of course!) and complaining at the butcher's next door that the *filet* had too much fat, wasn't *mignon* enough to be served.

At the bottom of Mouffetard stands the ancient church of Saint Médard, squatting in a hollow spot. The service is over. I step into the shade of the side alley behind a group of a hundred or so, where I shake hands with Gabriel Mergui, Pierre Tambourin, and embrace my other colleagues from Genopole. We are mourning Valérie, 46 years old, the manager of the business division. She died of sequelae to a glioma operation: an oedema. Accuray might have cured her, but our instrument isn't yet deployed in France. Gabriel and I start to discuss plans for their new biotech fund; we walk up the street again.

Over at ESA, scientists are reorganizing, one more time, and have no idea what to do with the space station. Our own project, which would answer the question, is 'approved'—but unfunded. Thanks to Kepler's Law, the ISS and its inhabitants keep spinning around.

Tomorrow morning, I leave Paris after nearly a month. The air glides easily into the lungs, scented with the messages of spring. There's freedom, bright ideas, tantalizing food, books and books, all those interesting signs. It's a trap, of course, but it does force me to ask harder questions about our American projects.

Hummingbird. Sunday 5 June 2011.

Back in California, alone at the low point of Sunday afternoons. If it weren't for Flamine holding me to this reality, I might have left the US by now. My children, as tenderly as I love them, no longer need me. My work in research is done, used and recognized. The world waits for resolution of its crises, from rebels in the Middle East to the budget impasse in Congress. Is this only a breather as world industry recovers? Big firms are flushed with billions, but unemployment remains high, and families try to hold on to homes they can no longer afford.

Hummingbird. Tuesday 14 June 2011.

The weather was so fine this morning that I walked all the way to the Zuni restaurant where my favorite physicists had invited me for a very private lunch. They wanted to tell me about their progress. Their experiments in Louisiana show actual fusion; they have a complete framework for a propulsion device under both their Type I and Type II gravitation, which may even explain UFO behavior in flight.

Hummingbird. Thursday 16 June 2011.

This evening, I called Hal to tell him about Whitley's latest, somber manuscript and my own review of Brenda Denzler's abduction survey. I made it clear that my real interests did remain with the active data, including my planned return to the Uintah Basin in a few weeks. Let everyone assume I'm working on old legends and obscure ancient sightings!

Hal is still convinced that deeper levels of the information will 'soon' become accessible, giving me some hope.

Hummingbird. Sunday 19 June 2011.

In a message from Kit, responding to my complaint that we never disclosed our criteria, or even agreed on a conceptual framework, he wrote: "We're behaving as if the fundamental construct is that the truth of the phenomena is based upon one or more veridical technologies. I personally have always been 'mostly' persuaded that is so, and I'm very happy with the way things are 'mostly' going. The part that really disturbs me…is that as a group we have eschewed active discussion about the most dramatic observations on the Ranch…and deferred to others (Bob, mostly, as the owner of access) the leadership to bring them up to us."

Kit goes on to note that the observations are "most often sensory, mostly visual, seldom singly auditory, occasionally olfactory, and increasingly generating fear. MRIs show significant inflammation and sometimes cell hyperphasia. *Ergo, they are not endogenously psychiatric.*" His conclusion: "I propose that we gather. We pay our own way. We prepare what you've given us the leadership to understand we need and define, in writing, a conceptual framework."

Hummingbird. Wednesday 22 June 2011.

Whitley's manuscript for *Solving the Communion Enigma: What Is To Come* mentions the scary prospect of human mutilations linked to UFOs. There's a case in Pennsylvania where Whitley hints that the coroner (a woman) was removed from the investigation. The event took place on 2 August 2002 and involves a 38-year-old hunter found dead in a wetland a short distance from his home near the Susquehanna River. A MUFON investigator, attorney Wayne Gracey, spoke to a woman who'd seen a silver disk in the area at the approximate time. There's another case, the suicide of a Marine Corps Intelligence officer, in what some people thought was a staged event. She'd been profoundly involved in an official project that left her pathologically depressed; all evidence of her death had been expunged, but rumors surrounding the case linger.

Checking into the circumstances of the medical examiner who left that position suddenly, in 2008, an hour-long autopsy of an alleged alien seems confirmed, but nobody would comment on the extraterrestrial issue, or who signed a paper entitled 'Gate 8' about an incident last year: mysteries on top of rumors, useless in the end.

Hummingbird. Saturday 25 June 2011.

I was walking in Golden Gate Park when Dick Haines called. He'd just left a hush-hush Washington meeting set up by Leslie Kean and introduced by John Podesta, ex-Clinton White House chief.

The audience included people from AFOSI, Congressional staffers, NASA folks... Among the speakers: Haines, Nick Pope, Colonel Walter deBrouwer, Dick Henry. Professor Sillard had cancelled and was replaced; others are held anonymous. Neither Greer nor John Alexander was there.

Hal knew about the meeting, aimed at creating a low-key government clearinghouse to handle data. He didn't go, but John Peterson did. Airspace threats were one of the topics. "It was a first-order waste of time," John Alexander said, "someone was just re-inventing Blue Book." But I wonder if it wasn't a trial balloon for a mechanism of wider information release, or a step to disclosure.

If it was, the opportunity was missed.

Hummingbird. Monday 27 June 2011.

At an Open Science session at the Institute for the Future yesterday, there was an unfortunate incident involving Dr. Seth Shostak, co-founder and senior astronomer at the SETI Institute. As we spoke about radioastronomy, I said the idea of a big radio telescope at Stanford had been initiated by the Pentagon for its own classified use, apart from academic work.

Dr. Shostak protested, staring at me: "This is yet another yarn that *some people* keep inventing, just to look clever!"

That was awkward: my mind spun around. Exposed before my stunned colleagues in the room, I began doubting my own sources but David Pescovitz, whose fingers had been racing over his keyboard, found recent entries on Google confirming all the facts: In the late 1950s, desperate to map out the Soviet anti-aircraft defenses in the days of the 'ferret flights,' the Pentagon got the splendid idea to catch the Soviet radar signatures reflected off the moon when it passed over the Iron Curtain, hence the secret project.

They only needed the dish for half an hour at a time, but they never officially told the professors what they were doing, or why they were so willing to pay for their toys.

Hummingbird. Thursday 30 June 2011.

Horrible news: Colm tells me that Bob Bigelow's twenty-year-old grandson Rod has died. I cannot put words around this kind of news. Bob raised him after the death of his older son in the early 90s. This is just too sad. Rod's sister has a keen interest in the business and may be the one to run it, while Bob's younger son never showed an inclination to do so. There's a pattern here: Trammell Crow's business is now run by his daughter, his other children losing interest.

New projects are taking priority here, especially the second exploration trip to Utah with Frank Salisbury.

Hummingbird. Thursday 14 July 2011.

A long conversation with Dr. Green, at his request. He was relaxing in Michigan and was happy, having just announced that he would be a father again in January. He wanted to alert me to the fact that I might be contacted very soon for another phase of research:

"Your ears may have been burning, because the things we discussed two months ago have been maturing at a rapid pace."

He intimated that a deal existed to upgrade our small group to the next level: "The reason we haven't been able to get together is that the time wasn't right," he went on. "Jim Lacatski's files and computer have been sequestered. All redundant material has been burned. But the new process is far along, and you may be called for a briefing. We don't all know the same things, and we don't play the same roles."

We went back to the situation in Utah. He's found eight people who'd been sick at the ranch, the disease extending into their families. It's now recognized that the phenomenon can be harmful, as I and others said for years, but I doubted the human origin Kit now suspects. He left me with a note of caution: "The people who are helping us now know where the Holy Grail is, *but they don't have the Holy Grail...*"

We also spoke of 'interference': Communication one kilometer away by Radio Frequency into the auditory cortex of humans was operationalized at Los Alamos as early as 1978. In another study, RF has been used for instilling fear and hallucinations at the Air Force labs since 1975. Finally, specific targeting to the hippocampus and amygdala for fear experiments existed at least 15 years ago.

Are such tools involved at the Sherman ranch?

Hummingbird. Monday 18 July 2011.

At a fine outdoor dinner in Palo Alto, the Institute for the Future (25), its staff and Board, honored Kathi Vian and me as Distinguished Fellows. Our friend Diana Hall had agreed to be my date, so that made for a great evening. Bob Johansen presented me with a huge, circular 'witch board,' a splendid version of the Ouija board.

Hummingbird. Wednesday 20 July 2011.

Taking the morning off, I found Golden Gate Park clear and serene today, with a long line of tourists and visiting families snaking around the lawns. I rented a city bike and did my modest 10K, then I spent another hour rowing around the lake. I have done it many times, but I cannot shake the beauty of that place, always enjoying the majesty of the ancient trees reflected in water, poignant as ever.

In the afternoon, as I was working at home, Bob Bigelow called and we spoke for an hour of his projects, his view of NASA as the last Shuttle returns to Earth, and the phenomena at the Ranch, where he confirmed that five security personnel have witnessed paranormal events: poltergeist and levitation.

He's convinced the orbs respond to mental processes, so I was wrong about my earlier assessment.

About NASA, he's still upset: "They gloss over the cost of the Shuttle ($1.5 to $2 billion per flight), and they ignore the danger that China may claim the moon in a few years."

Hummingbird. Saturday 23 July 2011.

A man I don't remember meeting sends me a note about a UFO landing that took place three years ago near Annecy, in daylight, close to hiking trails. The witnesses are said to be on-site officials managing road repairs. They saw two occupants next to a disk-shaped object. Many pictures, physical traces on the ground. A plane was sent from Paris to pick up the material. The analysis reportedly showed evidence of scandium. The CNES was not informed, and the witnesses were reassigned to other locations. It seems the data remains classified. But why scandium?

Over the net, we've been arguing about the significance of ancient sightings. Challenged (fairly) to be more formal, I answered: "I think the phenomenon relies on a technology that transcends time: our ideas about it are very wrong. I hypothesize there's a constant principle in the universe that can operate across various epochs.

"After all, think of the wheel: the principle of that shape hasn't changed in 10,000 years. If I had to put my money on a particular hypothesis, I'd say we face a phenomenon of non-human origin, with *sui generis* characteristics unrelated to our primitive technologies. However, there's a human element that's either aware of it, or part of it, and that bothers me no end."

To which Eric answered: "This is exactly what I think, too. And I can attest to the fact that our ideas of time are in fact very wrong because our psychological, perceptual, or mental sense of time has totally muddied up the way physicists try to understand and model it. There's a crazy dichotomy between the nature of time in quantum field theory vs. the nature of time in general relativity theory vs. the nature of time in thermodynamic theory. I was able to address this…in the section on Time Machines in Chapter 15 of my AIAA book. (26)

"The published works of physicists Paul J. Nahin and Paul Davies and Richard J. Gott, and a few others, also inform my thinking. Recent particle accelerator experiments and metamaterial experiments have shown spectacular results previously thought to be mathematical fictions routinely tossed away by theoretical physicists solving wave equations: the laboratory demonstration of negative frequency and time reversed wave motion."

Eric concludes: "I can sum it up: *time does not flow; there is no, and cannot be, a global chronology.* There is no unique cosmological arrow of time because it is defined locally by the different interacting particles and gravitational fields." A brilliant, fundamental statement.

Hummingbird. Wednesday 27 July 2011.

The world economy is suspended by the outcome of a political battle that displays Washington at its most corrupt, callous and incompetent. Technically, it has to do with the raising of the debt limit, but underneath boils an absurd conflict of ideologies.

Economic reality rests on the plight of tens of millions of average Americans who've lost jobs and homes. Morgane, who had to sell her house at a loss and now rents a small place outside Santa Barbara, tells me about an elderly woman who lost her real estate and then found out that Bernie Madoff had squandered her retirement money.

Many pension funds are devastated. The banks have lobbied the government into supporting them with billions of new money that now sits idle in their coffers. Similarly, major industrial concerns have large cash reserves, some of them bigger than the treasuries of an average country, yet that money sleeps overseas, with tax-free interest. In the meantime, there's no investment and few new jobs.

Later the same day.

An interesting article in *New Scientist* (27) exposes the devious behavior of many academics, including some great names.

— When Einstein gave a lecture at the Carnegie Institute of Technology in Pittsburgh in 1935, he made his 7th attempt to prove that $E=mc^2$ and "it contained the same flaw as all the others."

Max Planck had already pointed out Einstein's error in 1905. He'd applied the rules that governed slow-moving bodies to a situation involving fast-moving emitters of light. Other mathematicians had come up with the real proof, yet Einstein never conceded their priority.

— A worse fate befell Indian physicist Chandrasekhar in 1930 when he foolishly trusted Eddington who'd followed his work, then had made sure he'd speak after his presentation. Eddington stood up and called the paper "stellar buffoonery." Chandrasekhar was recognized with a Nobel Prize in 1983 but by then he had left England because of that humiliation and settled for a career in obscurity in the U.S.

— Galileo was wrong when he argued before Pope Urban VIII that the tides were a sloshing effect produced by a combination of the Earth's rotation and its movement around the sun. Three decades before, Kepler had correctly shown the tides were linked to the moon.

Galileo's calculations would have led to a single tide per day when everyone knows there are two, yet he went ahead and published his false proof in his *Dialogue Concerning the Two Chief World Systems* in 1632. Now, everybody gives him the credit!

Hummingbird. Thursday 28 July 2011.

Another blow: the death of Hilary Evans, one of the true scholars in the paranormal field, a collector of folklore and images (he ran the Mary Evans Picture Library), and a very kind, helpful gentleman.

A letter from Barbara Fisher (friend of Howard Rheingold) of Athens, Ohio, observes that "while these mysterious beings or objects do seem to cloak themselves in the cultural trappings of the time and place in which they appear, I cannot help but also note that sometimes individuals in a group have very different experiences at the time."

She goes on to describe two sights of an upside-down silver bowl in broad daylight, and an erratic light in the sky that kept changing color, followed by visitations from beings of light in her bedroom. Later in life, she and her husband moved to southeastern Ohio, on old farmland: "rolling field, and a small, wooded area at the top of the hill." The land was strange: "People saw and heard strange things, including aerial phenomena (stars moving in a circular formation) and glowing humanoids in and out of doorways in the woods....

"Which leads to one of the points in this letter: by your influence on other minds, is it possible that the mystery will now shape its manifestations accordingly? At least among the readers of your book? I found it interesting that you had tried to entice manifestations of the mysterious on a property that you and your wife bought in California."

She doesn't know how close she comes to what happened there.

Hummingbird. Friday 29 July 2011.

Turmoil on Wall Street as the debt crisis demands constructive decisions of which Congress seems incapable. A reasonable, ho-hum bill proposed by the Republican majority in the House has been torn to shreds by their own right wing (the populist Tea Party) while the Democrats under Harry Reid keep posturing around half measures that would do nothing to solve the country's long-term problems.

A side note: At the Institute, someone pointed out that it was the discovery of America that destroyed the European feudal system, because intelligent serfs could suddenly restart a free life somewhere else. If so, we need a new America now, another frontier to free us.

Hummingbird. Wednesday 3 August 2011.

Yesterday I climbed the hill for a sad pilgrimage: George Kuchar is in the hospital with a generalized cancer. We will soon lose this gentle artist. His twin brother, Mike, was there, distraught, along with Jennifer Kroot who helped with documents, donating the film collections (Harvard wants them). I was recruited to witness his will. I walked back through the park where Chinese kids were playing, the City splendid as ever.

Hummingbird. Friday 5 August 2011.

Flamine and I climbed up to Grace Cathedral in a rare break of sunny weather between onslaughts of the fog. I followed her to the center of the labyrinth, took her in my arms, and asked her to marry me. I treasure her presence, her intelligence, and her grace.

But what a strange world! People's trust in institutions has been shattered from Paris to Washington: economic realities are melting along with the stock market and rating agencies just sanctioned America's negligence by slashing the AAA note of the US Treasury for the first time in history.

Hummingbird. Tuesday 9 August 2011.

Contractors came over today to review my plans for renovation. As I showed them a list of paint colors that Janine had set aside, half a dozen books suddenly fell flat on a shelf with a loud bang. The title of the top book was *Evidence of the Afterlife* by Dr. Jeffrey Long. It had been her idea to redecorate the library. I'm still shaken.

Concerned with the silence in Vegas, I launched the idea of spinning out the Capella data warehouse as an intermediate step to a full classified study. At the present time it's idle, although some 800 cases with medical significance have been turned over to BAASS.

A supporter of the BAASS project called two days ago. "There's new activity to push," he said cryptically, and "interest by others" to see it pushed. So, perhaps there's still hope to reopen the process?

Flamine and I attended a wonderful soirée *chez* Rob Swigart with Jim and Dorothy Fadiman, and James Spencer, who used to work as an editor at SRI. It was a warm evening in beautiful Menlo Park.

Hummingbird. Wednesday 24 August 2011.

Steve Jobs, in very poor health, resigned his position at Apple today, and I just heard that Budd Hopkins died on Sunday. Perhaps his passing will allow a reassessment of the abduction phenomenon.

A series of breakthroughs kept secret for the last few decades could account for much of the data reported by modern abductees: (1) Radiofrequency into the auditory cortex in humans a kilometer away was operational at Los Alamos as early as 1978; (2) It was tested for instilling fear and hallucinations at the Air Force Human resources lab in 1975; (3) Targeting the hippocampus and amygdala to induce fear was done at least 15 years ago. Stealth started flying in 1970 and laser-mirror invisibility on suits and blimps existed in 1980 so it seems possible to account for some post-1970 UFO weirdness by assuming a very highly classified, possibly rogue human source.

Hummingbird. Friday 26 August 2011.

An expert within our circle writes cryptically about the Iron Mountain Report. He claims it was an Active Measures move related to the Soviet interest in paranormal and UFOs. It included 'salting' some real and some sanitized material in the MJ-12 papers. The fiction operation was expensive but not very expensive, he said: "most fabrications start low and stay low and cheap, but they are tough to get approved." Now Whitley Strieber writes that he, too, wonders if he's been missing an important factor: secret manipulation…by humans?

Hummingbird. Saturday 27 August 2011.

In the mail yesterday, a pathetic letter from Joël Mesnard, begging me to come and visit him in Poitiers again. He's disgusted with the circus ufology has become in France and the US: "a proliferation of craziness, phantasies, along with unverifiable documents." He goes on: "That small universe has become intolerable."

In San Francisco, The Coming Home hospice on Diamond Street is a former convent, so I joked weakly about that when I visited George Kuchar this afternoon, finding his sense of humor intact. There were pictures of his many friends on the wall, and he'd just enjoyed a birthday party, but the tumor had reached the spine. I tried to entertain him with gossip as he drifted into unconsciousness. This generous fellow won't be able to keep up the battle much longer.

Hummingbird. Friday 9 September 2011.

My physics friends have moved their antigravity experiments from Louisiana to a university in Texas, following the physics professor who is giving them access to his lab. Unfortunately, the equipment they had to use was too small for their test, so it produced some arcs that interfered with the process, a few wasted weeks.

They thanked me for my critique of their paper on advanced propulsion, which they've re-done. I challenged them about the unclear microgravity effects on human biology in their vehicle.

Hummingbird. Saturday 10 September 2011.

A very sad day. I drove to the hospice on Diamond Street, hoping to see George Kuchar one more time, and learned he'd died on Tuesday. I'd hoped to show him the nice photo of him I'd taken at Spring Hill. One always hopes there will be one more day, or hour. The manager told me it happened peacefully, with Mike and a nurse by his side.

Very little filters out of the group at BAASS nowadays, probably because very little goes on. Colm works at the aerospace office and Capella is maintained, but there's little interest in field investigations, except for the follow-ups to brain and physiological effects with no feedback to me: a frozen world again.

Hummingbird. Thursday 15 September 2011.

A flurry of email exchanges continues in our group. I raised the deeper issues and got a response from one of our Advisors: "I agree that our subject is open to many interpretations. Having collected over 40 years of information on the subject, some of it open, other classified, *and some of the information coming from alien sources,* I came to the conclusion which Jacques expressed many years, that the aliens are messengers of deception. Therefore, interpretation of their behavior is difficult.

"In addition, the cover-up by governments and the mocking of those people who investigate the problem make serious research very hard. We can only understand it if we put it into a much larger context."

He went on: "My interpretation of the information available to me is the optimistic interpretation [but] I have received *information from [other] sources* which would give a much more pessimistic outcome to the whole UFO problem, from a human point of view...Jacques pointed out the amazing lack for *independent sponsorship* to support research in this subject, which is so central, not only for understanding human history but also for the future of humanity."

I commented: "The situation appears increasingly disturbing as we get deeper. We've assumed that we were prevented from meeting for reasons of security. We may be wrong. Perhaps we were getting too close to something. It is that 'something' that I'd love to discuss."

The exchange led Colm to remark that "both the essay and Jacques' email appear to be implying a rather high level of granularity to a control system. The implications are disturbing."

Later the same day.

John Schuessler has joined the discussion. The apparent control system, he said, "intrigued me since the early days. Were we being shown a glimpse of the future? Were we being trained by some force (intelligence) that we didn't understand? Why were some people being given false information that led them to become true believers, *and then were easily discredited*? How could this happen, back in the slide-rule days of our early space program?"

Now I hear that one of our colleagues "is keen to pursue the idea raised by Jacques about unexplained interference in human affairs. If we could determine which agencies (human or non-human) are responsible for these actions, it may teach us something important.

"Therefore, I suggest that the group organize a preliminary meeting, maybe before Xmas to allow Jacques *inter alios* to present, discuss, and more fully explicate the nature and severity of the 'something' in specific relation to interference aspects—perhaps we've been prevented because we were getting too close to something—and discuss our findings in regard to such a strategy."

The Board considered four reasons such a conversation would be important. "We spend too much time talking about the 'Coffee Cup' aspects, because that is easier. We spend almost no time discussing the original and seminal ideas of Jacques…which is, after all, what originally captured us in the early '70s. The constructs of *Messengers of Deception* and *Invisible College* are as relevant today, as then. We have forgotten them to our possible peril…"

I wrote back that "we need a fact-based study, not a hypothesis-driven study," adding " In that respect we never implemented what we had set out to do at NIDS, namely (a) examine the entire UFO field in the way a turnaround business team would look at a failed company—and make the necessary personnel and strategic changes, (b) make a list of "what it is that needs to be explained"—we haven't even done that for the Ranch, (c) make a list of researchable vs. non-researchable issues, and (d) eliminate old theories we now know to be wrong—and all the baggage that came with them."

There's *"a lot of harm"* in the recent unclassified data, with hospitalizations, people savaged in their homes, endocrinological damage... blood cell anomalies: tissues affected by the blue orbs.

In the meantime, my venture projects are moving forward. I got 80% support from our investor community to establish a Liquidating Trust and continue our work on the active companies.

Hummingbird. Saturday 24 September 2011.

I'm over 72 years old, the age when my father died, but I feel healthy and every day brings me closer to Flamine, busy with the sale of her Paris apartment and her resignation from two jobs, ready to join me.

My daughter invited me to dinner last night at Westlake Joe's, a popular place in attractively foggy Daly City. It was a joyful occasion, Diane and Rebecca adding to the pleasant conversation.

At the end of the week, I will attend the second DARPA workshop in Orlando, curious to see where they'll take the Starship project. Ariel Waldman, a bright young woman who currently works part-time at IFTF, is the keynote speaker. Douglas Trumbull and Eric Davis will be there; a controversy has erupted over the speed of neutrinos in a CERN experiment. I will be there to take notes and learn.

The next big point on the agenda will be our meeting in Austin, set for November 8th. The number of cases of witnesses who have been "injured, burned or killed" in UFO encounters now exceeds 900, but some of my own cases haven't been added yet.

Hummingbird. Monday 26 September 2011.

Striking leftist victory in France. The socialists have conquered the Senate for the first time in the history of the Fifth Republic. They inherit a troubled system, even if France is not quite as economically fragile as Greece or other southern States, but it's only a question of degree now, and of how fast crises could escalate.

Orlando. Friday 30 September 2011.

Jill Tarter of SETI and Pete Worden of the Ames Research Center were on the same flight as me, on the way to this interesting NASA workshop.

This morning, David Neyland explained the DARPA plan for space and Pete gave the NASA pitch, after which Ariel Waldman stole the show with her plea to "start hacking Space."

I introduced her to Douglas Trumbull and the three of us happily compared notes and travel memories. (Fig. 8) Over a pleasant outdoor dinner, Douglas explained how he'd become disgusted and scared of Hollywood after the death of Nathalie Wood in 1981. "I knew I was next," he said. "Some execs were sabotaging the movie as part of a $15 million scam. They may have targeted Christopher Wolken, too. That world is deeply corrupt and dirty. I just had to escape. I moved to a farm in the Northeast, and set up my operation there."

Orlando. Saturday 1 October 2011.

Eric Davis gave one of the best physics papers. I am learning more and more about the Galaxy Zoo, Spacelog.org, York Time, negative energy, payload mass fraction, Alcubierre drives, brane cosmologies, Chung-Freese metrics, and invasive brain-computer interfaces that sound scary. My main interest was in the biology of micro-gravity, a major opportunity that can derive from space studies in the next ten years. I also attended a talk about the ISS life support system and a presentation by Athena Andreadis who reminded us that for Native Americans the stars are celestial campfires where our ancestors await us for storytelling...That would be very nice.

Douglas and I went to dinner with Richard Dell, founder of Dell Computers. His interest stemmed from communications he once received from two balls of light. He calls them "The Patrons."

Doug later pulled me aside and suggested we work together. Using his new technology of movies at 120 frames a second, he wants to develop six episodes for distribution on smart phones and tablets.

In the meantime, the US economy dives into recession as Europe continues to struggle. Instantaneous computer trading has taken inertia out of the system, generating perturbations; witness the collapse of Bank Dexia.

Hummingbird. Monday 3 October 2011.

A wonderful surprise: As I got home, I found a beautiful bicycle next to the kitchen, a very nice gift from my daughter who rebuilt and repainted it.

Fig. 7: *Dr. Salisbury and Junior Hicks: oil company HQ, Utah, July 2011
Note the extent of oil fields, just South of Skinwalker Ranch*

Fig. 8: *Discussing technology with Douglas Trumbull and Ariel Waldman,
Orlando, Oct. 2011*

My old SRI friend Dick Shoup, who now runs a private research outfit called the Boundary Institute, came to see me at IFTF this afternoon. We spoke of 'futures impossible,' the limitations of computer logic, and retro-causality. We agreed that a deeper formal foundation was needed to make sense of precognition. Dick believes the observed events are due to correlation, not to an actual transfer of information into the past; two random sources get entangled in the future, thus appearing to have communicated in the past.

We also discussed random number networks (Peter Bancelin, Paris) and the *Laws of Form* as defined by Spencer Brown. He said it took him a long time to understand the book; the key is in the differences between objects, not the objects themselves. I invited him to my physics presentation on Tuesday: three 'impossible' scenarios for the Institute's ten-year horizon project.

I made some coffee and heard of the death of Steve Jobs on the news.

Mabillon. Sunday 16 October 2011.

In Paris again, family plans: Flamine and I had lunch with Olivier's future wife Claire and her parents today, in the well-heeled Seventh Arrondissement where they live. More privately, we've begun sorting through treasured books Flamine will bring with her to San Francisco.

On the train returning from Caen. Thursday 20 October 2011.

A cab driver has told me that Mouammar Kadhafi had been killed today in Syrte. Will it help? After Tunisia and Egypt, another country on the map can be repainted in shades of spring, at least until new regimes emerge and reveal their true colors.

Will it be liberty at last, or a shaky return to hardline Islamic abuse? And Libya, just another tribal revival, fueled by oil? France remains shaken, and a turn to the left gains credibility behind Hollande and his promise of a 'normal' presidency.

Mabillon. Saturday 22 October 2011.

Lunch with Philippe Guillemant and his *compagne* Laurence, a chemist with Sanofi-Aventis. He explained his ideas about time, and I gave him my notes on coincidences, including the Balzar case (28).

In *New Scientist* (8 Oct 2011, p. 42) I read that Carlo Rovelli at the Center for Theoretical Physics in Marseille "has rewritten the rules of quantum mechanics so that they make no reference to time." He states there is no such thing as time at the fundamental level of nature.

In his view, Quantum Mechanics only needs to describe how physical systems evolve relative to other systems. Time is in our minds, not in the basic physical reality. Among those who disagree with Rovelli, the magazine cites Fotini Markopoulo of the Institute for Theoretical Physics in Waterloo, Ontario, Canada. He argues that time is fundamental, and reality's basic ingredients are quantum events *ordered in time*: "From them, space, gravity, and relativity emerge at larger scales and lower energies."

Yet others, like philosopher of science Dean Rickles in Sydney, say that "what we think of as time emerges from some deeper, more primitive non-temporal structure." I identify with that last view.

On the train returning from Poitiers. Sunday 23 October 2011.

This morning, I had the pleasure of taking Flamine to visit Notre Dame la Grande. The streets were empty, all of France frozen before TV sets because of a rugby match in New Zealand, the finale of the World Cup. Joël Mesnard came to pick us up. On the way to his place in the country, we stopped to see the field behind his house, where seven witnesses in two groups saw a low-flying object on Saturday 27 June 2009.

Among the topics we discussed privately: A man named George Metz, who is investigating the claims of 'Roro,' the nickname of a French contactee. No records at all can be found about the man's life in France, while he claims he was a prisoner of a group of aliens. Metz has also found French cases of very odd large animals with a behavior like that of the Skinwalkers in the States. Now I've become a target for a campaign of lies. Gildas Bourdais, a nice man who picks up disinformation rumors, has heard that "the DIA had taken me to a place where I inspected alien bodies and saucer hardware…" I don't get upset any more, because I understand the hoax has nothing to do with me. It's like Ummo, or the Serpo hoax: the story, when it gets picked up, could expose a hidden network, unrelated to anything we do.

Flamine's apartment in the Marais. Thursday 28 October 2011.

Much work is getting done: a presentation with Philippe Richard at ONERA yesterday and a follow-up with ESA with Alain Dupas to discuss science programs aboard the ISS.

On my way back I bumped into my neighbor professor Bokias, who invited me for coffee on the Boulevard Saint Germain. Always the ladies' man, he pointed out a French actress from the sixties who was walking by. She smiled at us as she passed, carefully made up, in a fancy hat and dark glasses, like Garbo. Bokias doesn't go to Greece anymore, disgusted at the waste and despair of his native country. The Chinese will emerge as the winners from the economic slumber of Europe, he said, ready to pick up the debts left by our wasteful politics.

My daughter and Rebecca arrived in Paris today, ready to celebrate my son's wedding, but I have to fly back to the States in a couple of days.

Hummingbird. Thursday 3 November 2011.

There was a presentation on space medicine at UCSF today, so I went back to the Parnassus campus with heavy memories. The impressive structure rises on the hill, with a seven-level garage underneath. The talks confirmed the fact that *humanity wasn't ready at all for a mission to Mars*. Dr. Nick Kamas, a psychiatrist, was the main presenter. Well-informed, he never overstated the data, but he too glossed over some key problems, like the relevance of gravity's impact on the genome.

Hummingbird. Sunday 6 November 2011.

Dick Haines forwards a message to me from Captain Jose Americo in Brazil, based in Sao Conrado. He states that Operation Prato never ended. It simply changed names when Colonel Hollanda told his superior, Brigadier Protàsio, that he'd maintained contact with six creatures from one of the objects.

Colonel Hollanda never intimated anything like this to us. Gevaerd and his group are trying to gain access to the 30 hours of movies the team had shot on 8 and 16 mm. That film does exist, from my conversations in Belèm, but it's held at a high level of secrecy. Bill Calvert, who will come see me in January, doubts that Hollanda committed suicide.

Austin. Monday 7 November 2011.

A nostalgic flashback: That gray sky of Texas, low and wet, reminds me of earlier days—happy careless days full of hope, when we lived on Duval Street, not far from here.

Eric was at the airport to pick up John Schuessler and me. We were all hungry, so we stopped for lunch at Bone Daddy's, a coffee shop where the scantily dressed waitresses are University of Texas coeds making an extra buck. Eric spoke of his research, arguing that the new 150 kW lasers could explain the injuries created by the chupas in Brazil. But can they lift the craft and fly them around, without any noise, or exhaust?

John Schuessler, like me, has doubts about the Bentwaters case. I asked him about the extensive study of UFOs that McDonnell-Douglas did secretly 30 years ago. It never led to any propulsion breakthroughs, so Boeing is said to have thrown away the files when they took over the company.

This is an emotional reunion, so deep have become the ties among us. The group is aging gracefully. Dr. Green remains elegant, his closely trimmed gray beard reminding us of Sean Connery. George Hathaway wears glasses and has an impressive white mustache.

John Schuessler, who has listed publications about UFOs, found 2,700 books and 2,300 videos between 1950 and 2011, just in English. Also 10,000 issues of periodicals and some 250,000 articles!

Over dinner at the Roaring Fork, I gave an overview of the Institute's Futures Impossible project; Bob Bigelow discussed biology experiments in space and the scary state of US banks. He spoke about the excitement of little kids who visited his aerospace facilities. In contrast, older teenagers spent all their time texting on social networks and seemed bored by the whole experience. So, in disgust, Bob has discontinued these public education activities.

In a recent analysis of the DNA of cryptids, they turn out to be hybrids with human mitochondria.

Austin. Tuesday 8 November 2011. Hilton Garden Inn.

We had our first group meeting today to expand on my control system concept, at an airy research compound of the University of Texas. George added the problem of interference and surveillance, reminding us of the concern about the possibility of contact with bio-robots. What are the fingerprints of the control system?

Another important player, an independent outlier named Jerry Miller, has said "this is the most complex thing that can be comprehended." Eric Walker and William Colby have said the same thing. So, what's the appropriate scientific method? I asked. The threat is global, immemorial. It is related to our origins.

Austin. Wednesday 9 November 2011.

Having recalled that a contact at the Joint Chiefs level once stated *there was indeed recovered hardware*, but so advanced it was useless, someone said: "It's the equivalent situation of giving Leonardo da Vinci a garage opener, he could never figure it out." To which I replied that if you also gave him the garage, he could figure out how to open it, even if he didn't grasp the circuits. *We have to look for the garage.*

"Is the Utah Ranch the garage?" someone asked.

Of course, there was more discussion: "If there is such hardware, it must be held as treasure," Bob said. In the case of the Joint Chiefs, we were told *the hardware was anomalous, not made on earth, not human, and held somewhere in the black budget structure.*

We noted that U.S. Title Code 301 makes special access programs that are waived and unacknowledged impervious to Congressional scrutiny. It didn't escape us that it left some room for rogue entities holding on to the intellectual property—and the material. We agreed that the proceedings on this meeting would not be published, given their sensitive nature.* Later discussion included a review of videos showing a small orb flying *inside a secure room* at the major facility. In another case, an infrared camera captured a softball of light moving in a controlled fashion. This is real, not a graphic effect.

On the way to Bergstrom airport, the conversation returned to the issue of data capture, so I offered a copy of my latest research proposal. I

---

* The meeting was set up under Chatham House rules, so I will only mention events at the Ranch, which have been widely reported in the last 13 years, together with their extensions into the Skinwalker Ranch TV series. Colm's insistence that the phenomenon mimics our military and intelligence environment is profound.

made it clear I'd only build the organization with committed five-year funding, near a major University, and with the ability to hire people with proven 'fire in the belly.'

"Who should fund this?" They asked. My answer: those most exposed to risk from an unforeseen threat from the phenomenon, the big insurance or re-insurance firms, the large hedge funds.

Hummingbird. Friday 12 November 2011.

Beyond my presentation on information physics, and Hal's relativistic theory, the new elements in Austin had to do with biology and coincidences, which dovetail with my personal interests. So far, nobody is aware of my anticipation of blue orbs in *Sub-space* (those had never been reported at the time, 1961) or the bending of spoons anticipated in *Dark Satellite* two years later. I am comfortable with the team ignoring these episodes on the mythopoeic side. Only Jeff Kripal, and Douglass Price-Williams before him, had picked up on it.

Hummingbird. Tuesday 15 November 2011.

A new project has popped up, initiated by a group of Russian investors in connection with a technology center in Skolkovo, a Moscow suburb. They talk about the birth of a free, new Silicon Valley, working with the West... Has Russia changed, freedom at last? Open for business beyond the expensive leasing of their spacecraft to us?

Dimitri Manakov, a young Russian with US degrees, has contacted Maurice Gunderson. "There's interest in starting both a nuclear energy fund and an aerospace fund," he said, "two areas where Russia and the US have long cooperated."*

Not so fast...Maurice will be on his way to Moscow soon to assess how serious the project is.

---

* It is ironic, for instance, that the titanium for the body of the SR-71 spy plane was bought from the USSR.

Mabillon. Saturday 19 November 2011.

As soon as I arrived in Paris, Flamine moved in with me, following the successful sale of her apartment. Her sister and her husband brought some furniture and her books, after which we all had a happy lunch at Les Editeurs. I feel tired, of course, but happy we're together, grateful to enjoy some of her precious memories, such as a small desk that belonged to her mother, the beautiful artist named Luce de Montety. Then her sister Isabelle, with her energetic sons, brought us more family souvenirs, and her father also paid a warm visit.

Brussels. Monday 21 November 2011. The Sheraton, Place Rogier

I always rediscover Brussels with pleasure, although much of the city is currently hidden by construction barriers. We got to the Bozar building just in time for rehearsals, but the staff of TEDx was late, which gave us an opportunity to meet other speakers: Rudy Rucker, John Shirley, David Brin, and Hasan Elahi. Peter Fenwick, a British academic in a suit and tie, who reminds me of Peter Sturrock, will speak after me on 'the future of dying.'

Brussels. Tuesday 22 November 2011.

My TEDx presentation today on the physics of the future (entitled "A Theory of Everything Else") was little more than one more brick thrown into a big pond, but it should be fun to watch the ripples. (29) Speaking before me was Paddy Ashdown, Member of Parliament and a James Bond figure, with observations on the modern world where enemies become linked *because they share a tragic destiny.*

Miko Hiponen described how society restricts liberty: "The issue is not privacy vs. security," he said, "but Freedom vs. Control." Under the guise of fighting terrorism, dictators and police use security software to flush out anyone who threatens their power: the FinFisher in Egypt, the Staat-Trojans in Germany.

"We used to recoil at the extreme tactics of the Communists, recording the unique patterns of typewriters, but our laser printers are identified through invisible yellow dots on every page. We must watch the government watching us—any government," he concluded.

In the same vein, one speaker named "Identity Woman" described our future as Participatory Totalitarianism: "Today's creepy is tomorrow's identity," she said, adding: "We used to have one identity in one context, but such distinctions no longer exist. Large-scale data mining does to society what large-scale mining did to the environment: destroying it in swaths of devastation."

I felt drained after my talk, so we skipped the cocktail. We paid a visit to the Grande Place, defaced by construction in the slow drizzle.

Reviewing my notes, I am struck by Patty Ashdown's eloquent talk about the shift in power in the world. When multinationals move up from their regional roots to the global stage, they go beyond the rule of law, evading regulation. Certainly, governance will follow, but treaties are slow and precarious, like the useless Kyoto climate talks.

Mabillon. Friday 25 November 2011.

Graham and Maurice have met over dinner with Dimitri Manakov and his wife, both graduates of US universities. Skolkovo, where the new funds originate, is the future site of a Russian technology center aiming to emulate Silicon Valley. (Yes, I know, one more! They have no idea...like the French!)

My research must consider the information structures evident in synchronicity. Jung and Pauli should be my guides, except that Jung thought synchronicities were significant. Perhaps they're useful in therapy but in my current model, most of them mean nothing at all.

Hummingbird. Saturday 26 November 2011.

On the plane I finished reading Céline's *Voyage au Bout de la Nuit* (*Journey to the End of the Night*). Somehow, I'd never been able to read that book, which I rediscover with amazement at its sharp, dark humor. When he leaves a sweet harlot he has loved in Detroit, for instance, he writes: "I kissed her, Molly, with all the courage I still had within my carcass. I felt pain, real pain, for once, for the whole world, for me, for her, for all men. That is what we look for through life, only that: the greatest sorrow possible to become ourselves before we die."

Hummingbird. Friday 9 December 2011.

When I got up this morning, the light over the City was so perfect that I became absorbed in the spectacle of the white buildings merging with the sky from Alcatraz to the Golden Gate.

Frilly bands of blue-white haze kissed the curving hills draped in their folds like vestals saluting the dawn, a perfect ritual of mystical adoration, time suspended in a magical moment, and the sea perfectly calm, as if in expectation of some divine visitor.

One of the UFO witnesses we discussed in Austin was assassinated in Pasadena two weeks ago. He worked at an aerospace company, had expired clearances, and expressed concern that he was targeted.

Hummingbird. Monday 12 December 2011.

Maurice flew back from Moscow on Monday having met at length with the three organizers of the proposed nuclear industry fund. The discussion, ably supported by Dimitri, was a basic tutorial in venture capital: The Russians had trouble understanding why they couldn't simply keep central control of all the companies in which the future fund would invest, so we're not getting anywhere! Maurice also met with Sergey Zhukov in Skolkovo. The anti-Putin demonstrations that have shaken Moscow now spread to other cities in Russia.

Fort Bragg. Wednesday 21 December 2011. North Cliff Inn.

My room overlooks the rocky Pacific shore and the harbor channel. The only sounds are of the surf and the hungry gulls and cormorants, and the foghorn at the entrance to the channel. I picked a quiet place to read Jeff Kripal's remarkable study of *Mutants and Mystics,* and to rest after so many frustrating days of complicated interaction with the Russians, but the truth is that I need this break.

Hummingbird. Saturday 24 December 2011.

Christmas Eve, spent alone for the first time in my life. Olivier jokes gently about it: "That gives you time to clean up your place," he writes, knowing that Flamine lands tomorrow.

Messages keep pouring in, prompted by my talk in Brussels. In this respect, this year was bracketed by that impactful presentation and an equally strong statement in Riyadh. Can we build on this next year?

Hummingbird. Friday 30 December 2011.

Flamine emerged from customs, tired from administrative hassles, the sale of her apartment, and the relocation of her library. We huddle here, reassembling the pieces of our existence, loving softly in poignant embrace. Another year of worldwide turmoil is about to start while San Francisco surrounds us with grace and a coolness that invites long walks, visits to little shops; a precious time, suspended high above the landscape of our wanderings, and the uncertain bustle of the planet.

Part Twenty-Two

LONESTARS

6

Hummingbird. Thursday 5 January 2012.

In late afternoon, the hills of Marin County glisten with the pale earth tones of dry brush and the mauve tints of the setting sun on grasses long deprived of rain. Flamine and I enjoy this autumn weather, the land that breathes quietly, and the sublime landscape from the top of the road at the Institute of Noetic Sciences, south of Petaluma. (1)

Tonight, I gave a lecture there with Jessica Utts and Dean Radin in attendance, as were Sergio Lub, our genial organizer, and Marc Ian Barasch from Green World. Also there was Stephen Bassett, an associate of Dr. Steven Greer. The audience was shaken by the idea of a future physics of no dimension and no time, until they remembered that Ed Mitchell, the astronaut co-founder of the Institute of Noetic Sciences, had an experience akin to Samadhi on his way back from the moon.

Hummingbird. Saturday 7 January 2012.

As we reviewed the Russian investment project in Skolkovo, my partners and I recoiled before the bureaucratic complexity on both sides. Washington demands export licenses for every trivial bit of technology, defeating innovation. It gets so bad that Bob Bigelow, with all his influence, was forced to send two armed guards to Kazakhstan to keep a sharp eye on an aluminum stool US diplomats decreed was "an aerospace product!" when he launched his two space station prototypes from Russia. We could never fund a startup company that was forced to pay good money for such absurd requirements. In contrast, Bob was impressed by the business and technical savvy of the Russians as partners, which encouraged us.

Hummingbird. Saturday 14 January 2012.

This is a precarious time. Yesterday, during a long phone conference with Moscow, we told Dimitry Manakov about the bureaucratic obstacles in the path to our hypothetical technology venture fund with Russia. We're ready to give up, but he still has hope and serious contacts.

Later, Flamine and I walked the labyrinth at Grace, and today, for a relaxed interlude, lunch at the Presidio café before watching the 49ers beat the Saints. The fact remains that my next moves, in a world seized with confusion, are undefined. European projects are stalled. The financial rating of France has been degraded even as it struggles to choose between an impulsive Sarkozy and a shaky François Hollande.

America, meanwhile, watches incredulously as a ballet of ineffective Republican politicians, led by Mitt Romney, hope to challenge Barack Obama. Unemployment is high.

Fred Adler, who called me back last night after I left a New Year's message, encouraged me to pursue medical investments, a clean area where he's done very well. Over 80 years old now, he continues to work with bright young scientists from the US and the Technion in Israel. He knows Romney (2) well; they were co-investors in Staples, the office supply firm; but he doesn't see a quick recovery in the US.

In France, as the magazine *Marianne* notes, "The political class has degraded into a world totally devoid of inhibitions. In the background are off-shore accounts, bags filled with banknotes, retro-commissions, envelopes containing money, armament contracts, oil piggybanks, trips and palaces subsidized by major construction firms."

Hummingbird. Monday 16 January 2012.

A visit by Saël today. He mentioned medieval Middle Eastern manuscripts, his research into the Kabala, and Islamic Occasionalism. I told him of my own struggles with Averroes and the Sylphs tradition, with the idea of a universe "created from moment to moment."

I have reviewed my progress with a study of coincidences; the key is how they provide closure. I've compiled nine such incidents in anticipation of a presentation at the conference of LoneStars (as we jokingly call our science group) when it gets together in Vegas in March.

Hummingbird. Sunday 22 January 2012.

My inquiries about the sightings near Palmdale have led me to a certain retired vice-president at Lockheed, a colleague of legendary aircraft designer Bob Rich, founder of the Skunk Works. They were once in charge of studying materials picked up from crashes in Nevada. Old story, he said.

The method is simple: The Air Force would call them as cleared contractors, indicating where "the objects" were. They would go out with trucks, gather them up, and study the materials. For some reason, funding halted and the stuff was stored away, like the Ark of the Covenant at the end of *Raiders of the Lost Ark*. For a while, as the rumor goes, it was held at Class 3 storage facilities near Chatsworth. There was a contract to maintain the secure lockers. It didn't support science.

So why was the research stopped? Here again, there's a straightforward answer: the military brass often shows a fundamentalist bias, while events like saucer intervention smell strongly of evil influence. But why not expose it then? I can only speculate about the answers.

Until November and December, we thought BAASS was close to re-opening our project. Bob begged his contacts to help create an appropriate SAP (which only the CIA or the Air Force can do at this level) to restart the quest. This effort was defeated by two things: first, Harry Reid conveyed a garbled request to Leon Panetta, and this led nowhere; second, a similar request supposedly went to Tara O'Toole, head of Science and Technology at Homeland Security, where deputy secretary Jane Holl Lute and Karyn Wagner became involved, along with a SAPCO manager.

They couldn't initiate a program, but they could recommend it to SAPCO, the body that oversees special access programs (3), so they held a meeting and hit a wall of three NO votes.*

Was it poor experience with Intelligence? Or the feeling that the administration had nothing to gain, given the high risk of exposure? Or mistrust of intermediaries viewed as shaky? Did Bob's approach ruffle some feathers? Or simply that DHS, alias Homeland Security, had just suffered a huge budget cut? Possibly, a combination.

The SAP structure was ready. The principals would be briefed and 'personnel' like Colm and me were on the schedule for briefing. One manager hoped to restart it "with a link to the people who know about the biological material." That would have been very nice indeed and scientifically logical, but it didn't happen.

---

* The Defense Intelligence Agency had funded BAASS in Sept. 2008 with a $22 million, two-year renewable contract to study "UFOs and their effects on humans." But no money was extended after September 2010. The project was given a 3-month no-cost extension to submit about a dozen reports by December 31, 2010.

Later the same day.

I've been contacted by a family that runs a large milk collection business in a small town north of San Francisco. They saw a "smokey, metallic color disk" over their home at night on 9 June 2009, leaving an area in the sky completely black. They've also taken many unusual photographs right in their yard, showing light balls, filaments, and odd luminous shapes, some of which we couldn't explain. I've examined the digital camera used and tested it. Some of the pictures, admittedly, may be caused by moisture on the lens, but others are unexplained.

Odd reports continue at the Ranch, too, affecting the staff. One scientist experienced poltergeist at his home, with a gray being in a bedroom, along with randomly opening drawers. On one occasion, two witnesses heard a giggle from a little girl, but there were no children in the house: The phenomenon is merging with classic hauntings.

Hummingbird. Saturday 28 January 2012.

On Thursday Flamine saw a falcon just outside our window and called me. He was clinging to the balcony, unafraid, lording over Pacific Heights and the white fog in his regal maroon robe. He looked at us coldly but didn't budge when I slid open the glass door.

Today we saw four more, in wide fast passes over the Van Ness corridor, cutting the air like razor blades, re-sculpting the sky in wide, raging curves, with savage angry wings. A breeze through the blue and silver drapes.

Hummingbird. Saturday 3 February 2012.

Visit by my French colleague Jean-Philippe Boige, who saw three unknown lights in Lyon at age 14. He brought news of the financial world: French Venture capital is in some decline. Big funds struggle.

Recent exchanges with the other LoneStars: I take the opportunity of the forthcoming meeting to raise issues of research planning for Capella, which can move in many directions from the base I've constructed. My concern is that there are as many concepts about Capella as there are members in our group.

A major neglected question rests in the religious implications. I reminded my colleagues that while all of us, given our backgrounds as physicists-engineers-biologists problem solvers, tend to gravitate to factual, reproducible clinical data, both the public at large and government continue to focus on spiritual and social-control issues. If physical evidence ever comes to light, either as exotic hardware or alien life, the disclosure would be the beginning, not the end, of an overwhelming series of questions—contrary to the simple-minded optimism our group assumes.

A breakthrough would only be a partial, temporary closure and probably would not answer any fundamental question but raise new ones. To me, that's why a serious push at data collection, screening, and validation on a professional scale is mandatory, in anticipation of obtaining hard physical or biological data, not afterward.

I don't mind being kept from the core group, even though I believe I could make contributions. It all reminds me of Jim McDonald ignoring everything I said, back in Project Blue Book days, with his superb way of rushing into the China shop. That attitude never works, it's a first-class failure point.

Hummingbird. Saturday 4 February 2012.

Touring the botanical gardens at Golden Gate Park with Flamine (the weather a succulent blend of morning moisture and blue-sky spring), we struck up a conversation with one of the 'birding' ladies, who recognized my name. Putting down her binoculars, she said she was the sister of superb writer Larry Collins. Sadly, she added that he had died. As we walked along the flowery path marveling at the hawks, the blue jays, the ravens, and the finches (and a dozen other species our smart guide recognized), I recalled my meetings with Larry.

She commented that it was a pity people weren't able to follow what they were best at. Larry spent his later years in dull Intelligence work and regretted it. His later books (such as *Maze*, for which he'd interviewed me and others from SRI) didn't meet with success.

Now a correspondent of mine, a test engineer for robotic tape libraries who was inspired by my talk about information physics in Brussels, writes that "we live in a block universe; dimensions and information are artifacts of consciousness." He concludes there must be *intentional transmitters* in the system, with competing agendas.

"I am the closest transmitter to myself," he notes in an amusing analogy to hardware networking. "This practice is the psychic equivalent of a beneficial local ping host."

A couple of other striking letters have arrived, one from Daniel Melvil in Prilep (Macedonia) calling my attention to early 19th century paranormal observations, long forgotten under the rubric of vampirism. Then Andreas Schneider from the Academy of Media Arts in Cologne writes: "You were the first person to use the expression 'virtual community'…you started to discuss the social meaning of computer networks before many community networks were launched." He wants to pursue the idea to logical conclusions.

Hummingbird. Sunday 5 February 2012.

A stunning entry from Kit: "Related to a previous note from George and Jacques in this thread, I think that the infectivity in a psychological sense, the disease in a psychiatric sense, and the epidemiology in a social sense are related to the initial insult. So, we have conversion disorders of teenage girls, religious fervor in Evangelical men, and UFO hysteria in sub-socially adept 30-somethings. But when there are real orbs, real craft, real RF injuries, and real creatures and mutilations…the memes that are psychological fade…and we're left with age- and sex-independent probands, nuclear family exposures, and a highly localized model of infection."

I loved it when Hal reacted with an ironic comment: "So articulate you are, Yoda of the medical arts!" To which I added: "Not only articulate, but meticulous as well!"

Since our very first LoneStar workshop, I've seen how impressive the MRI work was, but it wasn't clear to me that all patients suffered from the same issues; some only had flashes of unexplained terror.

Now for a Board meeting at Triformix, in Santa Rosa, where the hills turn a tender green, ready for succulent Spring.

Hummingbird. Friday 10 February 2012.

We had a happy visit today with Bill Calvert. I found my old friend a bit slower and more reflective, but as smart and well-informed as ever. His life in Galena (Illinois), where he runs a coffee shop and Brazilian import business is filled with friends and conversation. One of the 'coffee hour' regulars is a man named Leonardo, who claims he used to spy on Henry

Dakin's group in San Francisco for some agency (because of Henry's interests in 'parallel diplomacy' with the Soviets).

We spoke at length of the late Colonel Hollanda Lima in Brazil. A short time before he died, he told the Belèm story on *Fantastico*, a local TV show like our *Sixty Minutes*. He was planning a book about the Amazon sightings, co-written with a woman reporter, but he died in October in a supposed suicide scene staged to suggest sexual strangulation, a typical way to discredit his testimony.

Sr. Gevaert's release of the documents falls short of everything we knew, although it impressed the BAASS team. What about the missing documents? Colares reports went to Brasilia and were redacted there, higher reports locked up. Bigelow's two expeditions made a big splash. Very conspicuously, three of his people attended a meeting in Sobral, again triggering rumors about 'US imperialism' and manipulation.

Anyway, the damage is done, and little learned.

We spent the rest of the afternoon talking about psychic healers (Arigo and Dona Cicera Maria da Silva Almeida). Puharich had gone to Brazil, supposedly under a NASA grant (!) to study Peixoto Tinto, but Brazilian authorities stopped him. Only then was he redirected to speak with Arigo. Dona Cicera used to claim she worked under the control of aliens visible to her, and a female figure that was an emanation of the Virgin. These beings did not walk but "floated" as if sliding on a wheeled platform. She did her surgery *using their body*, not hers, she told Bill in private a year after his visit with Beverly. She described them as classic UFO beings (or classic Sylphs?) in shiny suits. She operated near Neopolis, on the Sao Francisco River, also Cristianopolis, and Umbaùba, close to Bahia.

She was already gifted as a child, healing animals, but the "Lady" taught her some special words when she was twelve, supposedly "transmitting the power of God" to her. Her mother simply concluded she was possessed. She left her family, upset about incipient bribery around her healing powers. Dona Cicera died at 33, but Bill proposes to track down her assistant, Carmelita, who was only 14 at the time.

Hummingbird. Thursday 16 February 2012.

A sharp exchange has taken place, as Eric compared the Catechism to a hoaxed narration. Kit replied from the mountaintop: "As a graduate of an Episcopalian seminary, and as a life-long scholar of comparative religion, and an Anglican Catholic, I now feel moved to witness to this group that I disagree fervently...It would require the same degree of delving into our personal experiences of receipt of Grace and Discernment as we otherwise grant to the much more narrow subjects of psychic information freely received, existent knowledge of the Collins Elite, coincidence and intuition."

This outburst left some of us in devout silence and others, like me, in consternation. What happened to our friend to shake him so? In an earlier message he was reading *Final Events* by Nick Redfern, to whom he gave an interview: "It terrifies me. I believe every word of his investigation, including the historical underpinnings. For what we are discussing, it is critical." He's sending me a copy.*

Kit had been invited to join the Collins Elite in 1982 and declined. There are rumors about the members, but do we need more rumors?

Hummingbird. Friday 17 February 2012.

Typical San Francisco Street scene: As I admire some ornate books in a storefront, a man approaches me. He's recognized me as the author of *Invisible College*, his "favorite book." He puts down his bags, shakes my hand, tells me about God, and gives me a copy of his manifesto about "our integral wholeness." His name is Wendell.

As I try to follow his discourse, a crazy character comes into view: a middle-aged Chinese woman who watches us from a few feet away, shuffling and talking to herself in some incomprehensible singsong where we only catch the word "Obama."

She displays her front teeth and goes on, dancing in her sneakers and

---

* Dr. Green may have felt (justifiably) upset because the classified proposal for a new, comprehensive study to continue and expand the work of BAASS (including a new computer science effort I might help design) was secretly declined when the Department of Homeland Security "directed its immediate termination" on February 10, 2012.

This took place after much discussion and many reviews in Washington. Existence of the proposed Special Access project, whose classified name was KONA BLUE, "not to be declassified until April 25, 2036," was kept secret as well as the date of cancellation until February 5, 2024, when a letter to that effect from Homeland Security Acting Deputy Secretary Kristie Canegallo was sent to Kathleen Hicks, Deputy Secretary of Defense.

green socks, gesticulating with her hands gloved in cheap panther-like fur. She goes back into her own parallel universe. Wendell blesses me.

Hummingbird. Tuesday 20 February 2012.

On Sunday, St. Mark Lutheran hosted an exquisite concert of Monteverdi's music (*Madrigals of Love and War*), played and sung by the Magnificat Ensemble. The sky was clear but the City was windy and cold, so Flamine and I rushed along to the church, staying on the sunshine side of the street.

Hummingbird. Friday 24 February 2012.

Exceptional weather, suitably warm. Flamine and I were married today in San Francisco City Hall, under the great rotunda, by Commissioner Mary Ortega. Our closest friends were there: Zoila Muñoz and Marta for her, Diana plus John and Michèle Forge for me, with Rob Swigart in a suit and tie and, most preciously, Catherine and Rebecca. After the ceremony we walked the labyrinth at Grace Cathedral. John Forge had invited us for a drink at his penthouse atop Nob Hill, and we ended the day with dinner for ten at the Fairmont that reminds Flamine of a castle in Austria with its domes, columns, and soft blue paintings. Olivier and Claire sent gorgeous white roses.

Hummingbird. Sunday 26 February 2012.

Last night we drove over to Berkeley to hear Saint Matthew's *Passion* by the American Bach Soloists, Jeffery Thomas conducting. Zoila and Marla were there, an added treat.

I have mixed feelings about the upcoming LoneStars meeting in Vegas. It hasn't been prepared very well. Colm and I have suggested to dial down the discussions about databases because everybody was too ready to jump in with their crazy miracle 'expert' solutions.

Hummingbird. Tuesday 28 February 2012.

The German American Business Association held a debate tonight among experts in computer security who'd come to attend an RSA conference. The subject was chilling, between engineers from Siemens and Infineon who certainly knew about the Stuxnet worm recently used to attack the Iranian nuclear centrifuges, and an American audience aware of recent attacks against power plants.

Howard Schmidt, special assistant to Obama for cybersecurity, impressed us with figures from the Intelligence world. Satellite imagery today represents 7,000 petabytes per orbit (the entire YouTube collection is only 1,000 petabytes, and Facebook one tenth of that). There's also a new 'Analyst notebook' program that handles unstructured data at high speed.

I get ready for Paris, where everyone waits for the elections to determine who gets reductions to their budgets. The world hangs on hopes of recovery, and a great deal of fear.

Douvres-la-Délivrande. Sunday 4 March 2012.

At Annick's charming house, resting from the flight, I slept in the pink bedroom amidst family pictures and many paintings, including a couple of framed *gouaches* my mother had given her. We're both aware of a burden of grief nothing can lift. Driving under the drizzly sky to reach Yvetot-Bocage, I went to place my flowers on Janine's grave, grateful she had visited this tired world and graced my life.

Annick has decided to spend all her time here in Normandy from now on, enjoying her home. Reading my Journals of the 1970s, she enjoyed reliving our conference in Brunville with Aimé Michel and the lively banter of Pierre Guérin. "All that is gone now," she says.

UFOs are treated by the media as a matter of curious irony. Perhaps that's for the best; they're not ready for the implications.

Mabillon. Monday 5 March 2012.

Warm phone conversation with Joël Mesnard. He recommends that I reconnect with Dufour in Nice, such a skilled investigator. There are new cases in the region, along the very isolated Vallée de la Tinée. One man who lived alone in a trailer confronted strange beings looking through his windows. They beat him up and fled. He couldn't be carried to a hospital in time and died on February 3rd. The body was a mess, the insides reportedly mashed up, and blood from every orifice. The coroner wouldn't bother with the remains. A woman doctor signed the permit, and he was promptly cremated.

Mabillon. Thursday 8 March 2012.

A bright sun puts a handsome glow on the cold towers of Saint-Sulpice church, free from the scaffolding that imprisoned it for years. France holds her breath, scared of the Greek financial debacle.

On Tuesday evening, crawling up along the Paris suburbs, the train that brought me back from the Genopole stopped at station after station along the ugly eastside towns: Juvisy, Maison-Alford—a landscape of empty lots, piles of discarded vehicle carcasses and junkyards. The gleaming towers of Paris adorn the northern horizon as a stranded humanity survives, humiliated and exploited.

I spoke to Maurice Gunderson again. The concept of a new energy fund may be viable at RosAtom; would EDF join us as investors?

Hummingbird. Monday 12 March 2012.

So good to drive up from the airport after a 12-hour flight, ring the bell at my door, and find you there, smiling...A few days to recover from Europe, then we fly to Vegas for the LoneStars II reunion at the company's space facility. I have four presentations ready for them and high hopes to see progress, since we'll have an honored guest, the first time I will see the two tycoons together.

Las Vegas. Saturday 17 March 2012.

Flamine had never seen Las Vegas, so I shared her amusement amidst the lights and sounds of the city. Things are still depressed here, many buildings empty, in the lingering economic malaise.

Eric picked me up, along with Hal and Kit, for the drive to the expanded facility in North Las Vegas where the group was given a tour of the spacecraft factory: displays of future bases on the Moon and Mars, and a model of the future large craft, a complete laboratory on three levels. All that still waits for major contracts and a new version of the Space Shuttle, but many theoretical problems are solved, and Bigelow has negotiated a waiver of ITAR regulations. (4)

The meetings took place in Bob's conference room, up from the 'mission control' facility his engineers have designed. We began with an infrared video of an orb floating in a locked room dedicated to radiation studies.

There was a cobalt source on the table. One could see that light source circling above, as if inspecting the room.

A summary of Capella was requested, so I'd prepared a few slides. I didn't want to launch into a discussion of databases, but that didn't stop someone from interjecting: "Why don't we just buy a Google appliance and throw everything in it?"

This is the kind of comment that drives me to despair. We can already search for text correlations in the Capella file. Flat searches on unscreened Google data would be worse than worthless. Similarly, another member "has heard of crowd-based research" and wants to distribute everything to web groups for sorting, another odd concept since 90% of UFO information is inaccurate or biased (or not relevant to the specific problem). But the prospect of systematic scrubbing and a few years of serious research on data validation is lost to them. *Nobody wants to face the hard work.*

The next topics were under the privacy rules and involved new forensic cases of very high strangeness, possibly using non-ionizing radiation. In the same section, Colm picked up the discussion about blue orbs and recalled that on the Utah Ranch, in 1996, the Shermans had a two-year history of being subjected to psychological and physical harassment by such objects. House lights dimmed as orbs passed by, and three dogs were reportedly incinerated, as already reported in public.

Las Vegas. Sunday 18 March 2012.

We resumed the conference at the intersection of Alien and Human technology such as implantation of sounds and thoughts into brains at a distance, and super-empathy (reflection of a human's emotional state). Yale University has even recovered sounds a ferret had heard, from the animal's own brain. Eric went on to describe invisibility cloaks made of metamaterials, nanophotonic crystals, or plasmonics.

Our guest stated that the UFO subject, on a political level, had an impact similar to the atomic bomb: once recognized, none of the previous geopolitical scenarios will apply any more, everything will need to be re-thought from scratch. *I completely agree with that view.*

"We have to consider an element of permission and control," said Bob Bigelow, "but time is running out. We'll soon be an off-planet species, even as our reptilian brain and our religious beliefs continue to drive our thoughts."

Kit was more negative: "We face a discontinuity," he said, echoing my private thoughts. "We're no longer experts in anything."

Later the same day.

The last substantive discussion (apart from my presentations, the Sonoma photographs and my catalogue of coincidences) had to do with the Collins Elite, the secret group of religious fundamentalists who keep blocking UFO information because they suspect a Satanic origin and nature to the data. The recent book *Final Events* by Nick Redfern chronicles the thoughts and actions of this shadowy quasi-government group. As noted previously, "the book is replete with Christian theology, eschatology and demonology and it makes the case for the existence of a portion of the US military that believes in an imminent Armageddon, winnable only if the world can be forced into an American fundamentalist version of Christianity." (5)

We know there's a cabal. It has tried to recruit people since the 1970s. One was Alonzo McDonald, White House Staff Director for President Jimmy Carter, linked to the Episcopalian church, who attended breakfast prayer meetings. Topics about demons would come up, and about people possessed.

The meeting ended awkwardly. As we were ready to return to the hotel, Bob pulled two key members into his office and expressed bitter complaints. More pleasantly, John Schuessler made it a point to thank me for my presentations. He was happy, he said, that I'd opened up futuristic perspectives rather than rehashing old history.

Hummingbird. Tuesday 20 March 2012.

We flew back on Monday night. We hardly had time to unpack and rest, then it was time to drive to Palo Alto for Peter Sturrock's birthday celebration. I was happy Flamine could meet him, as well as Federico Faggin and a dozen other guests at Peter's elegant retirement home on tree-lined Byron Street.

We read poems and limericks, traded news about our current occupations, and enjoyed Peter's conversation. I had brought Natalie's pictures from Sonoma, so we spoke again about orbs and their effects. Some of her pictures are still unexplained, especially the large red beam.

Hummingbird. Thursday 29 March 2012.

I had a horrible night on Sunday, after another episode of asthmatic bronchitis that brought me to the ER. I keep thinking back to our meetings in Vegas and cannot shake the frustration that replaced the initial glow of good comradeship. There's a lingering sense of *naïveté* in the theories we discuss and an absence of progress when it would seem we have all the tools to relaunch research.

Lunch at the Rosewood with Maurice Gunderson, relativist Slava Turyshev from JPL (6) and Sergey Zhukov, director of the Space Technologies and Telecoms Cluster in Skolkovo, a jovial man, fifty-ish, in pink shirt and mauve tie. "You look like a man who loves the sea," Sergey told me as he looked at me across the table, catching me by surprise.

"Not really," I replied, "but my ancestors were Normans, so my Viking roots are not very far away."

We laughed and the jokes continued. I confessed I got seasick easily and would rather glide down the Volga in a canoe.

"Actually, we're much more interested in submarines," said Slava.

I could travel happily in a submarine, I thought, a good compromise; submariners only get seasick when they surface. Was that a foresight of collaboration? Or a Russian warning? I didn't ask.

Hummingbird. Friday 6 April 2012.

Lunch with the Twins yesterday, at Cibo. Their papers continue to be rejected, but they improve them steadily. The latest version even predicts the mass of bosons and the three regions of probability for the Higgs particle. Their experiments (now moved to Dallas) generate high power output. We spoke at length of the experimental setup and its dangers. They were relaxed, happy to meet Flamine.

Coincidentally, Eric Davis has just sent me his assessment of the work of Claude Poher. "I think he's detecting the usual electric current and magnetic field leakage from his thruster," he writes. "If left uncompensated so as not to 'pollute' the thruster environment, then these will produce an imbalanced electromagnetic force."

My visit with Claude had already taken this into account, so Eric is jumping to conclusions. "I had a lot of trouble with his papers, which explains why no one talks about them," Eric continues. "His universon field and related propulsion theory is bad. He makes a large number of hand-waving, and sometimes circular, arguments and assumptions. General Relativity doesn't need another field of particles to produce spacetime curvature: It is produced by matter."

I've heard Peter Sturrock say the same thing. Eric goes on: "There's a flaw in his universon field, in that it has an anisotropic interaction with any accelerated matter that it encounters. This violates momentum conservation. Why does any matter have to be accelerated, if Poher needs the anisotropic interaction with the universons to produce net thrust for propulsion?"

Later the same day.

Phil M. came over for lunch today, a dynamic man semi-retired from Naval Intelligence and now assigned to analysis in a unit that supervises major contractors. We spoke of the lessons from World War Two, the entanglement among world powers, and the survival of powerful German financial interests around the world. If there's collusion among some leaders of industry, does it extend to controlling frontier subjects in physics? Are those areas connected?

One of his friends knew McGeorge Bundy well. He swears Bundy, as Kennedy's national security advisor, totally rejected real UFOs.

Phil had several stories about UFOs seen by military officers he knows. Their reactions surprise him; *many simply forgot the incidents until he raised the issue.* There's data about images analyzed by experts, showing obvious unidentified sources (some in infrared) classified as "spurious," and no follow-up. The analysts had never heard about UFOs. I told him about Redfern's reports on the Collins Elite. Phil supported the book's observation that the military (and especially the Air Force) was full of fundamentalists with very short imaginations, raised on a diet of Bible stories.

This evening, some progress at last: a series of messages from Sergey Zhukov, care of Dmitry, including a letter of support for our international aerospace fund, together with an implementation plan. There's a target date of 15 September to begin operations.

Hummingbird. Sunday 8 April 2012 (Easter).

We're spending the day at home, shuffling around, checking email, touching up projects, making phone calls, and watching the first episodes of *Rubicon* Phil has given us. Annick called from rainy Normandy and put Olivier on the line. My son has radical new plans: quitting his job, buying a house in the country.

Now I receive a report on two encounters over the Mediterranean in November 1983. The squadron, based in Athens, had two missions flown out of there and back into Athens. Mission One was a "Combat Sent" mission collecting technical ELINT for long term purposes. Mission Two, the following day, was a 'Burning Wind' mission collecting tactical ELINT for real-time use. Both aircraft were RC-135s flying under RIVET JOINT. Both were intercepted by a UFO.

The Mission One event occurred as the aircraft was headed south between Cyprus and Israel. The unknown vehicle placed itself in 'right formation' on the starboard side, *close enough that the crew couldn't see the whole thing from a single vantage point.* The edges appeared to scintillate, according to the primary witness. He described the surface as composed of "panels over inlaid panels" that were rectangular and translucent, with edges that would disappear or dissolve or go out of focus as one stared at them. They were luminous *like opal, with a hint of violet.* He observed no markings or symbols and no bumps, gashes, or openings.

My correspondent adds: "I asked one guy if the surface was reflective and appeared metallic, and he said no." The crew detected no RF emissions from the vehicle and no interference, although they were still getting normal RF readings throughout. Per standard operating procedure for these missions, they were in real-time SATCOM communication with NSA in the U.S.

Two F-14s were launched from the USS *John F. Kennedy* but the "vehicle" departed before they reached the RC-135. (Similarly, during the Mission Two event, three British F-4 Phantoms were launched from Cyprus and again, the object left before they arrived.)

Upon landing, the crews were initially debriefed by their own Det-IN (non-aircrew intelligence personnel assigned to the squadron). When they turned in their material it was placed under armed guard, an unusual procedure. Then three officers interviewed them alone.

That same evening, an anonymous debrief team arrived in an Air Force

C-130 with tail markings from a base in Germany. It parked next to the RC-135. The members were "white American males in ill-fitting civilian suits (sic!)." The witnesses were debriefed again, individually, by two interviewers from the anonymous team. They did not appear to be scientists or engineers. Their questions were mundane, and they didn't seem to know much about the phenomenon. They just asked what it looked like, if it had engines…

When they were ready to leave, the anonymous team had the witnesses read a prepared statement indicating that their own organization, designated only by a 4-letter code, was taking custody of all material. Each witness had to sign a non-disclosure agreement promising not to divulge the 4-letter code.

"*This violated the normal chain of custody for such material*, demanding transfer by SCI courier to the 544V SIW at Offutt AFB, and from there to NSA at Fort Meade," said my friend. "No one in the squadron protested; the incident was never mentioned again."

In November 2011, he debriefed a second member of the squadron, who confirmed the very unusual, borderline illegal process.

Hummingbird. Saturday 14 April 2012.

A week ago, we had lunch with Steve and Linda Millard in the refined atmosphere of the Burlingame Country Club. Reminiscences of Paul Baran, extraordinary inventor and rough manager. Steve was behind him in every company, to keep the team together and find the early backers for the devices that emerged from Paul's fertile mind. When will the real history of networking science be documented?

In the evening, after our ritual coffee break, we've been watching all twelve episodes of *Rubicon* Phil had given us. (7) On Wednesday we took a pleasant drive to Sacramento during a break in the seasonal storms to incorporate the management company for the new aerospace fund, but Maurice returned from China with an attack of gout: *No more Vodka!* he says, but we'll be in Moscow before the summer.

Dick Shoup, Ed May, and Bernie Haisch with Marsha were at Russell Targ's 77th birthday celebration tonight. All the talk was about retro-causation in quantum mechanics. Bernie plans a proposal to John Templeton Foundation for a new summary of UFO data, a long shot.

Hummingbird. Monday 16 April 2012.

Lunch in Mountain View (at Baklava, Turkish food, prepared by Mexicans!) with Dick Haines, Larry Lemke, Brian Smith, and Ruben Uriarte (8). I presented my recent research, asking for help. Flamine was impressed by Dick Haines' warm, helpful personality: a conscientious leader as he summarized for us the work of NARCAP, his international contacts, and his analysis of the latest Chilean video, released by General Bermudez.

Larry Lemke has picked up a rumor that Homeland Security was getting interested in the UFO phenomenon because of its 'intrusion' aspects. Like objects going through walls maybe?

Hummingbird. Wednesday 19 April 2012.

The former director of MUFON, Jim Carrion, who once signed an agreement with BAASS to supply a flow of fresh cases, has blasted his former partners in a sharp memo. He's critical of the project's foray into Brazil, "trying to purchase UFO investigations and research from the major players." He says that the team "managed to spark anti-American sentiment… wanting information but not allowing it to be used in return by the local organizations."

Carrion concludes: "This anti-American sentiment was so strong, I could almost hear the chants of '*Yankee, go home*' as the crowd was riled up." I can't defend what the project did since I never knew the details. The teams misunderstood the culture—and the key events.

Hummingbird. Saturday 21 April 2012.

Bill Calvert has met with the investigator on our Midwest case. The father is now 80 years old and suffers panic attacks.

This morning, I voted for François Bayrou in the first round of the French presidential elections, as I did five years ago. Bayrou is an academic, not a proven leader, but support for him points to a desire for a more honest statement of the country's problems.

In the afternoon, Flamine and I drove down to Almaden in splendid weather, delighted with the lavish landscape of the Santa Cruz Mountains. We met Larry Lemke there and heard a fine lecture about the Almaden Air

Force Station, the 682nd Radar Squadron, and 'the Box,' a fascinating radar facility dismantled in 1980.

Hummingbird. Sunday 22 April 2012.

Last night, an urgent message from Wisconsin. A witness we were hoping to interview said he was ill (he sounded very weak), so no meeting took place. I suggested an initial exploratory trip with Flamine during which we would simply meet peripheral people, without pressure on principals. It would be an opportunity to visit the Effigy Mounds (9) and the banks of the Mississippi, retracing the steps of French explorers.

Hummingbird. Saturday 28 April 2012.

Today, my son will sign the documents for the purchase of his French country home, leaving America behind. Here I wait to speak with Dmitry Manakov who arrived last night. I am pleasantly tired after a long evening with Alain Dupas, his wife and son, and our friends Pascal and Anne Bouillon on vacation from New York.

Flamine and I have attended a presentation with Tim O'Reilly and a band of young programmers from Code for America. Two journalists, Paul Radu and Drew Sullivan, lead a group to track down corruption through the banking system, a dangerous but fascinating process, not unlike hunting crocodiles in the jungles of the Congo.

Now the CNES sends me the latest notes from the GEIPAN meeting in Paris. They reviewed nine cases. The first one comes from a pilot of an ultralight who saw a flattened sphere flying about 50m above ground, faster than his own wing, in 2008.

The second case, in 2009, involves a bright light, less than 100m from the ground. The third case, on 31 July 2008, is reviewed by Louange; witnesses saw a formation of three triangles moving east to west, with 9 lights, near Rambouillet. Others are in the same vein.

Hummingbird. Sunday 29 April 2012.

Concerned about silence among our once-busy team, I found out that BAASS was neither dead nor alive, still looking for a home. The plan retains credibility in Congress. The intricacies of the bureaucracy are the source of the delay. "Questions of timing, *and also of what could happen in the future*, determine who gets money."

"It's likely there will not be just a single contract to one project, as in the past," I was told. "Contracts may go to Dr. Green, to us, and to you at your company. Bob is busy with his space station; the old staff is gone." This means Capella is no longer maintained, a real shame.

We returned to the topic of Mrs. X who once claimed there once was a reverse-engineering project at TRW, called Zodiac. Her background is verified (including her badge number at Area 51). The Collins Elite was involved as early as the 1970s. She felt the topic was becoming too weird…there were twin aspects in the phenomenon: (i) a nuts-and bolts, spacecraft aspect and (ii) an ancient control system manipulating religious ideologies…"I've been coming closer to your view, in recent times," my friend said with reluctant laughter.

Hummingbird. Monday 30 April 2012.

Flamine and I enjoyed lunch aboard the Delta King today, watching the Sacramento River and the quiet old City. Dmitry was in Oakland last night for dinner with Maurice Gunderson, so we're able to review the negotiations with RosAtom, our most likely lead investor.

A recent blast by Gary Bekkum, of Starstream Research fame, author of *Spies, Lies, and Polygraph Tape*, has stirred up the web: "Gary believes he can prove that senior officers in Intelligence have, over decades, created *a set of memes that have unintentionally distorted reality about UFOs and remote viewing*." He believes that the CIA's loss of the RV program resulted in its transfer to an organization where it had become an operational SAP in 1995.

Another hot topic is the infiltration of the Intelligence community by groups with cultist agendas and new Catholic sects. General Reinhard Gehlen, of Hitler's SS fame, was a member of the Knights of Saint John, and emissary to the Vatican. He later recruited his entire network for the CIC of Allen Dulles. After 1945, Operation Paperclip, in Los Alamos…

United flight to Chicago. Saturday 5 May 2012.

I'm finally on the way to re-investigate the old Midwest encounters, retracing the steps of Allen Hynek. The story is just too hot to be written up, however, even after all those years.

Hummingbird. Monday 21 May 2012.

Facebook went public on Friday, at a record initial valuation of 104 billion dollars. SpaceX is at work on the Falcon9.
 Flamine tells me of her sadness, her distant mood; her mother died twenty years ago. I think of my own lingering pain, of my daughter and my son, on every trip to Palo Alto, as I pass our hill of Belmont, and become infused with that sadness.
 Yesterday evening we watched the solar eclipse, with a simple pinhole device, a reconnection of sorts with the greater world.

Hummingbird. Monday 28 May 2012.

On Saturday, popular authors John and Rickie Shirley had invited us to a garden party at their house in Pinole, a quiet suburb near Vallejo. Rudy Rucker, whom we'd met in Brussels, was there with his wife, Sylvia, as well as half-a-dozen 'post-modern' science-fiction writers and illustrators: Terry Bisson (Hugo Award), physician Michael Blumlein of UCSF, Marc Laidlaw of *Half-Life* fame, artist Paul Mavrides, and Jay Kinney of *Gnosis Magazine*. Yesterday, Memorial Day, we had dinner with Diana Hall, watching the sunset from the Cliff House, foggy San Francisco and its magical overtones.

Hummingbird. Sunday 3 June 2012.

Last week, a pleasant dinner with Luisa Teish at Bateau Ivre in Berkeley. Then today, in fine weather, we drove north to Novato to visit the Ravenhearts. Oberon and Morning Glory are wise and kind, many years in love and dreams of magic. Their collection of goddesses has expanded into a joyous temple. Today, a rare solar transit of Venus, the last one I'll be able to watch in this lifetime.

Hummingbird. Thursday 7 June 2012.

Splendid weather: the mild rains triggered an explosion of blue blossoms and flowery trees, red, yellow and purple bushes, birds everywhere. We powered off the computers and went to the lake.

Maurice returns from China today; new plans for space ventures.

7

L'Escoublère. Sunday 24 June 2012.

The castle hides its turrets and battlements among the solid farmhouses of Anjou. A vision of medieval grace, l'Escoublère stands like a time machine: a fortress with a wide moat, once used by local Chouans, with two massive towers guarding the bridge. A corner building on the side is a castle by itself, while the two-story home is covered in blue-grey slate.

Flamine's aunt received us with her husband. Both are medical doctors. They inherited the place from Flamine's maternal grandmother, Hélène de Montéty who practiced law in Paris and bought the property in a decrepit state in 1933. Inside, the quiet majesty of a country manor has been restored with a vision to keep the 14th-century fortress alive at the dawn of the 21st. The walls, furniture, and plumbing that suited the Du Guesclin warriors and the rebellious Chouans are impractical in the days of Airbus and the Internet. Or are they? Flamine and I slept soundly in the large bedroom overlooking the moat, waking up to birdsongs. Today is Flamine's birthday; we'll visit her cousin near Le Mans. Yesterday in Angers we admired again the tapestry of the Apocalypse.

Later the same day.

Shortly after we arrived in Paris, Flamine and I had dinner at Dominique Weinstein's home. He commented on the WW2 angle in ufology, especially from the German side. In spite of the 'evidence' that accumulates about Roswell, and of my friends' belief it was an ET event, I still wonder if the case wasn't part of an early biological/space experiment gone wrong.

There were massive balloon tests with humanoid dummies at the time. Were there also follow-ups to Nazi tests with radiation components?

The Germans had just been condemned at Nuremberg before a horrified world, so this was no time for a revelation that SS doctors were doing morally questionable work in New Mexico alongside our own scientists and military. The public and the media had developed a convenient fascination for saucers, so the crash story may have offered the best cover-up. After an initial controlled leak, every attempt at explanation and every claim of censorship and even every denial would actually reinforce the fake story, *all blind alleys*.

"Is that a possible theory?" I asked.

Dominique took another serving of Mady's excellent dishes and answered, "No way to be sure. Whenever there's a danger of someone digging up part of the truth, another 'official' explanation gets released, a layer of obfuscation: mogul balloons, dropped dummies, Stalin's mood..."

He gave me a copy of his latest pilot sighting catalogue, confirming that UFOs behaved differently when confronting military aircraft. Yet his analysis of the foo-fighters simply concludes that some German secret weapons didn't work very well. In the end, we agreed the weight of evidence was for Roswell to be a genuine crash; but a crash of what?

Paris was gray and muggy when we went to see *Margin Call*. A rainstorm on Wednesday didn't relieve the pressure, or the pollution.

Mabillon. Tuesday 26 June 2012.

The wonders of this area are the bookstores, the cafés, the bakeries, and specialty foods, yet our tastes are frugal; we've been living, breathing, and reading books. We make room for Flamine's collections in psychology and spirituality, throwing away yellowing tomes that had followed me. On my bedside table: Laura Huxley's poignant, insightful *This Timeless Moment*; Mikhail Bulgakov's *The Master and Margarita,* and Paul Claude Racamier's *Les Schizophrènes*.

Another gray day. I'll join Philippe Richard for lunch. The French news is about Europe, a backdrop for politicians' egos. The new socialist minister of startup innovation is a talented young woman with no technical background who's never worked in business.

Pleasures: the quiet night, walking with Flamine through Paris, or a dash across the street to our Thai restaurant; the next day, lunch with

Claire and Olivier, happy plans for their home in the country.

The physical reality of the phenomenon remains obvious (my latest field investigations: Soissons, Marley Woods), so is their ability at camouflage as airliners or geometric forms complete with a field of stars, when they don't simply manifest as blue baseball-size lights. We know that. But what really defeats our attempts at rationality is the control of witnesses. Aimé Michel had tried to unravel it; if the phenomenon can set the circumstances of the observation, we have nothing to go on but elusive traces in the ground and the occasional metal residue. Even the humanoids could be a lure, clever manipulation. *The extraterrestrial theory explains nothing.*

Yet 'somebody' is doing this, with access to seemingly magical tools. What value do I bring? Should I stay in the game? I do have the hard data...How can I use it best?

Mabillon. Thursday 28 June 2012.

On the heels of the triumph of François Hollande, supposedly a 'normal president' for a change, the French parliamentary elections have consolidated a government with a toolbox of expediencies, while a renewed racist fringe on the right captures the frustration of the downtrodden: blame immigrants, ostracize the Muslim... It works every time.

A dream last night: In the middle of a confused scene (a group of us were chasing a cat that kept changing colors), Janine came into my arms, tragic and poignant. She spoke tenderly, with deep regrets. I held her in sorrow. Then she was gone; I woke up.

Michaël Vaillant came over for lunch, with his big laptop and reams of statistics. He's conducted clever data mining of the CNES files to explore the sociology of the reports, something American researchers never seriously attempted. He believes there has been not one but a double series of waves, having reconstructed the method I used to extract the pseudo-random patterns in *Invisible College*.

Air France flight to San Francisco. Saturday 30 June 2012.

Yesterday, our last day in Paris, I had breakfast with Yves Messarovitch and lunch with Alain Dupas, who pointed out that Hollande, who rose to power against the leftist Barons and improvised a campaign to fill the gap left by the shattering of Strauss-Kahn's ambitions, had no obligation to anyone and thus could shake up the French system, if he found the courage.

Yves was stressed as he often is with decisions at BPCE bank, witness of the slow progress of Europe.

Willits, California. Friday 6 July 2012.

My son and grandson have joined us, so I'm taking Maxim on an excursion to the redwoods, an adventure aboard the historic train known as *The Skunk*, which winds its way up the coastal range across forty spectacular miles of deep forest on the way to Northspur.

The towering, ancient trees (some 700 to 1000 years old) display a standard of rectitude, the wisdom of beings that have survived the vagaries of our so-called civilization. They make me nostalgic about Spring Hill: *The trees are drawing me near*, as the Moody Blues used to sing, not long ago, so long ago.

Hummingbird. Saturday 7 July 2012.

Nothing like a journey through the silent, secret green world (aboard a quirky diesel-powered train car dating from 1935, a moving museum) to jolt you out of a disagreeable series of interactions over the Internet. Somehow, we came to discuss disinformation and the people who practice it. My friends wondered if 'Agent Doty' was leading us into wild chases and urged me to stay out of it.

When I asked who profits from manipulating UFO information, I couldn't get a straight answer; everybody started blaming others. It was good to run away to the clarity of the groves where the Noyo River plays on beds of pebbles. Maxim caught three frogs.

We took a side trip to Redwood Valley to show Flamine what was left of Spring Hill Ranch. Our neighbor Marie Cleveland, looking very frail, happened to be checking her mailboxes. She was practically blind, and confused, yet she gave us permission to walk on her land for a better look at our old ranch. Maxim couldn't believe we'd given up on such a beautiful place. I had to explain why we simply hadn't been able to keep the dream alive when the drugs ruined everything and the shooting started, even as the grand old trees called us back.

Red Bluff. Monday 9 July 2012.

We drove on to Mount Lassen, the southernmost volcano in the Cascades range. We happily climbed up the mountain, enjoying the gentleness of the weather near the timberline, the creeks bouncing across vast green pastures, icy mountain lakes of cerulean and turquoise, and patches of snow, some of them three feet deep, dotting the slopes. We drove back along the western side of the volcanic landscape studded with black boulders through hamlets lost in time in magical disdain of civilization.

Hummingbird. Friday 13 July 2012.

Robert Guffey's book *Cryptoscatology*, devoted to "conspiracy theory as art form," captures the spirit of our current era of extreme political falsehoods, polarized creeds, and violent movements. Guffey quotes William Burroughs: "Control is controlled by its need to control." Burroughs also remembered waking up when he was four years old and seeing little gray men playing around.

A professor of English at CalState Long Beach, Guffey deconstructs the current fascination with conspiracies: He goes back to the Values and Lifestyles classification invented at SRI; he explores the 1951 CIA-funded LSD biological tests in Pont-St-Esprit in Southern France, which may have led to Franck Olson's 1953 suicide; and he cites Col. Corso's references to "camouflage through limited disclosure."

General Nathan Twining once said: "The cover-up is the disclosure, and the disclosure is the cover-up," while Marshall McLuhan summed it up: *"Only puny secrets need protection. Big secrets are protected by public incredulity."* We have an example in the Project Serpo material (10). How can anyone take such garbage seriously? Perhaps we should, for another reason: Serpo includes classified material on two biological warfare matters, *leaked verbatim*. Parts of the releases were written by a staff officer who wasn't arrested, leading to speculation it was an Active Measures program, with unknown motives.

On 12 July Eric Davis wrote in support of the credibility of Roswell: "Recall that back in our old NIDS days (ca. Jan. & Feb. 2002), I twice interviewed 83-year-old Brig. Gen. Arthur Exon (USAF ret.) who met Pappy Henderson on the tarmac at Wright Army Airfield while his crew was offloading the crates of UFO wreckage that he flew in from the 509th in

Roswell. He showed Exon some of the wreckage with unusual metal Exon was allowed to hold.

"Exon confirmed all this to me. He flew many TDYs (11) to Roswell; not long after meeting up with Henderson to see the debris, he flew there again on a daytime Army flight. Exon immediately saw where it was, plain as day. He described it as something that had crashed into the ground at high speed and then slowed down as it came to rest. There was a very long debris field that included charring from fire. It appeared to have been scraped clean.

"On a later TDY to Ft. Worth, he spoke with Lt. Col. Tom Dubose (Ramey's chief of staff) who opened the classified safe containing Gen. Ramey's Roswell files. He pulled out *photos of the alien bodies*, photos of outside and inside the remains of the craft, and some of the debris. The craft was broken apart but there was significant structure left. Exon told me that *the bodies were not human, and their craft was definitely not of human design.*"

Then, Gen. Clements McMullen went into a rage over the phone with Ramey and ordered him to fabricate a cover story.

Hummingbird. Wednesday 25 July 2012.

In anticipation of "LoneStars-III" in Toronto next month, the group has reviewed the research issues. Actually, I should qualify 'the group,' because Hal and Colm don't say much, and John Schuessler only participates by forwarding press items. Most of the discussion was between three of us, and it became strained. Last week, Dr. Green took sharp issue with my statement that the data was incomplete and biased, and that scant analysis was being done.

"Why is it 'silly' to ask what the data means, where it comes from?" I asked, "Or the context of the injuries? Shouldn't we investigate the cryptids? Shouldn't we wonder, What's next?"

Kit called this morning, and we spoke for over an hour. We resolved our differences in interpretation of what each of us called "The Data" (which he called "fungible, precious and fractionable") and we went on to discuss the agenda for Toronto. Colm will have to take vacation to be with us. As for Bigelow, he won't be coming. Right after LoneStars-II, he let go everybody at BAASS except for Colm, now re-assigned to the aerospace company, marketing a plan to study space-borne diseases and bio-threats.

The phenomenon is hurting people. It is manifest and scary. The circumstances are evil, as in actual injuries. And what we've called "interference" is very real. As much as I thought the Collins Elite was wrong in its fundamentalist attitude, I wouldn't blame anybody for talking about demonology here, as in Marley Woods. The flaw in that 'explanation' is that it doesn't explain anything, of course.

Hummingbird. Monday 30 July 2012.

An agent for East coast financier Larry Frascella, who got my name from John Alexander, has invited me to a gathering in Bucks County, Pennsylvania: "Your gracious host is just a true believer." I responded that my interest centered on the science, not belief.

This afternoon, Flamine and I attended a talk by Amir Weiner, associate professor of Soviet history at the Hoover institution. This is the man who gathered the archives of the KGB captured in satellite countries from Latvia to Czechoslovakia, analyzed them, and got them transferred to Hoover. Weiner spoke of the "schizophrenia" of the agents: they had a very high level of professional ethics... "even as they committed indiscriminate targeting of their victims."

They thought of themselves as communist heroes, "the only honest people." Yet, *they had none of the imagination or creativity we often attribute to them.* They knew little, and cared little, about the society where they worked. They did know their targets well, however; when they returned into those countries after the defeat of the Nazi, they had sophisticated card indexes. It only took 200 agents two months to completely take over each country.

Their method involved interrogation ("Who do you know?") and fake trials to get new information, as well as fake organizations, false Zionist movements, and blackmail. They knew their data was bad; after Stalin's death, 80% of convictions were thrown out, two thirds of the informers were fired. In 1991 the whole organization collapsed under the combined pressure of economic reality, the information revolution, and the lack of a will to fight reforms.

Following his talk, I approached Professor Weiner to ask what he thought of the situation today, when any dictatorship—or simply any powerful element in society—could find out "who do you know."

"They can trace every one of your movements," I said, "from your smartphone, your Facebook 'friends' file, or Twitter stream, without the clumsy methods of the old KGB. Doesn't this imply control of the public? Couldn't it distort or defeat democratic government?"

He simply didn't understand the question. He said his kids spent all their time on Facebook and Twitter, and they were happy. There was no way to stop it, and he didn't see anything wrong with that.

Hummingbird. Tuesday 31 July 2012.

The second volume of John Fowles *Journals* contains an appraisal of Colin Wilson, who wrote to him as *The Magus* was published "to say how pleased he was to at last have a disciple in the wilderness." An error because Fowles, whom I admire greatly, is a writer of mysteries who doesn't really believe in mysteries, while Wilson definitely does.

Fowles goes on to say that he feels sympathy for Wilson's "terrifying earnestness and obsessive need to publish," and his "bewildering unclouded vanity: 'In fifteen years' time I shall be seen as the most influential thinker in Europe'! And reflected glory: 'We are the only two major writers in this country.'"

Fowles concludes, observing Wilson's cannibalistic tendencies: "Sooner or later I shall have to inform him that I am not his dinner."

His diary discusses the writer's life: "To be a writer it is not enough to jettison what is bad in you; the good also—the good you might have been, or done, in other fields—has to go too. It is certainly more difficult to get rid of this, and much more important."

Gary Ross, a successful writer-director from Hollywood, works on a new movie about UFOs and has read my books. According to his assistant, he wants to invite me into the production. They catch me at a time when I am unsure about the deeper nature of the phenomenon, its potential danger, its exploitation, and the circus the believers create. I must seem very arrogant, but I feel like staying away.

Now a nice note arrives from Flamine's uncle at the Château de L'Escoublère. He speaks of his pleasure at seeing us as a couple *tellement uni, presque fusionnel* ("so united, almost fused together").

Hummingbird. Monday 6 August 2012.

Berkeley in the morning, all freshness and congeniality: flowers in all the bushes, birds in every tree. Tennis players on the courts at the Claremont Club & Spa as I arrive with Maurice and August. Dmitry and two analysts have built a financial model for the financing of Danotek by RosAtom, but there are errors, and August spends days correcting the tables. Could this help launch our proposed fund? Sergey Zhukov is back in California, at JPL this time, to watch the landing of the Curiosity Rover on Mars, where it will investigate the geology of the planet and search for signs of life.

Flamine and I watched it too, last night, on a big screen, a magnificent demonstration of the progress of robotics and celestial navigation. One now feels that the mechanical part of the trip to the planets has been mastered, the biological part still in question.

Calling Bob Bigelow in Vegas, I found him fully in control of his agenda and objectives with Boeing, the United Launch Alliance, and NASA. He said he valued the fact that I had "both a science mind and a financial mind" and invited me to review his business plan.

Hummingbird. Tuesday 7 August 2012.

A new book by Milton Leitenberg and Raymond Zilinskas on the history of Soviet biological weapons deplores the CIA's 'horrendous mistake' in creating disinformation in the late 1960s about a clandestine program in the US: *This actually reinforced the Russians in their paranoia.* Commenting on this, Dr. Green stressed the strategic reach of the 'Active Measures' Program...The same program that may oversee the manipulation of the UFO memes and discourage open research by scientists like me.

Today, Maurice, August Fern, and I were among the participants when John Holdren, head of the White House OSTP (Office of Science and Technology Policy), invited some 30 managers, investors, and civic leaders to discuss measures in favor of entrepreneurship.

When the time came for questions, I rose to ask why we still enforce the ITAR regulations that the Obama administration had promised to review four years ago? Holdren answered that sadly he'd held many meetings on that issue, but too many power plays blocked every solution, even as everyone agreed that the specific regulations were obsolete, hurting American industry.

Hummingbird. Friday 10 August 2012.

Colm sends a note about the latest cryptid discoveries. Over 100 samples have now been gathered from multiple locations (15 states, two Canadian provinces) and subjected to electron microscopy, histology, and DNA analyses. The mitochondrial DNA is human, with several separate phylogenetic patterns, which means…what?

The researchers claim both double- and single-stranded DNA, and a mosaic of anomalous sequences "very distantly related to primates." Three were good enough for full sequencing, showing "highly similar creatures sampled at three different times in three different States." The data are embargoed by a leading journal but there is no guarantee that the discoveries will ever be published.

Hummingbird. Wednesday 15 August 2012.

Lunch yesterday (at the lovely outdoor garden of Arlequin, but in cold weather) with the author of *TechGnosis*, Erik Davis, and his wife, Jennifer. Mark Pilkington was there with an English girlfriend, and David Pescovitz of *BoingBoing* fame presided in his usual gentle manner. Mark had brought his laptop with the movie for *Mirage Men*, which he'll present at Sundance. It shows long interviews with Doty and Linda Howe.

The movie makes it clear that in a weird combination the US military (i) covers up the real phenomenon and (ii) works hard to convince the public that UFOs *do exist*. The scheme, as I claimed before, uses false (planted) information and is as devious as ever: When someone accidentally observes a test they shouldn't see, you simply "reveal" to them they've glimpsed an intractable UFO.

The weather has been characteristic of San Francisco: gray and foggy in the morning, turning to cool, windy filtered sun in the afternoon. I am half-way through the very sad second tome of John Fowles' *Journals*. Also today at Fields, the legendary esoteric bookstore on Polk, I discovered the *Occult Reliquary: Images and Artifacts of the Richel-Eldermans Collection*.

Hummingbird. Thursday 16 August 2012.

Flamine leaves early nowadays, to attend English conversation classes, preparing herself for re-entry into the busy world after her sabbatical in my crowded universe of friends, books, and saucer stories. Back in Moscow, Dmitry has been busy educating RosAtom about venture capital (an amazingly complex task, since they have some 125 separate corporate branches under their heavy bureaucratic umbrella inherited from soviet days), guiding them with Maurice's help in the art of investing. He's also trying to teach a framework for financial innovation at Roscosmos, which finds it hard to recover from a series of rocket failures. This includes a crash of their long-time workhorse, the Proton rocket, which destroyed two communications satellites on the very same night when JPL did the perfect landing of Curiosity on Mars, added humiliation for a once-proud Russian space program.

Flamine and I walked over to Kevin and Sheila Starr's for dinner this evening (12). Their apartment is an accumulation of books and souvenirs, including the many honors Kevin has received from Harvard and other institutions, and the National Humanities Medal. A painting of a California scene with a lake where he swam as a child dominates the living room—"bought on credit..."

Kevin's conversation is brilliant, combining personal recollections (Saint Sulpice, Catholic history, the Bohemian Grove he visited with Giscard d'Estaing, and cultural life in LA) and a display of erudition that ranges from the theology of *ressourcement* to the political life of California and the French classics. He keeps a portrait of Balzac on his desk for inspiration, and a manuscript page with furious, hand-drawn corrections by the great man, who died at 51 of working too hard with too much coffee, as Sheila reminds us: Poor Balzac!

Overall, a wonderful evening as the fog rose around us, arguments with Flamine about sacred music (Couperin versus the moderns?), and questions about the coming elections, the Mormons, the economy. Sheila is discouraged: "The government can't be trusted with healthcare, even social security is broke."

Fig. 9: *NARCAP meeting in Silicon Valley, 16 April 2012. From left: Richard Haines (NASA), Bernie Haisch (Lockheed), Larry Lemke (NASA), Flamine & Jacques, and Brian Smith (NASA)*

Fig. 10: *LoneStars-III meeting at G. Hathaway's Toronto lab, August 2012*

Hummingbird. Saturday 18 August 2012.

The San Francisco Choral Society produced *Carmina Burana* last night, a brilliant, colorful, visually and musically stunning performance. Carl Orff's German soul was troubled by the stupidity of World War I and driven into a universe of angry, lustful humor, rejecting conventions.

We enjoyed it from one of the boxes with a plunging view over the orchestra, the choir, and the dancers.

Toronto. Tuesday 21 August 2012.

The Crown-Plaza near the airport. The LoneStars have assembled here, minus Bob Bigelow and Colm, (finishing a big proposal to NASA for a life science center) and John Schuessler. The main attraction of this third meeting is George Hathaway's laboratory in Mississauga, where he tests materials and conducts research based on insights from Hal and on experiments on future devices. (Fig. 10)

George's lab is a large, modern facility in a technology park. It contains several airy offices, a small conference room where we sat all day today, and large spaces filled with measuring equipment, experimental setups, optical tables, ovens and presses, industrial-strength drilling machines, and even a large, insulated Faraday cage that can be used by several people with equipment.

George is at work on several projects, including tests of antigravity devices with gyroscopes and complex pendulums, an electromagnetic pulse experiment, and an analysis of materials allegedly picked up at Roswell, obtained through Art Bell and Linda Howe (unfortunately, with no real evidence of provenance).

We handled one sample, a lamellar metallic object about 5cm square and 1cm thick, composed of 20-micron magnesium and one-micron bismuth sheets. There is speculation that the material could transmit terahertz waves.

The first presentation expanded on the medical case studies. Many things have changed in the last three months, possibly in anticipation of greater disclosure. I remain cold to that idea. This is still too narrow.

I was glad, back with Flamine, to speak of normal things. We had a splendid dinner at Scaramouche after a tour of this large city. Discussion returned to the complexities of the research, with little funding and no access to the resources of Big Science.

Our sponsor told an amusing story to illustrate our predicament. "Two frogs once fell into a pail full of milk," he said. "One of them tried to swim, was not able to stay on the surface, and drowned quickly. The other one kept furiously paddling and kicking, so hard that eventually the milk turned to butter, and she was able to jump out. Which frog do we want to be?"

He also reiterated that he'd been strongly influenced by *Messengers of Deception*. As dinner was served, I was sitting across the table from George, so I asked him what kind of materials he was testing, in the furnaces and ovens we'd seen in the lab.

The material derived from a biomedical engineering professor at Arizona State University, working with a remote viewer.

Flying back to California. Thursday 23 August 2012.

The lab visits and presentations are over, but I found out that Kit was staying over for an afternoon flight, so we sat down after breakfast to review our impressions, and to discuss the prospects for a serious project that could replace BAASS, now in limbo.

We were all very clear about the recent history of Pentagon opposition and Washington shenanigans. The first attempt to set up a research SAP (which presumably would have access to the Holy Grail) had been denied *on religious grounds* by an assistant secretary after being approved by intermediate levels, including Homeland Security. However, new project preparations have gone so well that the necessary appropriations are included in the prospective 2013 budget.

"So, what's the next step for us?" I asked, with fearful visions of another delay that could make any recommendations irrelevant.

Kit leaned back in his patio chair, shielding his eyes from the rising sun, and said: "Just keep watching CNN. When you see that Congress has passed a new budget, instead of a series of continuing resolutions as it does now, you'll know the project is funded. But there's no interest in AI or in your databases: it's all about cryptids and orbs now, physiological effects, and propulsion theories."

Good luck, I said.

Hummingbird. Tuesday 28 August 2012.

Lunch today with Maurice (at Zibibbo) along with Gary Hudson and Eric Laursen of Hyperion, about the possibility of early funding of their new rocket company through RosCosmos.

I knew Laursen in the late 1980s when Peter Banks and I formed EarthData. Maurice told me that the Danotek deal was likely to close, increasing the probability of success for Radiant, the first fund.

Hummingbird. Wednesday 29 August 2012.

For Mark Pilkington in his *Mirage Men* movie, Doty leaves the impression that he's used in a "professional deception" role, and that no document about UFOs, Roswell, or alien abductions can ever be trusted now, *even a Presidential briefing*, given the extraordinary lengths to which official parties may have gone to twist them.

I should remember that those lies are legal...which puts Roswell itself in question again, naturally.

Hummingbird. Friday 31 August 2012.

As I review my notes of the Toronto meeting, the result is disturbing. The awareness that we're swimming in "active measures" (13) that distort what little truth filters from government biases rational enquiry. It disgusts me to witness a process that works in such disregard for human freedom and intelligence (lower-case i).

As Janine once said in a rare moment of discouragement, "We're already bewildered by the universe we inhabit; we don't know where we come from, or what happens to our consciousness when we die. So, why do we have to support governments that are biased, distorted, full of lies, and pointless secrets, leading good human efforts to senseless failure? What ugly end does that system serve?"

Even frank explanations of the Active Measures program don't make much sense. We're told that such measures are only built to trick "one small group about one small thing," but that obviously isn't true when Hollywood is fed a fake story and broadcasts it to the entire world as Trust in Technicolor.

The lesson of the Toronto meeting, for me, is to shrink from public debates, and to rely only on direct observation and personal contact with real witnesses in research. *All else is suspect.*

One consequence of the misinformation from Washington is that other countries no longer take revelations seriously when they come. In Paris I've seen senior people treat reports of American UFOs as laughable fabrications designed to hide the shenanigans of the CIA, and they were not always wrong. I also recall Simonne Servais' amused comment towards *nos amis Américains* in the days of Ira Einhorn and Comet Kohoutek, when an office in Washington kept feeding her garbled New Age notes, raving about Soviet psychotronic weapons.

Hummingbird. Tuesday 4 September 2012.

Reading again an old letter from Aimé Michel (whose middle name was Louis) written in the early 1970s, I rediscover this forgotten fact, which delighted both of us at the time: In 1620, when a nun named Elisabeth de Ranfaing was declared possessed by the devil in Nancy, two of the exorcists who examined her were Jacques Vallée, *curé de Notre-Dame de Nancy*, and Louis Michel, *Supérieur de l'oratoire de Jésus-Christ!* (14)

Las Vegas. Wednesday 5 September 2012. Hotel Luxor.

Flamine and I took a cab to Park House this morning, meeting Bob Bigelow at his principal office. Bob and I sat in the meeting room and he launched into his reasons for wanting to see me: "I want to tell you about my plans for the aerospace company," he began, going into some of the financial details. He'd prepared a very clear PowerPoint presentation, which we discussed at length.

I imagined how traditional investors would look at the company in the current context, trying to relate it to various categories of structures that can use advanced venture tools. My own work has involved various financial minutiae like the treatment of loss carry-forward, pass-throughs, lease blocks, and the value of naming rights for Space. I feel rather dismissive of NASA. Eight years after the successful flight of the X-Prize, no progress has been made to support space tourism, while Dubai now owns 30% of Virgin Galactic.

Bob knows a favorite Sicilian restaurant in town, so he drove me there in his Cadillac Escalade for an informal lunch and conversation. He asked what I thought of the marks found on people's skin. I now regret not following up with him. We did agree the cover-up must be the work of a supra-national group ("a Brotherhood," he said) rather than one single agency, or country.

Las Vegas. Thursday 6 September 2012. Hotel Luxor.

We visited the Grand Canyon today, in all the splendor of a dry, warm summer afternoon, with only a few clouds when we left Guano Point. We stayed away from most tourist attractions, avoiding the skywalk at Eagle Point, but we walked to the edge on our own, to absorb the magnificence of the tortured landscape. In the evening, we were still under the spell of the abyss with its changing colors, in the sparkling confusion of Vegas.

Hummingbird. Sunday 9 September 2012.

Reviewing the notes from some recent industry conversations and recalling the details of the most illuminating close encounter cases I have studied, it seems plain to me that the LoneStars, smart as these gentlemen may be, have wandered away from the actual depth of the phenomenon. But perhaps it's necessary to take this weird excursion to see the problem under new angles?

Bob thinks of approaching Canada, Japan, Sweden, or Brazil for investment and the leasing of his space stations. Today these countries can hardly fly on the 1300 cubic meter ISS, which they contributed to build, because three seats in every Soyuz are dedicated to Russia and two to the US, plus one to a service technician, leaving a single extra to represent the remaining 6.5 billion people on the planet... at $56 million a seat, soon to become $63 million to Roscosmos!

In contrast, Bigelow's BA330 (at 330 cubic meters) can house six astronauts. It features encrypted communications and a 25-kW power supply, with a hydrazine system for attitude control—a very modern facility in space that can be leased in 'blocks' at $25 million dollars. Each block provides one-third of the craft for two months, the equivalent of a full volume on the ISS. The taxi service from Boeing would carry two BA astronauts, cargo and clients.*

Elon Musk has identified 54 customers for SpaceX and will soon be at breakeven. On October 8th, he has a big launch for NASA, which pays him 1.6 billion. (Gwynne Shotwell is the president, much admired by Bigelow for her business savvy.)

Bob believes that Earth observation, as a market, will be saturated in six years. His potential clients are energy companies, pharma, and medical firms but neither Ariane (fairing problem) nor Proton (space is too short) can be used to deploy the BA330, as the Bigelow station weighs 22 tons and can only be launched by an Atlas-V or a Delta-IV.

NASA has wasted time and money on the 10-billion-dollar SLS, a 70-to-84 metric ton rocket, and on Orion that has limited capability since it needs a Delta-4 rocket to reach LEO.

Hummingbird. Thursday 13 September 2012.

Reading Jacques Blamont's reminiscences about his life in French science (15), I am amused by his remark that in 1954, while "flying saucers were the rage in Provence during our tests" (of a Zeeman instrument to assess the presence of sodium in the high atmosphere) "we never encountered a single one, although we were flying several hours a day among the UFOs…"

When he describes the details of his work, however, it is amusing to find out that he was "in the cockpit of the two-seater Meteor, an oxygen mask over my face, squeezed into the navigator's seat, I had to manipulate an instrument located in the nose, using its electronic equipment and batteries that squeezed me in my seat."

Hardly optimal conditions to detect unknown objects! The most disappointing fact, here again, is to find such a bright scientist who automatically assumes that any UFO would be a craft coming from space and flying like an airplane and among airplanes, giving no thought to alternative hypotheses for the phenomenon.

New York Times columnist Mike Malone gave a lecture this evening before a group of senior executives and investors, in which he spoke about 'the Great Inflection' from the point of view of Moore's Law and called

---

* As we now know, 13 years later Boeing is still unable to safely fly two crewmembers to the ISS and bring them back safely.

the Internet "the defining technology of our time." He pointed out that since 2005 we've added 2 billion consumers (including a billion animists who have never been aboard a plane or a car).

In Europe, by contrast, "they think an entrepreneur is someone like a shopkeeper, not a creative agent in the economy, so how can they survive and thrive?" In the US the next big changes must impact education, which is incompetent, inefficient, and corrupt, and television, which is irrelevant and silly...

In response to a question about the weight of the military on Silicon Valley, he said that the era of Pentagon influence lasted from 1956 to 1970 with many companies, like Hewlett-Packard, wisely rejecting the 'opportunity' for military contracts after that date.

Hummingbird. Saturday 15 September 2012.

Colm posts a summary of our conclusions: "In-depth research, unprecedented since 2008, has shown that the UFO phenomenon is a threat to human health and well-being."

He lists "three inter-related aspects to this threat recently emergent from a close scrutiny of the data" as follows: (1) Transmission aspects: It can "invade" households, family groups, and communities and, acting like an infectious entity, cause these small groups harm. (2) Brain lesions now documented in dozens of cases ... a characteristic footprint in the brain detectable by MRI. In a subset of these individuals, symptoms can slide into dementia and death. (3) Profound disturbances to immune cell number and immune function in multiple individuals, as measured by alterations in circulating blood neutrophil-to-lymphocyte ratios, sudden autoimmune antigen expression, and increased circulating liver enzyme levels.

Colm adds: "The emergence of a cascade of autoimmune symptoms in individuals and family members who've had CE experiences will be explored in future studies. Here the emphasis is in treating the immune system as a trillion-cell information processing organ, with ten times the capacity of the brain, which is capable of recording insults from CE/abduction experiences."

Hummingbird. Sunday 16 September 2012.

The Conservatory of Music offered a harpsichord concert by Anthony Newman today. I loved the pieces by Bach and especially Couperin (*Dix-huitième Ordre*), but others left me unmoved. A harpsichord should be heard at close quarters in the salon of a *marquise*, not in a large modern building with plastic walls and steel furnishings! My pleasure was to walk down to Oak Street, holding on to each other like kids in love, laughing at Halloween disguises in the new garish shops.

The recent statements by Colm and Kit trouble me. If we are saying—as they do—that a close encounter with the phenomenon at less than 100 meters can trigger brain inflammation with lethal consequences, it should be possible to find evidence in recent decades, selecting 1,000 witnesses from close encounters of the Seventies and Eighties who have since died, and research the cause. Looking back over the cases I know, I don't notice any correlation.

The cases referred to here come from individuals in the military and intelligence, who may have been exposed to unusual situations, other than UFOs. I do agree that *transmission* happens, in the form of household invasion highlighted by Colm; we've already seen this in the Brian Scott case, for which I have actual recordings. Why reject the value of detailed historical analysis? I remember our early group, with Kit dismissing *all encounters* as anthropological fantasies, abductions as fugues, and cattle mutilations as the work of predators.

Hummingbird. Wednesday 19 September 2012.

The latest *Lumières Dans La Nuit* (*LDLN*) magazine mentions that Edgar Mitchell has created a new organization, Quantrek, for the study of consciousness and zero-point energy. With the unrealistic ambition typical of such ventures, he hopes to raise an endowment of $250 million (?)

LDLN, always impeccably edited by Joël Mesnard, mentions that the *International UFO Reporter*, the publication started by the CUFOS group in the 1970s, has ceased to exist, victim of the Internet and the public's lack of interest for UFO quarrels.

Hummingbird. Monday 24 September 2012.

The music of Marin Marais touches me in a dark, intense way, my mind peeled away from local reality and its expectations. To my uneducated ear, the impact comes from the way melodic phrases trend down into long interrogations, as if hoping against hope to find a solution to human existence. Last evening, we heard a concert of French oboe pieces with harpsichord and viola da gamba (the ensemble *Les Délices*) that included pieces by Marais, his *Labyrinthe* and *Suite in D*, most profound. I fall easily under the spell of his existential quest, an aura of sadness.

Today is my birthday, the 73rd. Having now lived longer than my father and having left behind many colleagues and mentors like Fredrickson and Deléage, and others younger than me, I have reached a position where I should be able to stand and contemplate the order of the world, or at least that portion I have traveled.

This I have tried to do in vain. I'll continue to collate impressions and edit them as a memoir, but the wisdom and mystery of this life eludes me, even when Flamine reassures me with her patient love.

Hummingbird. Thursday 27 September 2012.

Catching up with a pile of issues of *New Scientist*, I am struck by an article on paleogenetics that discusses the course of the brain's evolution, pointing at two duplication errors: the first one about 3.4 million years ago, when *Australopithecus* started using tools, the second one 2.5 million years ago, when the genus *homo* split off from the *Australopithecus*. In both cases a single gene (SRGAP2) duplicated itself, inducing neurons to increase their connectivity.

There may be other gene duplications (about 30 genes have duplicated themselves since we split from chimps between 4 and 6 mya) but I find it remarkable that the emergence of modern man was not due to some demiurge or to a wonderful plan of Mother Nature, but to a series of duplication errors—the kind of faulty error correction that plagues our electronic circuits.

Hummingbird. Friday 28 September 2012.

I have a question about reverse engineering. An ordinary Englishman named Maurice Ward once invented a coating with extraordinary thermal properties: You could paint an egg with it and blast it with a blowtorch and the egg would be fully protected. He took the recipe to his death, refusing to reveal it to Cavendish Labs, Boeing, or NASA, and no one seems able to reconstruct how he did it. What does that imply if we ever 'catch the school bus'? What's the likelihood that we could even find the function of any piece of UFO hardware? Especially in a compartmentalized lab where cooperation is censored?

This triggered a lively discussion on the properties of materials and novel abilities to test them. We discussed implants, referring to a professor in a department at the University of Michigan who has reviewed the biology and the metallurgy of implants removed from abductees ("about eight in the past year, and several dozen in the past 30 years"). The laboratory examines atomic-level properties, applying advanced atomic force microscopy, rapid spray ionization for nanoparticle molecular property analysis of surface and deep layer matter, and SEM. Carbon-fiber nanotubes are said to penetrate neuronal cell bodies and axon hillocks, in numerous samples, from *before* nanotubes were common.

Hummingbird. Saturday 6 October 2012.

We've had some arguments about crop circles (Eric doubts my hypothesis about a stealthy aerial device equipped with microwaves and lasers), but I've also heard the formations were the result of "an SAP consortium of British, French and US-AFOSI."

Tomorrow, Flamine and I fly to Paris together, with plans to renew family ties and visit with friends but we will update our field research from Nice and Vence to Avignon, Orange and Sisteron, the most active sites. Guillemant is on my list, as well as Roger Corréard and Jean-Claude Dufour, a refreshing prospect away from our own American shenanigans, too often stuck in shadows.

Mabillon. Tuesday 9 October 2012.

Flamine, her head on my shoulder, enjoys the luxury of sleep after our long flight and the quiet night. We feel privileged to have this shelter.

In a message to the LoneStars, Eric Davis comments on current theories: "Hal, your model is only first-order 'classical' and you know that spacetime physics has to account for quantum effects because it emerges from something more quantum-fundamental. The same goes for general relativity theory which your theory emulates; GR doesn't accommodate quantum phenomena. It's hard to figure out a testable hypothesis based on observed UFO encounters until I can spend time with Jacques hashing out cases and details." (I'm working on that.)

Eric goes on: "Information is somehow manipulated to produce the manifestations that we observe during close encounters, and *information appears to be more fundamental than spacetime*. Quantum effects have been directly tied to information. So, advanced propulsion and other things could be tied to quantum effects."

Given that, I thought, shouldn't we put some emphasis on the data warehouse, in the hands of information professionals, not "temps"?

The group has reviewed the work of a certain Colonel Hennessey, in charge of UFO information for the Air force, from the 60s through the 90s. NIDS interviewed him in 2009. He told them about intrusions in several Northern Tier bases in the 1970s but "had no clue what they were."

Mabillon. Wednesday 10 October 2012.

Evening with Monsieur de Margerie, the CEO of Total, at Cercle Interallié. He spoke to us of Government efforts to help small companies; a pointless exercise, he said, defeated by bureaucrats. The only visible action is the creation of yet another super-layer of financial bureaucracy, very safe, encrusted in pointless elegance...

I know who these people are; we've admired the lace of their *pochettes* as they passed, and we suffocated in the trail of Old Spice.

Mabillon. Friday 12 October 2012.

The first person I wanted to see again in France was Dominique Weinstein, who has the only copy of my FS-3 draft. He understood why I wasn't eager to bring it out. Dominique told me that Dick Haines' recent presentation before Yves Sillard, François Louange, and the new head of GEIPAN went well, but attendance was minimal, only two private pilots and no airline participation. Even Alain Boudier stayed away. Sillard provides prestigious support for the project, but his ability to help is limited. The bureaucracy of CNES even misplaced an important letter from General Bermudez in Chile that offered cooperation.

We spent much time discussing WW2 again, an unfinished topic. Dominique gave me *The Young Hitler I Knew*, an account of the early Nazis by August Kubizek. Chapter 11, entitled "In that Hour it all began…" dramatically confirms Hitler's mediumship. He wanted to destroy Christians as well as Jews. German subs took nuclear bomb prototypes to Japan in the closing days of the war.

It strikes me to see how WW2 remains current in the mind, so full of unfinished business, while everybody seems to have forgotten the long war in Iraq and now disregards the conflict in dusty Afghanistan.

Given the poor economic outlook in France, Alain Dupas and I have given up on the idea of a business book; what we want to say would fall on sterile ground. He quotes the Chinese: "If you sit on the shore and watch the river with patience, eventually you will see the corpse of your enemy floating by…"

Rochecour. Saturday 13 October 2012.

First visit to this gorgeous property, of which my son is justly proud. It is close to L'Aigle, the French town well-known to astronomers for the train of meteorites that peltered it on 26 April 1803, forcing the Academy to recognize that "stones could fall the sky" after all, as the foul-smelling peasants had been telling them for centuries.

A magnificent willow tree greets you when you drive into the stately courtyard. We arrived in great style, Olivier having acquired a 1979 Rolls Royce Silver Arrow from a car lover in Scotland. Claire, pregnant with their first child, prepared a quiche and a brioche we enjoyed with Maxim, his buddy Matthew, and Annick.

Rochecour. Sunday 14 October 2012.

We walked up to the village today, taking advantage of the last hours before the rain. Those country roads remind me comfortably of my childhood, of evenings when the air turned pale blue with the haze rising from the fields, when the countryside was quiet in the expectation of winter, the harvest finished, the trees losing their last leaves, and water gurgling everywhere in the noisy brooks.

The train returned us to Paris in the middle of a cloudburst, but Flamine and I were so happy to be together, back to the city, that we didn't mind the rain. Tomorrow we leave for the Midi.

Nice. Tuesday 16 October 2012. Hotel Suisse.

Flamine guides me through the old town she knows well, a labyrinth of narrow streets, away from the affluent displays and the opulent yachts. We enjoy the cafés where we read the French papers with their haphazard commentaries about the upcoming US elections. Then we walk again, amused at the ramshackle buildings, the windows with half shutters, and the Italian-style plazas with their lively fountains.

This afternoon we hope to track down Pierre Beake, long-term expert on the paranormal events at Col de Vence.

Later the next day.

Colm has been researching the anatomy of empathy in neurodegenerative disease. He has proposed to "tap into the immune system at a level where it is now feasible to decode the 'generative grammar of the immune system,' (the idiotypic network of antibodies) to identify a UFO event."

In an important, long email response, Kit teaches that "the immune system is the only memory system that is independent of cognitive distortion, delusion, illusion, and false memory. With PTSD, there are now antibody T- and B-cell markers that can be decoded to show if what is adduced cognitively is (in some cases) 'real' or false. These include traumatic shock proteins, putative exposures to certain chemicals such as amines from fires and burns, electromagnetic NIEMR/RF, claimed exposure to 'Aliens' and much more."

He goes on to assert that it's possible to design a specific panel of immunochemical tests tailored to claims of CE-V. The name for the rampant increase in PTSD is *alexithymia,* which leaves changes in the amygdalae when imaged with MRI. He hypothesizes that the syndrome we encounter results in loss of empathetic responses that can be measured with the new tools. Alexithymia is a dysfunction in emotional awareness, social attachment, and interpersonal relating, characterized by a scarcity of fantasies, with overly logical thinking…I don't know what to make of it.

Vence. Wednesday 17 October 2012.

An address to remember: Auberge des Seigneurs, hidden away in the medieval streets of this fortified town. From the Col de Vence one can see the lights along the entire coastline from Saint Raphaël to Antibes and all the way to Nice and Italy beyond, a sublime sight.

We saw Pierre Beake at his home in the hills of Nice. He gave us some advice about the reputed hotspot of the paranormal at the top of the 'saddle,' where cameras have been known to blow up inexplicably: Mesnard once gave Pierre Guérin a camera that suffered a similar fate. Yet few UFOs were reported, far away lights.

"We're simply wallowing in ignorance," I tell Flamine as I review the latest messages from the LoneStars.

"Ignorance may be good," she answers wisely. "It leads to innocence, the doorway to contemplation."

L'Estachon. Thursday 18 October 2012.

In Forcalquier, where we stopped for a quick lunch, people looked lost and tired. Pensioners with nothing to do but smoke and drink the local *pastis,* faces prematurely aged. In an environment where people retire so early, where a well-meaning but poorly managed 'social net' encourages middle-aged workers to stay idle and where the young suffer from unemployment, the emptiness of life takes a toll.

Philippe Guillemant and Laurence greeted us in their converted shepherd's home. As we looked over the fantastic jumble of rocks beyond Saint-Geniez, an area whose beauty had struck me in earlier trips in search of Theopolis with Roger Corréard, Philippe pointed out it was the least populated place in France: fewer than two inhabitants per square kilometer, and plenty of mysteries. We visited the 11th century Chapelle du

Dromon, with its interesting 5th century crypt, contemporaneous with the Roman Proconsul Dardanus, an erudite Christian ruler of the region who wrote to St. Augustine.

Our guide was a farm woman who crudely spoke of the Roman builders and rebuffed a few tourists who commented loudly about the massive boulder, relic of an ancient Mithra cult, embedded in the chapel's wall. The group took pictures and went away. Left alone with her, I quietly asked about local traditions, and she revealed herself to be very knowledgeable.

"Have you seen them?" I asked. Suddenly, she wasn't the rough farm woman anymore. Her eyes lit up as she replied cleverly, respectfully, "They recognize each other, Monsieur. It has to do with the heart."

Avignon. Friday 19 October 2012.

We left Philippe and Laurence after lunch in Sisteron. We drove along the fine roads of the Lubéron, leaving the dramatic landscapes and the telluric forces of mysterious Saint-Géniez. We stopped in Ménerbes, where Flamine's relatives own a villa, on our way to the Rhône valley. She has many memories of the place, a peaceful wooded enclave on a hilltop.

On the TGV from Orange to Paris. Saturday 20 October 2012.

We spent the day with medieval scholar Yannis Deliyannis in Orange, one of our co-authors in *Wonders*, reviewing ancient woodcuts of celestial phenomena, and some marvelous books.

We walked through the old city. Here again, the Roman presence is palpable, all around the enormous theater whose wall blocks the sun over part of the town. This is the site of one of the medieval engravings Yannis has found: a phenomenon of the sun (striking for the time, but explainable now) over the top stones.

Mabillon. Tuesday 23 October 2012

Lunch with Dominique again. Like me, he worries about the rise of the Far Right throughout Europe. In France, even some Jewish groups are moving closer to LePen's Front National in reaction against Muslim harassment, a curious development.

French jails are turning into Islamic universities, even 'graduating' non-Arab French recruits, ready to become radical agitators.

Mabillon. Wednesday 24 October 2012.

Messages are flying back and forth among the members of our venture team as Ivan, our main (but ineffective) contact at Rosatom, hopes to set up meetings next week with former Prime Minister Sergey Kiriyenko, who served under Yeltsin. If the meetings happen, I will have to postpone my return to San Francisco by one week and join Maurice in Russia.

Hummingbird. Wednesday 31 October 2012. Halloween.

We're back home. Nothing new came from Moscow. Suitably grey skies here, with rain on the way from the Pacific. Our hummingbirds dart in and out of the deck foliage. They hover next to the window, and peer at us.

I plan to restart operations while Maurice is in Moscow with our slides, but he says the progress is as glacial as a winter in Krasnoyarsk.

I tracked down two new cases of skin marks I found in France. The marks are not all identical. If somebody was testing a directed-energy device, aiming it at random strangers in casual situations like waiting for a bus, the marks could develop overnight as skin inflammation. Or is it just an urban legend, triggered by casual use of hair dryers?

Hummingbird. Sunday 4 November 2012.

Family implications of home interaction with orbs are scary. This is in line with the history of interaction at multiple sites, including Col de Vence in France. The experience with one family is striking: As seductive as a day-to-day involvement with these phenomena may be, we should never forget they're part of a larger, complex globally unfolding pattern. In terms of my research, two issues arise: (a) How to engage further with the phenomenon *in a way that protects human subjects,* and (b) What does it mean in terms of reflexivity, as Colm and I have asked?

We recently learned that orbs had attacked a family, starting with the son in his bedroom, and made him very ill. He was essentially 'beaten up,' black-and-blue marks on face and chest, made semi-comatose, and developed a high fever.

There have been at least eight other orb incidents with this family, and four other people associated with those who'd visited the Utah ranch. This afternoon, Colm wrote with more recent news: "I've just had a 45-minute conversation with the father. Last night's incident with the yellow orb is the third one in the past few weeks. Right after the Saturday evening incident, he went out and took several digital photos with his iPhone, *but nothing showed up*."

Hummingbird. Thursday 22 November 2012 (Thanksgiving)

There's new information about my friends' physics research: The Twins have completed another series of experiments using the range of frequencies I'd given them based on UFO observations. The new results were very good, with effective fusion and high ratios of energy gain. What was achieved comes close to the famous "WOW Factor" Bigelow was hoping for when BAASS was created, but it's taken them six years... The money comes from savings, can't last forever...I suggested to discuss funding with their business manager. The team's papers continue to be rejected by physics journals, which they (rightly) take as an indication of successful progress. I think there's an acute need for a re-examination of all our data, as I proposed, and a real database service to support the various disciplines.

Dr. Green believes my 'old data' is as irrelevant as some aboriginal tale in New Guinea. I answered testily that his own field of medicine would still be practiced by country barbers were it not for the insights that have come from well-structured case depositories!

Indeed, back in the mid-1950s, the natives of New Guinea who died of kuru gave modern medicine fundamental cues to the nature of scrapie, the reality of prions, and the imminent threat of an epidemic of dementia, even in this beautiful America of ours.

I added: "If we face a control system, the hidden reference signal may not be at the physical or biological level but at the cultural level, in which case some of the observations about societal effects are profoundly relevant."

Kit replied that my observation was "totally off the mark." He insisted: "Looking at the whole picture, I've concluded that's a waste of time. It's too fizzy. No way to discern any reality."

This is where I disagree and feel frustration with the way the LoneStars practice so little sharing of detailed, essential facts.

NICAP used to deny UFO landings until the late 1960s when my published evidence, based on Aimé Michel, shattered their reluctance to admit them. Abductions became the rage, with John Mack throwing away his previous work in psychiatry under the great urgency of a supposed alien invasion.

Jacobs has his Hybrids, Whitley his Visitors, and Leir his Implants. Bigelow has the Ranch, and its unprecedented secrets. Each thinks he's inaugurated a new era that pushed everything else into the dustbin of irrelevant research. Now we have an interesting study of brain lesions, acting the same way. I don't doubt that the diagnoses are relevant, but what do we know about the encounters? Who, where, what, when? Not enough is seriously disclosed, in context.

I'm reading John Fowles' novel *The Magus* again. Here is someone who truly understood this confusion. What a book! And what a beautiful language, that 18th century English tongue! I'm forever grateful to Kit for leading me to that book.

Hummingbird. Monday 25 November 2012.

The fog plays curious tricks today. We woke up inside a cloud of dark grey wetness but it soon cleared over the City, leaving us in sunshine, surrounded by a crown of thick, bright rolls of white. Soon, they tumbled from the hills and filled the Bay, obliterating Alcatraz, filling the horizon from Berkeley to San Jose. On days like this, the skyline appears as a mirage seen through a light blue lens, vibrating softly as if to levitate us into the sublime California sky.

Hummingbird. Friday 30 November 2012.

Multiple projects in anticipation of a new trip to France. Our major problem has been the slow pace of the Russians, struggling to set up their investment program. Maurice and Dmitri have tried to educate RosAtom about venture capital, but it couldn't even execute the easy purchase of Danotek, the Michigan motor company that needed Russian magnets. Now the company has closed, the deal lost in complexities of clumsy institutions inherited from the Soviet era.

I wasn't involved in that deal, busy enough with selling Triformix, where my goal was to make sure the employees found secure new jobs before we closed the doors, so I haven't flown to Moscow yet. Maurice and I have become skeptical about everything Russian.

Mabillon. Sunday 9 December 2012.

A two-day stopover in Paris before joining Maurice, Christofer Dittmar, and Dmitry Manakov in Moscow on Wednesday night. Someone named Alexander Nikulov will meet me at Scheremietvo.

Paris is cold, it will freeze tonight. Socialists are firmly in power, their rightist opposition having disintegrated. Ecologists squabble more obscurely than ever, and the political center, weakly attempting to rally behind Borloo, has no spine or muscle. This leaves a core of newborn 'Hollandistes' who repeat everywhere that "innovation is a good thing," blocking every tool that could release it.

Mabillon. Tuesday 12 December 2012.

An old memo has surfaced in the form of an SNIE (Special National Intelligence Estimate), perhaps a draft, dated 5 November 1961, entitled "Critical Aspects of Unidentified Flying Objects and the Nuclear Threat to the Defense of the United States and Allies." I am puzzled by it. It doesn't even read like an authentic memo.

Moscow. Thursday 13 December 2012. Radisson Slavenskaya.

My guide was an engineer with an energy company. The day was cold (minus 10°C), the sky empty and blue; a fine sunset made me forget the biting wind. The flight had lasted three and a half hours, but it took us just as long to find Alexander's car in the maze of the airport and to drive into the city, even with the help of a smartphone he was consulting. All roads into Moscow were hot orange on the screen, enough to make you regret Communism and its empty avenues.

Avoiding the freeways, we drove through villages, past the massive factory of Energomash that makes liquid fuel rocket engines (primary customers, American aerospace firms) before joining a massive traffic jam that strangled the downtown area.

Nothing I remembered from my two previous trips (in 1966 and 1990) is recognizable, except for the massive buildings erected under Stalin. They enjoy new life as luxury hotels or major institutions, revived in light, their spires pointing sharply into the chilly night.

Along the Moscova, all is modern architecture and a jumble of advertising panels in neon light. At the hotel, food is plentiful and excellent, in proportion to the prices charged. All the amenities of an online network service are available. I'm told the old Rossiya hotel where I stayed on my last trip has been torn down; there's no plan yet to replace it, no great loss to the city.

Nikolov, who has just returned from two weeks of kite flying on the warm beaches of Vietnam, laments the passing of Old Russia, its picturesque villages in the forest and its orderly country life.

I met Maurice in the lobby, busy with his iPad. Christofer Dittmar arrived with his driver amidst renewed jokes about traffic jams.

Moscow. Friday 14 December 2012.

Four meetings yesterday, and our project moves ahead, following constructive discussions with the director of development for Skolkovo (Zhugov's deputy) at the World Trade Center. Later, a session at Roscosmos.

Those meetings take place in large, closed rooms with a kind of light that makes faces appear greenish gray, with frozen expressions. This is not peculiar to Russia, of course: I've seen those same greenish-gray, frozen businessmen's faces in France and in America. Only later, in the corridor, when the boss is gone and we pick up our coats and fur *chapkas*, do the friendly discussions begin.

The group leader at Roscosmos showed a sense of humor, fortunately, and even commented to us (after a long tirade I made about achievements of the startups we were backing), "it seems obvious that you do this because you enjoy your work!"

The meeting with Andrei, from RosAtom, was the only disappointment. An aspiring *apparatchik*, he set us back six months with impossible demands. He wanted control of all investments and full rights to buy our companies (at a price he would fix?). This would prevent other investors from joining and discourage entrepreneurs.

Dmitry reminded him that in six months the huge bureaucracy of RosAtom hadn't even succeeded in funding Danotek, a small company that our Michigan colleagues were ready to share.

In the evening, we met with Vladimir Boutenko, the director of Boston Consulting Group in Moscow, at a fine Chinese restaurant. He, too, promised to open some doors but was amused by our efforts. BCG has been trying to make operational changes at RosAtom for three and a half years, he said, with no visible success.

Moscow. Scheremietvo airport. Saturday 15 December 2012.

We had three more meetings yesterday, and a return to the Skolkovo offices to discuss Radiant with Denis Kovalevitch, then our main presentation to Igor Agarmizian, the CEO of Russia Venture Capital. To our surprise, he turned out to be a former astronomer from Saint Petersburg, where he computed ephemerides in his youth, so we had an immediate, friendly rapport. As usual, Maurice was brilliant and impressive, the embodiment of authority and deep knowledge.

Taking advantage of a break at lunch, I drafted a conciliatory letter to Alexei, who had been so abrasive the day before (my colleagues, including Maurice, would have preferred to wring his neck). Some of the demands from RosAtom, I wrote encouragingly, would be legitimate "if they were framed in line with venture practice..."

How do you say, "tongue in cheek," in good Russian?

I recommended going over his head to Kirienko, but the former prime minister was unavailable. Instead, we had dinner with his deputy, Andrei Vasilievitch Dorefeyev, clever engineer, savvy entrepreneur, and cigar-smoking oilman with a vast grasp of the energy field and industrial relations in the real world.

Dmitry engaged him in a long, passionate discussion in Russian over dinner, from which I only understood that he might be able to short-circuit the bureaucracy for us. (I focused on my Beef Stroganoff, which was a lot better than what they serve in Paris.)

The women we've met in Moscow were svelte, elegant in disarmingly simple ways, a combination of helpful efficiency and good humor. They smiled at my confused questions (I am often confused here). The men, in contrast, are rather scruffy or dry, focused, silent, and withdrawn, until you can lead them into personal discussions where they reveal their souls. Bureaucracy is overwhelming, pressing on their burdened shoulders.

The wind pierces my body through my coat as we run from building to building, but the Russians have a secret: they escape the weather by rushing into delightful cafés where there are always tables on the second floor, once you have set aside your fur hat and your heavy armor. There you can get a hot bowl of borscht, or chocolate, or coffee with great pastries, and rebuild the world with friends, forgetting the time that passes, the hated bureaucrats.

Beyond the fences and barricades at all the worksites, you can see the splendor of an old city still trying to emerge from the squalor of communism. Traffic is frozen; Dmitry keeps his iPad on top of the steering wheel of his Lexus, in a hopeless attempt to discover some thin green lines on the crowded map.

In contrast with the rapidly mutating monster city, this airport still offers passengers the stale smell of Soviet drabness. A few shops sell shiny souvenirs and overpriced alcohol, but there's no place to buy the international press or even a paperback book. The VIP lounge itself is a plastic desert with no newspapers. People stretch out on the floor outside the door and just fall asleep. I will be in Paris for lunch.

Mabillon. Sunday 16 December 2012.

Yesterday, the baptism of my grandson. Today, Christmas carols at the American cathedral, with Flamine. The various threads of messages among the LoneStars have taken a strange turn: The roots of the Collins Elite started in the 70s and originated at Fordham. An examination of that University's Trustees from the late 60s gives an interesting mix: many linked to senior Intelligence officers including some 4-Stars, CIA directors, and Ambassador Alonzo MacDonald.

Mabillon. Tuesday 18 December 2012.

Last night I wrote a Christmas card for my neighbor Christian Bokias and was surprised, when I delivered it next door, to meet a young man who said he was his son. The old professor is not doing well and may not be able to come home. His son has been emptying the studio, packing several thousand books *pêle-mêle* into green garbage bags.

A sad ending for a very brave, humble man we loved.

Mabillon. Thursday 20 December 2012.

Flamine resumed her professional work in her Paris office yesterday for the first time since our wedding. I happily joined her for lunch at Saint-Lazare amidst swirling crowds, madly shopping.

Yves Messarovitch confirms to me that France has fallen back into recession. If the Hollande government continues with its anti-business measures, fiscal confiscation and aimless expansion of the bloated public sector, France's debt will become so large that the country will resemble Spain, economically crippled by mindless borrowing.*

Mabillon. Friday 21 December 2012.

Dominique Weinstein and Mady were here for dinner last night. He doggedly pursues the research thread that begins with Borman's escape from Berlin at the fall of the Third Reich. Like Phil in Washington, both of us are concerned with the survival of Nazi projects in science, finance, and business.

Perhaps my friends cling so much to their 'veridical' data because they must survive in a world of mirrors. I am not certain their data are as veridical as they believe, although there's no doubt some of the unfortunate patients are dying under puzzling circumstances.

Here again, where is the specific link? Where, how, and when were they exposed? What else was happening in their lives?

I've heard people argue that deception is legal when disseminating official tales presented as fact. If so, it's no wonder many scientists, aware of such 'active measures,' conclude that ALL supposed UFOs are nonsense, a State-condoned farce.

Mabillon. Saturday 22 December 2012.

A mild rain falls on Paris, setting up a mood of subdued expectation for the holidays. Many people have already driven away to the country. Last night, on a visit to Flamine's father, her whole family assembled for an early exchange of gifts and good cheer.

---

* As we publish this book in 2024, the French budget shortfall is reaching 300 billion euros.

Rochecour. Tuesday 25 December 2012.

Yesterday we took the train to Le Mans, where I showed the cathedral to Flamine with renewed pleasure at visiting the chapel with the Singing Angels, then we drove on to the home of Philippe and Kathryn Favre, where the *Grande Halle* now features gothic windows, 22 meters high. Philippe's friends May Kay Butler and Eagle Sarmont (a former Skunk Works engineer who came six months ago with his plan for a space elevator) were there for the Holidays, so dinner conversation kept swinging from fourteenth-century architecture to the take-off characteristics of the SR-71.

We drove home at midnight, as quietly as we could, the kids asleep. This morning the celebration started around the tree, we opened up the gifts and laughed; we told silly stories and ate too much chocolate. I felt grateful to see my son happy.

Hummingbird. Tuesday 1 January 2013.

Back in San Francisco, watching the peaceful City, sorting memories and pictures, all we have to do is witness the silliness and hypocrisy of Washington's "fiscal cliff" and huddle in fear of lasting uncertainty in Europe.

With all the cards and the Good Wishes and the phone calls out of the way, Flamine and I find ourselves in a nostalgic mood. January will never be a happy month for me. Paradoxically, the splendid California weather contributes to sadness; the sunshine fails to match our subdued feelings.

Letters and messages. Tom Tullien writes that CUFOS no longer maintains an office, moving files to Mary Castner's basement and Mike Swords' home. The remaining members—Don Schmitt, Mark Rodeghier, Eddie Bullard, Jerome Clark—met with him last August, sharing stories about Roswell, but their magazine, *IUR*, has ceased publication.

Hummingbird. Thursday 3 January 2013.

Bob Bigelow called me this morning, returning my good wishes. We commiserated on the slow pace of Congress, no decision on the nation's budget will come before spring. Bob is worried about the planet, engulfed as it is in conflicts that could mean annihilation, and he held firm that an announcement of ET presence would "be the mechanism to change human condition towards realism…" I hope he's right. In the meantime, no significant step is being taken towards sane, professional analysis of the data.

Hummingbird. Monday 7 January 2013.

Someone floats a rumor (one more!) that the White House has moved away from 'disclosure' and now leans towards 'confirmation.' Is someone feeding fake revelations in the media to gage reactions? I'm told 'confirmation' will be rolled out next year. I laugh as I hear that, and my friend says: "Of course, that's only what I've been told…"

Tom Tullien sends me a book called *X-Descending* by Christian P. Lambright. It picks up the history of Paul Bennewitz again, emphasizing the reality of the films he shot, lights and lenticular objects over Kirtland Air Force Base which he could see from his rooftop, and which could be prototypes for directed energy test targets. He then dissects the grand operation of intimidation, misdirection, and eventually mental destruction mounted against poor Bennewitz, including details I hadn't known when I met him (16).

Later the same day.

Colm has intervened decisively in the discussion of potential disclosure. He wrote:

"*Disclosure/confirmation hasn't happened since the 1940s, is not happening now and will not happen. Anything that might be released in the future in the guise of confirmation or disclosure will be a distorted, manipulated, diluted facsimile of the truth.*

"Confirmation-disclosure on UFOs will probably approximate to Doug/Dave's 'disclosure' regarding crop formations, or Ken Rommel's 'disclosure' regarding cattle mutilations: Nothing to see!"

I read this with total agreement. Colm adds a comment based on my control system analogy:

"My own hypothesis is that one of the thermostat functions of the Control System is to keep hidden knowledge hidden. It is kept ONLY with individuals and very small groups that expend a lot of effort to hunt such knowledge. When the temperature rises and there is a danger of hidden knowledge being disseminated to the public, the control system is activated and the knowledge becomes distorted, diluted, discredited etc. Active measures (bidirectional mimicry) play their part in this mechanism."

In an email message, Jean-François Boëdec, recuperating again from prolonged hospitalization, summarizes the French Air force's project Simon Goulart in a single phrase: *"On roupille!"* ("We are sleeping!). "The idea of deploying advanced electromagnetic stations and using the Helios satellites in hopes of catching UFOs in flight was well-intentioned, but unrealistic." Paris, back to normal.

Hummingbird. Sunday 13 January 2013.

John Shirley and his wife Micky came over for lunch today along with Jay Kinney of *Gnosis* magazine and *Anarchy* fame. It was refreshing to discuss spiritism and science fiction with gifted writers. They had come over to attend services at the Spiritualist Church next door.

In my latest conversation with Kit, he mentioned that the BAASS project had "been terminated *for cause* three years ago."

I'd never known that, and I need to find out what the cause was: Breach of security in DC? Encroachment on other programs? The group strongly believes there will be some sort of confirmation soon, and it will confirm an alien presence. At least that's what my friends are being told by "people they trust" in Washington, supposedly close to the White House. There's a belief the public isn't capable of comprehending the issues: "They don't have a 'right to know' if that knowledge is going to make them sick. The truth could be released on a slow glide path but there is a chance of another agency ('similar but different') picking up where BAASS stopped."

Hummingbird. Friday 18 January 2013.

Funeral mass for Andy Lipinski. Stephen was there, as well as my friends Mike Palmer and Greg Schmidt. The eulogy mentioned Hubert's historian grandfather in Poland and his tragic execution.

I read a *Skeptical Enquirer* article that borrows widely from my *Forbidden Science*, attempting to bias passages that may cast a doubt on Hynek's rationality. Yet any attempt to throw mud on his memory will backfire: He emerges as a visionary scientist, sometimes disliked by colleagues for his free spirit, honesty, and creativity.

Hummingbird. Sunday 20 January 2013.

Steve and Linda Millard were here for dinner last night. He's a strong believer in psychical research, based on his personal experiences and on the continuing studies of friends like Federico Faggin. Steve's group has a link to the Intelligence community, but it may be cancelled if Congress decides to make deep cuts in the federal budget, as seems probable.

This morning, I spoke to my brother in Paris, as clear-thinking as ever. He only complained that both he and his wife were "slowly getting older." I told him we might as well do it as slowly as possible. There was mutual pleasure in exchanging a few words, news of my son's new life, his children.

This afternoon, a private concert at the home of Paul and Béatrice Gomory. One of the musicians, David Tayler who played archlute and Baroque guitar, asked me if I was "the well-known astronomer." Tayler is another recovering astrophysicist who's read my books!

While the sunset played on the windows of buildings along the Bay and shadows crept across the entire City, they played Uccellini (*La Bergamasca*), Bach's *Sarabande* from the suite in G Major, pieces from Henry Purcell, Dandrieu, and Couperin (*Les Barricades Mystérieuses*), and finally Corelli's *Ciaccona la Virginia*.

While reviewing *Forbidden Science 3* (which still leaves me feeling a little sick of all the machinations), I spoke to a friend "back East" about Paul Bennewitz, and how Bill Moore misled me. The reply was: "How was [Bill Moore] misleading you? Bennewitz DID eavesdrop on classified links; he didn't realize what they were—downloaded Soyuz transmissions of biometric data our contractor was receiving."

Hummingbird. Thursday 24 January 2013.

Lunch with Peter Sturrock, who returned to me the binder with Natalie's photographs. He was evasive when I asked him whether he still followed the subject: "I've tried to do what I could," he replied, "but my colleagues in science don't seem interested."

Instead, he's been absorbed in the annual variations in the decay rate of radioactive elements, which he believes are linked to the flux of neutrinos from the Sun. This relates to his theory that the core of the Sun is rotating at a different rate from the convection zone, a novel idea. Here again, his colleagues in physics disagree.

Peter is as puzzled as I am about Natalie's pictures, taken with three excellent digital cameras. He continues to believe that the "orbs" are unexplained. He took a picture at a party where a vivid red orb showed up above the head of physicist Ron Bracewell, who died shortly thereafter. When he looked up the image again, the red color had impossibly turned to a faint yellow, both on the printed photograph and in the digital file, which I've seen.

Hummingbird. Sunday 27 January 2013.

We continue to debate the allegations in the book *X-Descending* by Chris Lambright, who re-opens the tragedy of Paul Bennewitz and the roles played by Richard Doty and John Lear. The author suspects that both were in Laos at the same time, at a secret CIA base supposedly known as "Site 5," a fact I don't have the ability to verify.

Hummingbird. Friday 1 February 2013.

Ingo Swann died in New York last night, about 10:30. I had spoken to him on Wednesday, and I also spoke to his sister, Merleen, who lives in Los Angeles. Ingo's speech was heavily slurred, almost impossible to understand. He knew he was dying. His close friend Mark, a very spiritual man, tells me he thought he was able to perceive Ingo "on the other side," where he was fine.

Hummingbird. Friday 7 February 2013.

Henry, an old contact from 'back East,' is in town for a conference. Flamine and I had dinner with him and another friend with his wife in North Beach. I'll spend time with him tomorrow. The conference was devoted to instrumentation in medicine.

He mentioned a proposal from EarthTech to a larger company, to expand on the reports done for BAASS. That contract never materialized, but it led my friend to becoming acquainted with someone with up-to-date knowledge of the crashed UFO hardware.

Hummingbird. Saturday 9 February 2013.

Henry and I had plenty of time to talk today as I picked him up at his hotel in the morning and drove him to Palo Alto for a visit of the Stanford campus and Hoover institution.

Our conversation filled many holes and opened some new areas, especially in physics. Henry had recently spoken to a knowledgeable scientist who described how "the government" would call his company with a request to go pick up something in the desert, yet when they got there, they often found ordinary junk. Few pieces were of interest when tested in the lab. At the end of the project, there had been no breakthrough, so about 1988 or 1989 they were asked to store away the material indefinitely, or so was he told.

In those studies, the hardware was made up of common elements, but the structure was unique, unlike any sample from any material from other countries. "You have to look at the atomic level," one source was reportedly told, "but even then, it made no sense. This was made by a science we don't have, using techniques that may not be invented for decades." Of course, that's what Marcel Vogel had told me about the pieces tested at the best IBM labs.

Thus, there was a community of a few dozen people who had some knowledge of the program, none of whom had a complete picture, although high-level executives appear to have been shown photographs of entire craft. All those people are now older than 60 or even 70, and their collective knowledge will soon be dissipated.

What I didn't know was that the whole domain was not under the responsibility of the military. When the Air Force and the CIA were created in the aftermath of WW-II, the CIA was given control over any piece of foreign equipment falling over the US, with the Air Force providing security, retrieval, and storage. The CIA (whose special budget was a line item in the Air Force's black budget) then parceled out the material to various research facilities. The contractor never had more than a few objects, up to a cubic foot in size, possibly from the skin of a craft. And they never had access to bodies, although they were sure they existed.

People in such projects have a lot to say about science. Modern knowledge is built upon principles that are mere assumption, not fact (for instance, Einstein's "weak gravity equivalence principle" states that inertial mass is equal to gravitational mass, yet that conjecture hasn't been tested).

This extends to climate change, one scientist said: Spacecraft observing the Sun find that solar weather "influences the atmosphere to a greater degree than human activity." Even if true, no reason to aggravate it!

Bigelow's facilities are no longer available for classified work, so I won't have access to the paper on directed energy. All our other reports are unclassified. Senator Harry Reid (who, along with Joe Lieberman and the late Inouye from Hawaii, strongly supports the effort) went to Leon Panetta, then CIA director, to request release of the material. Panetta denied access for fear of political exposure. The project was shut down in November 2011. Detrimental rumors and ugly lies were planted about some individuals to destroy their credibility.

There remains hope in a quiet approach through the Navy. The plan would be to hire one knowledgeable firm to work on the material. But here again the religious factors may interfere; 'Collins Elite' members believe that no one should be interested in anything that come from Satan. I said that if we're about to do battle with Satan, we might as well learn everything we can about his toys…

Glenn Gaffney, who serves as Director for Science and Technology at CIA since 2009, would probably raise another issue: Contracts are old, and "the stuff" should no longer be in storage, but it's unlikely to get returned. Our team would have to buy or lease a new building, and new clearances would have to be obtained, everything starting from scratch with no link to older programs, abruptly disbanded. The database manager, who'd moved his family to Vegas, is now an instructor at a new facility out of state. All others have moved on…So much detailed knowledge has been lost!

Henry and I stopped for a visit to Stanford's quadrangle and took the opportunity to call Joshua W., a medical student who'd described to me his observation of late June 2012 ("about the 27th at 10:30 pm") when he saw three lights moving together very fast in the sky. Again, what struck him was the unique *quality* of the light.

Over lunch at Café Vida, Henry told me that Capella lingered on some dead computers in Vegas. We discussed recent theoretical discoveries in Austria, implying that quantum entanglement is independent of space (something we already knew) *but also of time and of causality*. It means a new physics is required, confirming my speculations at TED in Brussels, with fundamental implications for the role of consciousness and spirituality.

More pieces of the puzzle are falling into place. I now understand why project leaders go directly to CEOs for classified support, ignoring the lower levels because they wouldn't know the real work. What Henry described also fit what I knew from conversations with Marcel Vogel and Art Lundahl, both of whom knew the material.

Hummingbird. Sunday 10 February 2013.

Thinking back, the LoneStars' efforts to restart the study of recovered hardware seems poorly designed. In archaeo-zoology you can rebuild an entire animal from a piece of jawbone, but that's made possible by years of careful analysis of many animals from many geological layers, climates, and ages. That's why the database work is fundamental, but this has been missed. Engineers believe one piece of saucer skin will tell them about lift, propulsion, even fuel and flight controls, as if they had picked up some Russian MiG.

My short note about the misinterpretation of Ockham's razor has been published by *BoingBoing* and is raising a minor storm. Now, plans for another series of meetings in Germany and Russia, trying new investment structures for business in space.

8

Mabillon. Tuesday 19 February 2013.

Wintry sunshine, sweet musings of France. I'm reconciled with the Parisians, having given up on their fear of innovation, their suspicion of anything too technical, their obsession with a false 'equality' that suppresses excellence...as if you couldn't have excellence in a Democracy, along with equality. We should take them as they are: lusty and bright, eager for superb meals around a good bottle and fierce pointless debates about the state of the world as they imagine it to be.

Parisians argue well, given their lack of authentic information about what goes on elsewhere. Their image of America is misguided, yet not totally unjustified: the US does look like a violent country of gun-toting folks with brutally dogmatic notions about a simplistic God, a people unable to enjoy anything not expressed in dollars or related to some passing celebrity. It's a caricature. Is France immune to that?

As soon as we landed, we were busy with family visits to Flamine's elderly father (bright as ever), to my son and his wife about to give birth, and connections with friends. I must prepare for the trip to Russia in one week amidst recent astronomical surprises: A 7,000-ton asteroid entered the atmosphere three days ago and blew up over the Urals, shattering windows, and injuring folks in Chelyabinsk.

On my bedside table is a book entitled *Encounters with Star People* by Dr. Ardy Sixkiller Clarke, an American Indian woman who is an anthropologist at Montana State University. It recounts dozens of unexplained, yet consistent experiences by American Indians all over the country, usually isolated on remote reservations in the wilds of Montana or New Mexico.

Mabillon. Thursday 21 February 2013.

Meeting with Alain Dupas: we discussed our prospects in Russia: He's well aware of the structure of the aerospace industry there, the centers of power, the aspirations of the young, and the ruthlessness of the oligarchs, good background for my trip on Monday.

Gérard Forgeron, a friend of Flamine and CNRS researcher, came over for dinner last night. He is a recognized expert on hermetic history. Long acquainted with Africa (six years in Mali) he deciphered for us some of the puzzles that surround the current war against the Jihad in the desert.

Most of our conversation was about Freemasonry and his recent interest with the Samothrace Lodge, which he created in Paris. He gave us copies of recent papers, notably a study of 'Démiurges' in Africa and Antiquity (in *Lettre d'Ile-de-France* of the Groupe de Mythologie Française no.88, June 2012).

We spoke of AMORC. He had joined them, initially impressed, but became disenchanted as I did when he saw their history was considerably enhanced. Spencer Lewis was sincere and bright, but his claims of initiation in Toulouse only ring true for Americans who've never been there. He told us that a current group leader caused a crisis, given the prejudices of the time, when he asked another executive to celebrate a personal ritual with him in the great temple of San Jose. He thought Bernard saw Rosicrucianism as a fulfilling career, but his claims of mystical visits to the Houses of the Rosy Cross were dubious. After the ritual incident, imperator Ralph Lewis replaced some executives. Indeed, at the time, such groups have been used, and are still used by various Services, notably in Africa. After the war, Washington injected money into many esoteric and religious organizations based in the US to further the image and impact of American ideals around the world. Freemasonry seems increasingly rudderless to me; it lacks an active line of research.

We alluded briefly to Rennes-le-Château, agreeing that the mystery was overblown. Abbé Saunière probably did find hidden documents about some influential royalist family (during the Revolution and beyond, everybody was hiding money and documents) but the site doesn't hold any deep secrets.

Forgeron is a tall, strong man with an attractive face, graying hair, a stunningly precise memory. We spoke too briefly of the relationship between some AMORC splinter groups and the Solar Temple massacres. Other references: Nazi records and the history of Paschal Beverly Randolph, the Fraternity of Eulis, Luxor, Clymer...

Later the same day.

A phone message says that my second grandson, Xavier, was born this morning in Paris. His mother is fine. I will see them all tomorrow.

Among my plans: a visit to Madame Servais.

Mabillon. Friday 22 February 2013.

Leaden sky. Intense preparations (through the web) for the Moscow meetings next week. I saw my new grandson today, at Saint Joseph hospital, a strong fellow with dark hair. It feels so good to see my son happy! And Claire, who glows with the pride of motherhood.

I just completed a revision of *Forbidden Science 3* that raises many questions. What was that secret business with Lear and Lazar, Bill Moore and Condor? What really happened to Bennewitz? What did he do to trigger a government plan to destroy his mind? Even if he picked up some encrypted signal, why did they have to crush his life? (16)

Mabillon. Sunday 24 February 2013.

Our first wedding anniversary. The weather is nasty, just above freezing. I went alone to visit my grandson again, to greet him on the planet and to kiss his mother. He's three days old, strong and lively.

On my visit to Simonne, we spoke of Valensole. She recalled going back in the mid-80s to see Masse again. She caught him after his game of *pétanque*. He was rough, non-cooperative, reproached her for not going with him to the place where he saw certain things…

She had a photograph of DeGaulle on her wall, with the dedication. He had first written "To Simonne," then scratched it, wrote "to Madame Servais" and added, "*Respectueusement.*" Coming from the leader of La France Libre, a rare tribute. She commented on Flamine's picture with me: "your eyes with the same peaceful certainty, the same confidence that only deep love can bring, the same strength."

On TV, a fine old interview of Aimé Michel. Then, Alain Dupas' advice: "Don't miss the beauty of the 'true' Russia, and savor the inspiration found in the countryside, the plain Russian people."

Moscow. Tuesday 26 February 2013. The Crown-Plaza.

Next to the Russian World Trade Center is a luxury hotel once forbidden to ordinary Russians. It's in the style of the Embarcadero Hyatt in San Francisco, an immense cavern with interior balconies for the first eight floors, a huge interior canyon, water jets, and glass elevators. But the finest feature is the eastern view over the Moskva lined with lights, and the rising moon sublime over the river.

We have several days of hard meetings ahead, with no time to bother taking notes.

Hummingbird. Wednesday 6 March 2013.

Our Russian meetings are over. Plenty of information and good contacts, but an amazing inability to make decisions. I flew back to Paris, and now to San Francisco.

We're ready for the fourth LoneStars meeting in Houston, and things are heating up. The uncovering of the Collins Elite by writer Nick Redfern has led Kit to consider approaching Father Boeche and former White

House Staff Director for Jimmy Carter, Alonzo McDonald, for background insight. I called him today, eager to clarify a few items that don't deserve to take our time face-to-face in Houston next week.

Kit and Hal have given up on seeking support under the second agency they've contacted. They're now trying to find a third one, but I pointed out to them that the very absence of a big, classified program, if true, would be the best thing that ever happened to this research. It would free scientists to investigate on their own ticket, especially for biological insight.

Hummingbird. Saturday 9 March 2013.

Yesterday Colm himself had an encounter in Las Vegas that reminds me of Fred Beckman and of the strange characters that appeared in San Francisco in the old days. He writes: "Forty minutes ago, as I was outside my home getting into my car, a strange-looking woman passed me by. She had very black eyes, a white pasty face, and wore very prominent black horn-rimmed glasses. She also had an obvious black shiny wig. She gazed at me intently as she passed."

He adds: "As I got into my car, I noticed that on the back of her brown jacket, emblazoned in large florescent white lettering was the single word: LONESTARS!"

Years ago, several Bay Area ufologists were involved with a woman who claimed to be an alien hybrid. She too had large black eyes, evidently wearing the kind of contact lenses used by sci-fi actors in Hollywood. That woman ('Wendy') was a multiple personality patient. She was handled by a manager at Ford Aerospace and my friends traced her to a detective agency in the South Bay.

In the case of Colm, I think the message is: "We know who you are and what you do." It's not especially ominous, just a reminder that they can find us, wherever we are. But it's also a warning. It could be from anyone: cults, the Collins Elite, and who knows what else...

Later the same day.

Flamine and I just walked back from a memorial service for Michael Toms, the founder and long-time host of *New Dimensions* radio. Michael was a wonderful man and a great interviewer. His 3,000 recorded programs are now preserved at Stanford.

At the Unitarian Universalist church, a few blocks down our street, some 200 friends of his came to hear Jim and Dorothy Fadiman, Charlie Tart, Steven Halpern, and Ralph Metzner of LSD fame pay tribute to him and encourage his wife, Justine, to continue his work.

Stanley Krippner was in the room, and Anadea Judith, Diane Darling, Sergio Lub, Dan Drasin…all important figures from the formative years of the consciousness movement in California.

It was like a tapestry of the explorations of the mind that began with the 1960 Summer of Love, down through the decades of modern American history, from the Hippies and Pagans to the scholars of modern parapsychology, 'enlightened' or not.

Houston, Thursday 14 March 2013.

The LoneStars have assembled again, all ten of us, with Flamine as a guest, at a local resort, superb this time of year.

Most of the discussion so far has had to do with abductions. I brought up the Perez de Cuellar incident in which UFO believers became involved through Budd Hopkins' efforts. Neither NIDS nor MUFON touched it. With the passage of time, I thought, we might get some clarity. My assessment of the use of hypnosis in all this has not changed. As practiced today, I believe it is mostly fantasy.

Houston, Saturday 16 March 2013.

George Hathaway neatly began the first session last night with a recap of our Toronto meeting, and we discussed how the Air Force had used Project Blue Book for social manipulation. We also discussed the SERPO story, with evidence of hoaxing by insiders.

After breakfast, impeccably served in the meeting room by silent waiters, I had the floor to present several recent cases I'd selected, including a follow-up to the 'lethal fog' episodes reported in Pennsylvania by Gerry Medvec. Curiously, these unrelated episodes occurred at the same exact latitude (39°39') although they were 168 miles, years apart. Colm and George recalled similar situations.

Next, I showed my slides from Col de Vence, stressing the similarities but also the differences with the Utah Ranch: no cattle incidents and no reliable report of structured objects in Vence, but there are many unexplained orbs, falls of stones, and two unexplained sounds that haven't been reported in Utah. The best witnesses are clamming up, disgusted both by the trash on television and the arrogant skepticism of scientists. But sloppy, invasive inquiries by ufologists didn't help build trust with upset families.

Following my three slide presentations, the discussion returned to the fundamental issue of censorship. Since most of us have spoken to high-level technical people open to the reality of the phenomenon, we're very curious about the roadblocks.

It's also very interesting to track new studies of the Sasquatch based on three genomes from three different animals, studied at the University of Texas, Southwestern. The mitochondrial DNA is from human females, not from lower primates, but only 1% of the whole genome has been analyzed. The male source is unknown. The SNP (single nucleotide polymorphisms) evidence is full of anomalies. *Nature* has refused the paper, arguing there wasn't enough genomic data. It was eventually published in a secondary zoology journal. My colleagues remain agnostic.

George asked me to expand on some statements I'd made in earlier messages to the group about the survival of German financial and industrial power, possibly along with some post-war Nazi concepts. I reminded them of the secret gathering organized in Strasbourg by Martin Bormann on August 10, 1944, (the so-called *Hotel Maison Rouge* meeting) where German industrialists like Krupp, IG Farben, Messerschmitt, Bussing, and others were invited.

The attendees were told that following the Allies' invasion in Normandy two months before, the Battle of France was about to be lost. They must immediately prepare a post-war commercial effort, relocate assets to friendly countries (Middle East and Latin America particularly), move cash to secure foreign banks, move intellectual property and key engineers, and initiate new contacts with foreign firms without attracting suspicion. As they did, the Nazi party would secretly commit to supporting them financially and logistically, putting submarines at their disposal, among other assets.

Jacques Bergier often told me that the surviving Nazi in Argentina and Brazil used the belief in extraterrestrials as part of their political manipulations, as did Lopez Rega under Peron. The ownership of modern industries that could be attributed to the expatriated German funds was anybody's guess, but it was most significant in chemistry, paints and dyes, metallurgy and automobiles. (Germany recently requested return of its gold from foreign banks, too.) On the other hand, I said the stories about Nazi flying saucers were not credible.

We ended with a round-table discussion of what we individually thought and planned, at George's request. Kit seconded the idea, reminding the group that "when NIDS and BAASS went away, we used the LoneStars as a way to continue as a group."

My notes have recorded someone's worry: "*My view of the phenomenon has changed.* I think the human race is in a precarious position. I want to make the world a better place, but *something is coming...*" The LoneStars are "the only game intown" in terms of doing something about this, he added: "We're like the lone sailor in the Titanic crow's nest."

When my turn came, I said I felt the same way, except that I'm the guy down in the engine room, without even a chance to see the iceberg coming, or to jump out after the collision, unlike others. When the laughter died, I added: "I'll continue to interview witnesses in the field. I like to meet them at their home, sit down at the kitchen table, and listen. I know how to assess the data. But at this stage in my life, I must also think about transmitting what I know, helping newer researchers. I devote time to revising and improving the study of ancient sightings, alongside our good group of experts on the Internet that has become larger since the publication of *Wonders in the Sky.*"

Privately, I reflected that it would take five years of hard computer science work to answer the basic questions, and I would know how to run it. But I now believe there's no real hope of a project in that direction.

It would make sense to develop new technology for Humanity. The group felt that an existential threat had been detected, and that we should stand clearly for an open investigation.

The time scheduled for dinner had passed, so George suggested we adjourn and continue our talks in the more relaxed setting of the bar. The Spring Break holidays had filled the resort with families and students who milled around or walked over to the pool.

The next morning, before packing our suitcases, Flamine and I walked pleasantly along the mile-long path that curved through the woods around the complex. The gnarled trees with hanging vines put a tropical touch to the place, flowers everywhere. Then, darker clouds started coming over, pushed by the moist winds from the Gulf.

Hummingbird. Tuesday 19 March 2013.

In the aftermath of the Houston meeting, questions linger. I'm concerned about the many steps that could be taken to throw light on the phenomenon. Our past efforts have brought little new knowledge about UFOs. They came in the form of poltergeist, lights, structured craft, cryptids, and the occasional injury to humans, but in the absence of correlation with a larger body of calibrated data, they teach us nothing, scientifically, about the origin, nature, and evolution of the meta-phenomenon.

Now another Russian oligarch, Boris Berezhovsky, former friend of Yeltsin but enemy of Putin, has died in exile in England at the young age of 67. This raises dark doubts again, about real changes in Russia.

Maurice and I have found two company projects that new funds could consider, one in superconductor manufacturing, based in Germany, the other in nano-metallurgy, but the changing attitude of our potential partners in Moscow is more disconcerting than ever.

Hummingbird. Saturday 30 March 2013.

Echoing our concerns, Maurice, August Fern, and I had a pleasant breakfast in Menlo Park and a frank discussion with Igor Agamirzian, head of Russia Venture Capital. It clarified the process that new venture funds will follow in Moscow, after the Cyprus financial crisis. Coincidentally, on my bedside table is the remarkable book *Dragonfly* by Bryan Burrough, the true story of the Mir space station and the good, ongoing Russian-US relationship in orbit.

At the request of Peter Beren, I just completed a foreword to a new edition of Jack Katz' brilliant comic book series, *The First Kingdom*. The plot is so like current theses about telepathic robots (cyborgs, in Katz' world) that the parallel is funny. Katz started drawing it in the early 1970s when he came to San Francisco.

Now I hear that Jack Houck has died and that Ted Rockwell is in hospice. The parapsychology community is in mourning, recalling the days when we were bending spoons and wondering why nothing useful came of it, not even a theory of metallurgical exceptions. Will UFO research follow the same path?

Hummingbird. Thursday 4 April 2013.

In the wonderful jargon of government, I have found that my 'single life remainder factor' was 0.11329. If I do the calculations correctly, Washington says I still have 9.33 more years, to die in 2022, age 83.

Life has limits, as Janine once reminded me, and they're cruel. A friend who recently spoke to David Jacobs found him discouraged: "He was sitting at home, spending most of his days watching TV. He said he'd been ostracized all his life by his colleagues at Temple University."

It would be sad to lose him as we lost Jessup and Jim McDonald, so I hope he'll find comfort in his supporters, but the forceful style and insults of his mentor, Budd Hopkins, have discouraged many people.

Hummingbird. Friday 5 April 2013.

To meet Jack Katz, as we did today over lunch at Hotel Mac with his agent Peter Beren, is to enter another galaxy, a world of power and intrigue in the undiscovered past, a universe of emotions and glory and occasionally despair, born of the Great Depression and nurtured in the shade and powerful exhaust of the modern rocket ships.

Jack is legendary, one of the very early illustrators of the genial American pulp, and he's still drawing his own fantasy, after a long career with the major publishers. The inventor of the comic novel, he's ever attentive to everything. I see his sharp eyes on me, penetrating my thoughts, until the pretty waitress comes over and he offers her the sketch he was drawing on the closest napkin.

Hummingbird. Sunday 7 April 2013.

In response to Bob Bigelow's request for advice, I wrote: "Your position on the 'access to space' issue is crystal-clear and will be enhanced when NASA announces the attached inflatable craft..."

Actually, Bob's initiative is sensational. He's brought NASA into a position of partnership in a wide exploration plan. I added: "My short-term advice is to carefully study the reactions to the disclosure campaign and the Washington pseudo-Hearings...This will be a test of the media's attitude.

"Absent a solid core of new scientific data, *which has not been assembled*, we may enter a period of confusion and pointless ideological battles between the various factions and the scientific community. There's no simple answer. My concern is that, once confirmation occurs, the public will demand quick answers and will be manipulated, once again, into a pre-packaged belief system."

Hummingbird. Friday 12 April 2013.

Steve Millard and Federico Faggin attended a talk I gave on "Futures Impossible" at the Association for Corporate Growth last night. Also there, Tania Fernandez from Burrill, and others from major companies and VC funds. The mood of that community remains optimistic.

Now it's time to board a plane to Paris and perhaps Moscow again. Today is Yuri Gagarin Day; politicians have joined Putin in the Amur region where a base is being planned, to replace decrepit Baïkonour.

Mabillon. Sunday 14 April 2013.

Another cool, gray Paris morning, later cleared up, morphing into a pleasant day of spring. The city is quiet, surprisingly silent. This is a brief transition before we take the train to Geneva on Tuesday.

The important news comes from Dmitry, telling us that RVC is ready to initiate due diligence on both of our current funds.

Geneva. Wednesday 17 April 2013.
Hotel des Horlogers at Plan-les-Ouates

This small establishment doubles as a lovely museum of clock-making, across the street from the Rolex factory. Glass cases in the hallways display the tools of the clockmaker's trade, the intricate machines and their precision parts.

After a couple of media interviews, I presented "The Impossible Futures of the Internet" at the TEDx conference, in the funky atmosphere created by Théo Bondolfi and his Ynternet.org Foundation.

The day ended with conversations on the lawn with science students and their teachers, full of ideas for new projects. We walked home in the sunshine and lost our way in the fields. There's construction everywhere, the new roads lined with opulent villas. Everything is polite and seems far from the worried world.

Cully, Switzerland. Thursday 18 April 2013. Major Davel Inn.

Théo's old friend, Henry Rosset, and his wife, Colette, drove us to Lausanne in their new Citroën, through a landscape of old hamlets and vineyards. Lunch at Yverdon-les-Bains where we visited the Maison d'Ailleurs, which holds writer Pierre Versins' overflowing collections of science fiction and fantasy. A young man provided a tour of the buildings filled with treasures of graphic art and visionary literature. In the evening, we met with Théo's group in the hills of Clairvaux to plan open-source hardware and do-it-yourself tractors.

Mabillon. Saturday 20 April 2013.

A cold Paris morning has turned into another spring delight, the streets illuminated by the sunshine reflected off the stately façades. After lunch with Olivier's family, we walked home through the affluent Seventh Arrondissement where all is sumptuously calm. Yet it only takes a few days to understand why many economic measures are inapplicable in France; it takes half a day to mail two packages.

My trusted friends don't paint a better picture: Pierre Tambourin had to cancel our meeting at Génopole and Yves Messarovitch is overwhelmed. Alain Dupas was the exception, a glass of red wine in front of him. We spoke of NASA and Bigelow, the capture of asteroids for fun and profit, and, again, our Russian projects.

Aboard the train to Valognes. Wednesday 24 April 2013.

Springlike weather, and a graceful view of the French countryside, all rolling hills and fat pastures on the way to Caen. I read a novel by Roger Peyrefitte, *Les Clés de Saint Pierre*. Peyrefitte is a must, each page a refreshing French lesson with his sharp, subtle style.

This train is a time machine: to the country of Janine and Annick, but also to those Norman names so familiar: You would ignore these little towns at your own risk. They gave the world Norman intellectuals who dared scrutinize the minds and the hearts of mankind from the shelter of their castles, in this land of milk and with the occasional shot of Calvados.

From Valognes to Yvetot-Bocage, the walk takes less than an hour in a pretty landscape, a delight as you reach the village beyond the castle at

Serpigny. Birds were improvising with such abandon I was carried again into a time of innocence, imminence, and beauty. I saw no one along the way until I reached Yvetot where an old man was tending his garden. His flowers spilled into the asphalt, so I stole a few myosotis to complete a small bouquet of wildflowers.

I was surprised to find fresh flowers on Janine's grave as I added mine to the small display. Annick later told me a friend had driven by. She also said she no longer went there; it was just too hard.

Mabillon. Friday 26 April 2013.

Who wouldn't want twelve weeks of vacation? That's the norm at *Enjeux*, the magazine managed by Laurent Guez, where the 35-hour rule leaves editors and staff with reserves of extra time they must use up on a regular basis. As a result, key people are always missing when the magazine needs a serious push, or when the pressure of business goes up—an inefficient way to work for everybody.

Not surprisingly, as an unintended result of the short-work rule, unemployment is at an all-time high. Flamine and I paid a visit to Antoinette, an old friend and long-term astrologer. She warned us about a dangerous *quadrature* between planets.

On the flight back to California. Monday 29 April 2013.

The new DSM-V medical reference estimates that 80% of people fifty years and older exhibit psycho-pathological disorders. Having 'medically observed' the LoneStars, Kit said the ratio applied to our group as well. This excited Eric, who wrote he "couldn't find his anti-paranoia pills and suspected Hal of having pilfered them." I pointed out that the seven LoneStars being over 50, an 80% rate of psychopathology meant 5.6 crazies among us, which implied I was the only sane guy.

The discussion of cryptids has taken an interesting turn because of Dr. Greer's latest ET suspected specimen, a desiccated fetus-like being from the Atacama Desert, has reportedly been confirmed by Dr. Garry Nolan of Stanford.

Hummingbird. Saturday 4 May 2013.

Two Air Force sergeants who were among the first responders at Rendlesham Forest now claim to have been hurt in the encounter with the unknown 'light.' They're unable to get their medical records from the Veterans Administration. We also discussed the Chinese experiments with viruses, and the spread of yet another strain of the flu. The Chinese were playing with genes while the Pentagon invented prions at Fort Detrick as a bioweapon... So, what else is new? Horse meat was still on the menu at some of the best restaurants in Geneva and Lausanne.

I only knew about Rendlesham from my early meetings with Jenny Randles and Colonel Halt who oversaw the base. I'd concluded early the whole case might be a security operation, which was none of my business, so I never pursued it. But the sergeants' testimony has come up, so the matter may re-appear.

Hummingbird. Sunday 12 May 2013.

Yesterday Flamine and I had lunch with Peter and Mary Banks, on their way to visit their son, and we had the Herzenbergs over for dinner at our home. Len and Lee came over with their son Rick and with Madame Gigi, the adorable little dog who follows them (or proudly *precedes* them) everywhere. They spoke highly of "our student Garry Nolan," to whom they provided an introduction.

I already knew that Garry had appeared on the UFO stage by providing a genomic analysis of the Atacama creature on behalf of Steven Greer's organization and his *Sirius* movie. But Lee added that he spoke freely of his wider interests. So, I called Dr. Nolan, who answered warmly, to set up breakfast with him.

Hummingbird. Tuesday 14 May 2013.

Lunch with the antigravity team, confident they can generate more energy than they are using. Last month, they performed over 30 consecutive runs with Q ratios (power out/power in) greater than unity. All runs lasted over one minute, the longest was 20 minutes. Those results are unprecedented in fusion research.

9

Hummingbird. Wednesday 15 May 2013.

"You're a brave man, Doctor!" I told Garry Nolan when we met for breakfast at Rosewood. Skeptical articles have greeted his interest in the possible extraterrestrial 6-inch-tall, desiccated entity from the Atacama, a shocking admission by a Stanford professor. He laughed and said he'd never shied from controversy.

Dr. Nolan is in his forties, with a neat beard and obvious energy in his step and his attitude to the data at hand, which includes the mapping of the genome of the Atacama creature and several striking videos posted on the web by a forest ranger in Oregon. Garry met Joe after researching his background, and he observed some strange lights during his visit to the location.

As a child, Dr. Nolan had the experience of seeing a face at his window (on the second floor, as did Janine as a child in Morocco). Later, in Portola Valley, he was aware of night visits by entities that stood around his bed and 'told' him to go back to sleep. He would wake up with unexplained marks and bruises.

We spoke of our various research interests, but I didn't probe into Atacama. Garry regretted his association with the case; he's now concluded the creature is likely human. The full analysis has yet to proceed with a study of the non-coding part of the genome; that's the kind of research Colm is most interested in pursuing as well.

We spent two hours over coffee on a terrace overlooking the San Andreas Fault. We disagreed on some points of future research, including what was visible on Mars. He left very upset at me.

Hummingbird. Thursday 16 May 2013.

Last night Flamine and I had the pleasure of meeting Kay Massinghill, the nurse with the passion for searching through older records of unknown craft. I asked her about experiences she might have had. Reluctantly at first, she spoke about seeing a strange shadow on the wall and a light figure in the air she took to be Jesus when she was four or five years old. But later, at 21, driving along with her first husband, she saw a bright light too close to the expressway to be the yard light of a house. When they reached the top of the hill, they suddenly realized it was a large hovering disk. She wanted to stop, but her husband, shaking, sped away as the disk flew up and vanished.

Kay added, self-consciously: "You're the first person I've told that story to, other than my husband." Such confessions always astonish me and make me feel a sense of great responsibility.

I just finished reading the fantastic book *The CIA's Greatest Covert Operation* by David Sharp about Project AZORIAN and the Global Explorer's Ocean voyage. I hadn't realized that the insane secrecy around the project, which continued for many years after the recovery of the soviet submarine was exposed, was designed to cover up *what the CIA didn't know* about the technologies it might or might not have picked up when part of the vessel broke and fell back to the ocean floor.

There's a parallel here with the UFO conundrum: it was critical to keep the Russians guessing. In fact, despite the weaknesses of some results, the soviets had to spend enormous time and money revising strategic procedures. There's also a parallel because what's classified is what *the government doesn't know* about UFOs and hope to learn by 'granting' clearances to researchers who may have stumbled on something...*They've classified their own ignorance.*

Hummingbird. Sunday 19 May 2013.

Jeff Kripal and Jack Katz came over for dinner last night. Over a simple vegetarian dinner with optional salmon and ham, we engaged in a long, colorful discussion. As usual, Jack Katz regaled us with many stories and theories; also, some stunning memories.

Jack, who is 85 and doesn't drive, had requested that I pick him up at the gated community near Port Richmond. Right away, he started telling me about his experience as a boy, one day of frozen weather in Canada when a vehicle with narrow slit-like windows 'floated into his body.' He told me again about his account, while in the service, seeing six disks diving above a Grumman plane the Navy was refurbishing at the Floyd Bennett Field in Brooklyn, and an officer instructing the men *never to report it*. Jack also believes that a UFO crashed in Canada in 1933.

Jack is a complex, tortured man with original ideas. He remains a true master at drawing the human body in all deployments of grace or strength, and a profound lover of music with a limitless memory of little-known classical composers of mid-century America who had to make a living in Hollywood.

I could see that Jeff was in turn fascinated, intrigued, and bored by the endless flow of Jack's monologue. Most interesting are his personal recollections of early ufologists like Frank Edwards in the days of the Long John Nebel show; he remains a friend of Stanton Friedman, who stays at his place when he comes west.

"You have no idea what kettle of fish you're dealing with, young man!" Jack warns me as we drive onto the Bay Bridge in slow Friday evening traffic. "Morris Jessup was killed, you know, and they stole the 500-page manuscript he was about to publish. It all has to do with religion. As for disclosure, it's never going to happen, and it's not even a good idea, the world isn't ready."

In enthusiastic terms characteristic of a much younger man, he spoke of his plan to complete *Beyond the Beyond* before his death while I chewed on his sublimely tragic theory.

Professor Kripal, who drove Jack back to the Cove after dinner, reflected warmly on our evening: "The ride back to his place was lovely. He shared with me a number of his stories/memories (it's difficult to tell the difference, no?) of getting zapped as a three-year-old boy and regaled me with conspiracy theories, which you have no doubt heard. I gathered that he gets much of his creative material from his dreams, which he reads as somehow related to that childhood encounter and subsequent incidents.

"He is actually a perfect example of what I was trying to explore in *Mutants and Mystics*, that is, an artist-writer who has had a number of anomalous experiences and then uses the genre of the comic book or the sci-fi novel to express and code them. He lives, as you know, near the poverty line. He showed me (all of) his paintings! It became comedic at some point, with me begging him that I had to leave and him inventing another reason for me to stay, it was very touching."

Hummingbird. Thursday 23 May 2013.

Colm and Kit arrived in Palo Alto today in anticipation of their first visit to Dr. Nolan's lab tomorrow, so we met for dinner at Scott's Seafood across the street from Stanford and spoke about our various concerns and projects. They had arrived early, so I found them nibbling on a bowl of fried prawns and arguing about biology.

I asked them about the status of our proposed new project, expecting the worst, but Kit was hopeful to see something new organized "within three months." He made it clear he was sick of the politics and the confusion: "We thought it would happen with DHS (Homeland Security) after we lost our primary home, and then with Arêté, but a third group is involved. I'm not sure I want to be part of it, and neither should you. You'd probably never have a chance to do the research you really care about."

"What's new at the Ranch?" I asked, turning to Colm.

"Mutilations have started again," he said somberly. "Our neighbor Garcia, the father, whom you know, has found a cow with the head stripped of flesh, and a cored rectum. Cases have increased."

We spent the rest of the dinner discussing Dr. Nolan's research. It should be possible to detect evidence of past injuries to the victims' immune system, assuming the case has happened in the last 5 years, he said.

Today the NASA-Bigelow press conference took place in Washington, announcing Bob's mandate to assemble a group of private companies and plan a return to the moon.

Hummingbird. Sunday 26 May 2013.

As time passes with no decision about a follow-up to BAASS, I feel increasing doubts. All of us have a different picture of it and a different idea of the sponsor. First it was going to be Homeland Security, then Arêté, then a Navy sub-entity. In the meantime, Washington has become messier: Obama fails to apply to his second administration the lessons from the first.

Even the space industry, lodged in the recesses of the military-industrial complex, has surpassed itself in deviousness. Many scientists leave their legendary labs when they no longer stomach the money-grabbing rules of the big contractors. Whatever the government or its industrial partners have in their vaults, the view of the phenomenon—and what should be done about it—is skewed and faulty.

As for Hal's theory of UFO physics, I first heard it in 1971 and it has yet to be demonstrated; other alternatives, still confidential, make equally slow progress. Physics is hard. I trust Kit's assessment of the lethal effects. He's also correct, in my view, when he says we'd be nuts to join yet another secret project where we won't be able to manage our own destiny and design the right research.

Hummingbird. Tuesday 28 May 2013.

My reading returns to Philip K. Dick's *VALIS*, a book that didn't hold my interest when I first bought it. I dropped it after Chapter 7, unimpressed with Dick's obsession with Christian symbolism.

Reading further now, however, I am struck by his anticipation of the physics of information. He cites Mircea Eliade: "Time can be overcome." He writes that his character, Fat, "developed a theory that the universe is made out of information. He started keeping a journal ... the furtive act of a deranged person." I know the feeling!

Returning to Eliade, he comments: "Time can be overcome, that's what it's all about. It has to do with the loss of amnesia; when forgetfulness is lost, true memory spreads out backward and forward, into the past and into the future and also, oddly, into alternate universes; it is orthogonal as well as linear."

He cites *Fragment 123* by Heraclitus: "Latent structure is the master of obvious structure." Then, in a remark relevant to our Midwest research, he writes: "What if a high form of sentient mimicry existed—such a high form that no human (or few humans) had detected it? What if it could only be detected if it *wanted* to be?"

I am most struck by his Entry 14 from the *Tractate*: "The universe is information, and we are stationary in it, *not three-dimensional and not in space or time. The information fed to us we hypostatize into the phenomenal world.*"

Then he turns on the TV: "Morons and simps appear on the screen, drool like line heads and waterheads; zit-faced kids scream in ecstatic approval of total banality."

Hummingbird. Friday 31 May 2013.

Peter Sturrock and I had lunch again at the Duck Club in Menlo Park. Peter now wears hearing aids, but his eyes are sharp and his intelligence is as clear as ever. He spoke of his amusing troubles with Shakespeare experts, who dislike his latest book, *AKA Shakespeare*. He has stirred up as much controversy with scholars of English theater as he'd done among physicists with his UFO interests, and he seems proud of it.

He asked about my current work, so I spoke about *Wonders in the Sky*, and the realization, when we come to the Enlightenment, that scientists of the day were very open-minded, accepting unusual phenomena with joy and creativity, nothing like today's fear.

"What happened to Science, Jacques?" he asked. "It seems that if you present today's scientists with an unusual phenomenon, they will immediately fit it into a conventional explanation, and if that doesn't work, they will drop it altogether. *But that's not science!*"

"What has changed between the Enlightenment and our current Age of conventional darkness is the funding structure," I said. "Any indication of truly independent research is often frowned upon, and another career is destroyed if it doesn't fit."

Peter still works (with Dr. Pappas, of the University of Arizona in Tucson) on the effect of solar neutrinos on radioactive decay rates. The latest information from a lab in Moscow shows different decay rates from two detectors monitoring Strontium 90.

This could be an interesting discussion with Roscomos on my forthcoming trip. Maurice and I are still waiting to hear from Dmitry Manakov about the meeting of the Russian board, where they may give us the green light to start our proposed venture funds.

The reason for Peter's call was elsewhere, however. He now believes that his isotope studies of the Ubatuba material have been insufficient. He even suspects that some of the institutions he consulted may have hidden the true results, so he's giving pieces of the material to several close friends for safekeeping, and he asked me to keep the primary ones.

Hummingbird. Monday 3 June 2013.

A three-page, tongue-in-cheek article I wrote about "The banker, the dinosaur and the little nomad" has appeared in, of all places, the Annual report of the BPCE bank. It contains a severe appraisal of the slow evolution of banking. Reactions should be amusing, since BPCE is now the second largest such establishment in France and its publications get into the hands of all major investors.

This afternoon, I drove back to Richmond to spend the afternoon with Jack Katz, who was in turn fascinating, charming, fragile ("you can't imagine how long I've waited for a friend," he told me) and full of his new celebrity: the first of five volumes of the *First Kingdom* has just appeared, so he's being called for interviews by admiring fans the world over. The volume for which I wrote an introduction is *The Space Explorers Club*. Jack made black tea for me, extracted many drawings from old cartons, including some precious originals, and told me about his passion for music, especially the forgotten composers of the 20th century, like John Alden Carpenter and E. J. Moeran.

Jack also told me more details of his childhood experiences. When he was five (in 1932, in Ottawa) he once felt very hot and decided to take a walk down King Edward Avenue. As he was walking along, a "thing" passed by him. It looked like a very flat car with slit-like rectangular windows, *floating a couple of inches from the ground*. It went up State Street, leaving the boy feeling hotter than ever, so hot in fact that he took off his overcoat on the way home and was scolded by his grandmother. People on the street had seen this; they asked him, "Did they talk to you? What did they ask you?"

Sometime later he was in New Jersey with his father when he saw a pear-shaped object spinning, with smoke around it, the color of metal. A seam opened up in the object and a being emerged, very tall with square eyes in a square face. It pinched the boy's hand near the base of his thumb, paralyzing him.

"That's when my ideas started coming," he said, "about the First Kingdom, a contact with the ancestors of Humanity." Curiously, he had tried to call his father's attention to the thing but his father "didn't want to be bothered because he was there on business."

"You see why they'd put me in the basement at Bellevue (New York's major mental hospital) if I told that story?" he joked.

John Alden Carpenter's *Symphony No. 1* was drowning his words. He made me listen to it for a while, and then turned it down.

"Music is intelligence itself," he remarked as he shuffled from the kitchen to his living room where the disc was playing. So, we listened to Moeran's *Overture for a Masque* as he assembled several books about religion that he wanted to show me, including *Past Shock* by Jack Barranger and *Breaking the Godspell* by Neil Freer. He also quoted the tales of Roger Zelazny, John Campbell's 1928 story *Twilight* (whose depressing ending got changed), and underlined the word <u>Dehisce</u>, his key to *Beyond the Beyond*.

Dehiscence is the spontaneous splitting of an object or a surgical wound along a natural line, or the opening of a plant at maturity. What does that have to do with the history of Humanity? He saved that story for a future lesson. "Transpermia was poisoned," Jack told me in conspiratorial tones as we sat at his little kitchen desk with our steaming teacups. "Now the Universe wants to heal itself."

Hummingbird. Thursday 6 June 2013.

Another blow to venture plans: the leader of Russia-VC, Sergei Gurieff, has been forced to leave Russia and flee to France. He was chairman of the Russian economics school and of a major bank. He was under increasing 'pressure' (which means threat of assassination, in Moscow these days) because of his involvement with the Yukov affair. There's no date for any meeting.

Chris Aubeck has had an unusual experience in Madrid: "I woke up around 4am in my pitch-dark room, near the door. The door was ajar and I had put a pair of shoes stopping it from opening. I noticed a small, intense light coming at the level of the shiny brass door handle (no keyhole). I said, drowsily, 'I know someone's there!' No reply. I noticed the light made a chair and some objects visible in some detail. I sat up, quite afraid, I thought, 'if this isn't real, I'm doing a fine job of getting the details right!' I don't recall if I saw the light disappear, but finally I lay back down to sleep."

Mabillon. Tuesday 11 June 2013.

The second day after landing in Europe is always the toughest: bad sleep, tired muscles, and a brain that starts and stops in fragile spurts. I did visit a friend at the venture offices of Idinvest yesterday morning, reconnecting with my French network. Then I arranged for our follow-up visit to Col de Vence next week with Alain Bauquet.

The press has been full of a new NSA scandal, with a flurry of revelations about the wholesale invasion of the privacy of US and foreign citizens as the agency sucks up data from every social network and search engine: The kind of situation I had tried to warn against, starting with *Network Revolution* in 1979, has now arrived.

Laurent Guez, at Les Echos, was surprised I'd never been approached to work in Intelligence, so I explained to him the difference between open research (occasionally funded by the likes of ARPA or CIA, like the development of the Internet) and the business of spying on people, which always disgusted me.

Now, a conference call with my Euro-America partners as we prepare to make a near-final cash distribution to our investors, quite a victory in today's troubled financial waters.

Mabillon. Saturday 15 June 2013.

Lunch with Maxim at Le Suffren. My seven-year-old grandson was calm and kind to me as we spoke of his interests in books. He recalled our trip to Mount Lassen and the ride on the Skunk. He wanted to see a movie, so we watched *Star Trek Into Darkness*, where big, special effects substitute for story. Our Paris home is changing as Flamine's books find their place here, a graceful inspiration.

Nice. Thursday 20 June 2013.

We drove up to Col de Vence again with Alain Bauquet yesterday, and today we went up to Roussillon-sur-Tinée to visit Jocelyne, who knew about the sightings there.

At the Col, Bauquet told us of the time when he and his family observed a large 'orb,' brilliant white, among some trees a kilometer or so away. *Two simultaneous photographs were taken*, from which I'm able to estimate a diameter: about 2.5 meters.

The road up to Roussillon-sur-Tinée is a dramatic two-lane highway that follows a large torrent fed by the recent rains, bouncing down the mountains in white foam and rushing noises. Jocelyne took us to a site where encounters with cryptid creatures have occurred. Her 18-year-old son came rushing back from the forest one night when he was confronted with a large grizzly-like creature that chased him all the way back to the house. Jean-Claude Dufour reports that the boy was traumatized and could hardly describe the beast when his mother put him on the phone, seeking official help. The house next door, which is empty and locked up, has been the site of hauntings since the 19th century, when it was investigated by Camille Flammarion.

Hummingbird. Tuesday 25 June 2013.

Today I finalized a long, important text drafted in Paris, intended only for Hal and Kit. It could terminate my association with what remains of the BAASS/DIA project. The letter begins:

As we wait for some decision by sponsoring agencies about the next step in UFO research, I have given serious thought to my future participation.... If a "real" project was undertaken, with the active involvement of the two of you, I would of course hope to be considered for it. Yet, what I have seen in the last 20 years or so, whenever I have been aware of related US government activities, doesn't lead me to presage anything of the kind.

Expanding on the reasons for my growing uneasiness, I tried to place some serious facts in front of them:

My analysis begins with the task I have conducted for BAASS under agency sponsorship, namely Capella.... Even the critical nature of a platform upon which other research projects can be erected is not understood by the sponsor.... An expectation exists that revelation of hardware and biological material will 'soon' happen, solving all extant scientific problems in short order, and making a long-term, careful analysis of global data "nice to have, but superfluous."

Such an expectation is unrealistic. If such physical and biological material in fact exists, it will be critical to correlate it precisely with underlying patterns in this eminently deceptive phenomenon, a correlation that can't be done reliably with current data. There's bitterness in these statements, which I can't hide. But I cannot contemplate spending the next few years playing silly games. This forces me to consider breaking with years of joint work. I no longer trust the process, despite my high regards for the participants.

Hummingbird. Sunday 30 June 2013.

Quiet days: lunch on the deck, a warm wind plays with the bougainvillea. A curious hummingbird flutters here and dashes there along the railing. Flamine and I delight in our closeness.

There's great excitement in San Francisco today: the Supreme Court has permanently opened the way to Gay marriage. This City, which renews itself constantly, witnesses one of its greatest celebrations. Other trends are not so happy: the *Lumière* cinema, a nice theater for art films and avant-garde, has closed, while Fields Books, the legendary esoteric bookstore on Polk Street, a century old, has moved into cyberspace, where I have lost track of it.

I love this City, its diversity and its moods, its moments of foolish enthusiasm and its renewed attempts to grasp beauty beyond the hard reality of every day. Part of me will always haunt these streets, these hills, from the heartbreaking Tenderloin to the sublime seashore, from the arrogant financial center to the guns of the Presidio.

Hummingbird. Wednesday 3 July 2013.

Last Friday evening we drove to Palo Alto to hear my old friend Dan Drasin talk about his documentary on electronic communication with the Beyond, now called Electronic Voice Phenomenon (EVP) or Instrumental Transcommunication (ITC). Russell Targ was in the audience, friendly and open as always.

The film reflected interesting work with little new material from the days when I discussed the topic with Father Brune and Rémy Chauvin. Researchers cited included Raudive (pioneer of the field) and Lisa and Tom Butler in Reno. I asked Dan if any physical phenomenon were noted along with the unexplained voices: the answer was negative. Good mediumship remains the best form of communication with such entities. Whether or not electronic signals represent dead persons is inconclusive, in spite of very odd coincidences among the noise.

Lunch today with Robert B., his wife and daughter. He's a 'retired' counter-intelligence agent knowledgeable about UFOs. The history of US Intelligence is full of grievous mistakes, I told our visitor; shortly before the Cuban crisis of 1962, Kennedy received an Estimate of the Situation stating: "We cannot find any set of circumstances under which the USSR would put missiles in Cuba…"

Later the same day.

I just read of the death of Doug Engelbart, at 88. Reactions are flooding in from the programming community in Silicon Valley and beyond. The French press (*Le Monde*) is hailing him as "the man who invented the mouse," which is not even wrong, but like too many journalists, they only see what in front of their nose, the obvious gadget that looks interesting on TV. Nobody asks *why* this man invented the mouse and the breakthroughs he achieved afterwards.

Hummingbird. Thursday 4 July 2013.

Dr. Garry Nolan came over for lunch today. He isn't mad at me anymore. He explained to me the work of his lab, building on the Herzenberg breakthroughs, and expanded his interest in UFOs: He had all the right questions. Can he get blood from Kit's witnesses, and start serious research on the effects of the phenomenon? The research on cryptids has led to premature publications, careers in question, and ridicule in the press. We were wise to stay away.

Hummingbird. Saturday 6 July 2013.

Egypt: another intelligent, warm and generous people, suddenly on the verge of civil war. The finest impulses of our spirituality form the roots of our most murderous instincts. Studying historical facts about religion, reviewing Kelsey Graves' old study of the role of 'Saviors' in civilization, I am reminded of the fact that no less than sixteen such would-be Saviors were crucified, *not counting Christ.*

Eight of them were born of a virgin, usually at Christmas, the Winter Solstice. Several of them had a father who was a carpenter, were of royal blood, and were threatened by a tyrant trying to kill them. They crushed Serpents. Their birth was announced by a new star and attended by shepherds and wise men from far away; one of them, Krishna of India, furnished the template for the Christians: he received gifts of gold, frankincense, and myrrh. By Hollywood standards, the story is a brilliant remake of the Krishna saga, twelve centuries later.

Hummingbird. Monday 15 July 2013.

Upon returning from Europe yesterday, Garry found himself confronted with the phenomenon before dawn. He'd gone to bed early at his home in the hills: "I woke up at 1:30 am for no apparent reason (I usually am an extremely heavy sleeper). My back was to the balcony. I felt I needed to 'turn over' to look in that direction.

"When I did there was a 'diffuse,' wavering light outside the bedroom window, bright enough to cast a shadow on the floor. The dogs noticed it as well but were not growling. It was not imagination. It was there for about 10 seconds, wavered, then simply seemed like it floated off. The balcony is 30 feet off the ground."

That experience, again, reminds me of that of Janine, as a child in Morocco.

Hummingbird. Saturday 20 July 2013.

Alain Dupas was in California for an AIAA propulsion conference, so I invited him to the monthly dinner of the Band of Angels, and we attended a two-hour methodology seminar at the Institute for the Future, led by Mike Liebhold. Yesterday, August and Maurice joined us at home to review the status of our Russian ventures.

Hummingbird. Tuesday 23 July 2013.

This evening, Senator Mike Gravel gave a lecture at IONS, north of Petaluma, about the politics of democracy and the special problem of UFO disclosure. It was good to meet a man who served from 1969 to 1981 and was responsible, among other things, for releasing the *Pentagon Papers* and for renewed efforts to end the Draft.

By a perfect California sunset, we had a potluck dinner under the arbor, so I had a chance to discuss several points with Gravel before his speech, notably about the wisdom of Congressional Hearings (he says they're a waste of time). He's convinced there is an ET presence. The problem is that the US is so bellicose it isn't mature enough to deal with it as a research opportunity. He also thinks that the aliens are far advanced in terms of energy, and their impact "will make a difference in our civilization." He also drew a difference between 'representative government' and real democracy, found only in a few nations like Switzerland. "We never had 'our' government," Mike Gravel added, "even in Philadelphia. *The Constitution is not a democratic document; it contradicts the Declaration of Independence that claims all men are created equal: why restrict the vote to male white landowners?!*"

In response to questions, he stated that in the event of Disclosure the elites would panic, but not the common people. And the government "doesn't know much more than you do…"

Hummingbird. Thursday 25 July 2013.

"Which century are we in?" I asked Flamine this morning as we watched the news with the feeling of being thrown back some four hundred years, in gothic shade. Item one was the announcement at Buckingham Palace that the Duchess of Cambridge had been delivered of a fine boy. Item two was a report on the Pope's travels, and item three was the abdication of Albert the Second, King of Belgium, in favor of his son Philippe.

Far down the list were 'minor events' like the civil war in Syria, the riots in Egypt, and the exposure of wholesale spying by the NSA on every citizen of the planet: phone calls, email, travel data. Small stuff.

Hummingbird. Wednesday 31 July 2013.

Dinner with Peter Sturrock and his friend Tita, a cellist. We met in Menlo Park at the Flea Street café. He entrusted me with more samples he hasn't had a chance to analyze. He reminded me of the bank break in; we agreed we wouldn't talk through email or the phone, and that the material would not be stored in any bank vault, wide open to penetration by the people managing the censorship.

The truth is that I am frustrated by the delays. Good science begs for support, as I've seen over the last two days at a conference in San Jose. Serving on a panel about microgravity, I heard about plant modification and mitigation of harmful effects. Those are topics where intelligent investment could make a difference for the whole planet.

Hummingbird. Friday 2 August 2013.

"There's nobody like me!" jokes Jack Katz as he rummages through a binder full of his latest drawings. In the background, his favorite music is playing at full volume for inspiration: early 20th century neglected composers who've been widely ripped off by Hollywood: Howard Hanson (*Lament for Beowulf*), Samuel Barber's *School for Scandal*, Arthur Foote, and Joseph Suk (*Serenade for Strings*), followed by Griffes' *The Pleasure Dome of Kubla Khan*.

Over dinner at noisy Hotel Mac, where a piano played in the bar 'for atmosphere,' Jack failed again to reveal all the secrets of the universe, but he did give me some details of his childhood experiences.

He was born in September of 1927, so the events must have happened in the 1931-1933 timeframe when he lived with his parents in Ottawa. The first sighting involved the strange vehicle that flew past him on King Edwards Street. The encounter with the very tall figure happened in New Jersey, where they had gone back to see his paternal grandmother. He also talks about a series of vivid dreams, and an episode when he was just a tot, hiding under a table; he fought hard against a doctor who tried to examine him. Jack has been married three times and claims he has 'known' over 80 women, one of his more believable boasts.

I point out that's less than one woman a year, a poor record for a new Don Juan, and he gets mad at me.

Jack isn't the only gifted graphic artist in the world, but his portraits are stunning in their authenticity. His rendering of the motions and anatomy of the human body is perfect, his work on portraits is inspiring. I am fond of his paintings, which remind me of Clovis Trouille: both loving of humanity in its depravity, our pitiful failure to rise above the gutter, yet so full of hope for redemption.

Hummingbird. Saturday 3 August 2013.

Lunch with Garry Nolan at the Stanford Faculty Club. He told me he'd met Greer through his correspondence with the man who shoots UFO videos in Oregon. Sadly, Greer continues to hype the 'alien mystery' of the small being of Atacama although Garry has proven it was human.

We discussed his lab, the Herzenbergs, and his experiences with cancer. Garry has gone to Arizona with Colm, driving three hours into the hot desert in hopes of drawing blood from one of the Bentwaters witnesses but the man's attorney denied access, and the VA refuses to release medical records, so the opportunity was lost. Colm says the crashes at Roswell, Kingman, and El Indio Guerrero are all genuine.

Stan Friedman called tonight, following a conversation with Jack Katz. He told me he'd made a presentation at the Citizen Hearing on UFO Disclosure but he didn't believe that sudden Disclosure was a good idea; it would start too many troubles around the planet. He also reminded me that Blue Book was closed in 1969 by General Bolender, the man in charge of the lunar excursion module.

Once again, I've called Hal and Kit to express frustration at our physical models. It's clear to me that the phenomenon can take over physical effects and human perceptions over any area for a significant amount of time (a notion I derive from unpublished but solid cases). I reminded them that "Those who take the physics of information most seriously (claiming that the physical universe is a virtual reality and that we live in an information matrix) find no problem with higher intelligences replacing any volume of space-time with their own version, where they can project displays and their own machines."

So far, I've resisted this model as too simplistic, extrapolating our digital technology to build a seductive vision of the universe as videogame. But it does account for the absurd reports we have on file: witnesses stuck on expressways where no cars pass by for 20 minutes, people going in and out of invisibility, parallel reality landscapes, and of course the synchronicities.

Kit responded: "I can explain the cases I've personally investigated in-depth (about 50, over the past 7 years): fugue state, acute; induced de-personalization/de-realization, sub-acute frontal lobe amnesia. For me it works, all can be explained without resort to complexity."

So, why should that be incompatible with a better physical topology? My questions were not at that level. Hal calmly told me my presence and contribution remained valued, reminding me of the extreme pressure, exposed patients dying. He's just as frustrated as I am. He's directed the compilation of 38 papers that resulted in no follow-up. "They're gathering dust, sitting in some big warehouse next to your Capella data," he tells me. "Yet I had key people willing to talk, if only we provided a SCIF, and an airline ticket."

Hummingbird. Sunday 4 August 2013.

Eric Davis has jumped into the discussion with a classic comment: "Kit, your reply to Jacques makes zero sense. We're talking about a model for reality in which EM/RF radiation is a byproduct of the universal information hologram producing projections that we perceive as EM/RF radiation."

The truth is that the group never considered the meta-reality of the phenomenon, whether in the anthropological context (what Douglas Price-Williams called Mythopoietics) or in the hard physical context, where Eric and I are trying to erect new hypotheses.

Hummingbird. Sunday 11 August 2013.

Senator Gravel and his wife, Whitney, came over for dinner last night, along with Garry Nolan and his spouse, Tim, whom we were meeting for the first time. Flamine had prepared a light dinner (Whitney was dieting) enjoyed with good humor while we spoke about Garry's research, Gravel's interests, and my experiences as an investigator. I gave them a demonstration of the database structure I'm proposing as I continue to refine Capella beyond what the BAASS project could do.

Hummingbird. Friday 16 August 2013.

Flamine just left for Paris, where she'll keep company to her elderly father. On the news, the tragedy of Egypt emerges in all its ugliness among the bloodshed of the Middle East. To stop a slide into radical Islam, the military has removed the first democratically elected president and slaughtered his supporters, thus making the problem worse, elevating the Muslim Brotherhood to martyrdom and creating the conditions for an even bigger civil war. I've long admired Egypt, where I found dignity and beauty. To see it reduced to such ugly hatred is very sad.

The next Vienna meeting of the LoneStars is shaping up badly. Bob Bigelow will not attend, and Colm won't come either, due to 'pressing deadlines,' and John Schuessler has back pains. I'm especially sorry not to see John, the most patient, informed, and analytical member of the group with vast experience in field investigations. I am intrigued by his feeling that there is *"a horrific threat level behind what we are seeing."* I do feel the same way, often.

Kit has pre-empted our discussion of historical trends by insisting: "I will try and argue…that history can paradoxically be the *enemy* of the future. Lessons learned, are not. The key ingredient is … to investigate abductions with modern tools in the eye of historical lessons. *NOT* using historical process."

Just a month ago, he was brushing off all abduction research as absurd, a product of false science and crazy beliefs. Now he's willing to forgive and forget: "We have no reason to re-educate anyone…we have but to move forward and address our interests with our modern approaches. Will we miss some stuff? Absolutely..."

From my point of view, we're missing too much stuff; the failure of the old approach left biases that will haunt us. The invention of new science has never erased the reality of bad old data. Just because we have modern techniques to analyze cometary gases doesn't mean it is useless to research the history of comet sightings, which give us precise orbital and essential periodicity data the farther back we go in time. How else can we be certain of a 50-year period?

So, Kit comes back with "a pair of abduction cases, to try and prove my point. In one, the person has steadfastly refused to be hypnotized, except by a qualified physician. Handled as an 'abductee,' the case would have died. So, you're right. The abduction is not what should be eschewed; it is the method of dealing with abductees."

He then concludes, aptly: "Years ago Jacques invented the criteria to accept and investigate the cases. We have forgotten, too soon, forgotten our own advice. We have reverted too often to historians seeking stories, and not scientists seeking hard data."

This puts me in a hard spot, despite the passing homage. Good stories are hard to get, too.

Hal and Eric will be there, fortunately, but Flamine's absence will empty a palace in Vienna of the elegant opportunities it promised.

Hummingbird. Saturday 17 August 2013.

Last night I drove across the Bay to meet Jack Katz again in Point Richmond. I found him energized, walking smartly to meet me at the gate. He said he'd run five miles along the beach the previous day, an admirable performance at 85.

Jack gave me a copy of *Legacy*, one of his old graphic novels, and an original drawing that he inscribed before we went off to dinner in Albany, where we had an excellent meal at Bua Luang Thai, one of his favorites. On the way he reminded me of the three basic urges of Man: *the desire to survive, the urge to procreate, and the urge to migrate* that is the basis of creativity and discovery, the latter only found in a few, exceptional people.

I asked him about his father, an itinerant craftsman and salesman who traveled widely, "working across seventeen States." Jack remembers helping him fix novelty items, like small toy birds sold to kids. Everywhere he went, their home was full of live music, with impromptu chamber concerts and groups of learned friends. He recalls leaning against the cello, feeling its vibrations. Jack doesn't think they were Rosicrucians or Masons. Yet one day in Redwood City an old man saw him: "You're Jack Katz, aren't you? I was a friend of your father. We tried to change the world, to create a better society. We've been around for centuries…"

Back in his street-level, one-bedroom apartment in a low-rent section of Point Richmond, he let me read through the original plates of the *Space Explorers' Club*, his major work. I found it complicated (and occasionally garbled by careless spelling) but the images were so beautiful and held so much power that I shivered at the obvious genius.

"Would you miss me if I took my life?" Jack said abruptly. In shock, tears in my eyes I asked, "Why do something so stupid, when you still have so much to tell?" He believes he was given an insight—perhaps during the incident with the strange floating vehicle he saw as a child in Canada—into the imperfect nature of Man, the flaws in creation that go back to the Big Bang. I am not too concerned about his suicide threat because he's at work on his ultimate opus, *Beyond the Beyond*, that should put a crown on *First Kingdom*, and there are rumors of a movie…and more… and yes, I will miss Jack Katz terribly when he dies.

Later the same day.

I just finished reading *The Tunguska Mystery* by Vladimir Rubtsov, recommended by Peter Sturrock, a well-researched book. Not only does it bring news of active, intelligent research in Russia (and the sad news that Felix Zigel, along with my friends Alexander Kazantsev and Alexey Zolotov, has died) but it re-opens the issue of what triggered the devastation in the Siberian Taiga in June 1908.

Now, a letter to Flamine: "I went to our usual restaurant on Polk Street for lunch. It's the second time the owner sees me by myself. I must have looked sad because he asked, 'Is your wife coming back?' I reassured him, but I don't think he's completely convinced."

She answers from Mabillon: "I don't want you to be sad; I miss you, I sleep badly. I'm relieved to know that I get back in one week."

Hummingbird. Saturday 17 August 2013.

Nice surprise this afternoon: two of the grown-up grandchildren of Annick were in town, so I spent time with them, my daughter, and Rebecca. The sky was clear in a mild summer day; on the Bay, three catamarans of the America's Cup race back and forth for fun.

The pleasure of Catherine's visit to Café Quetzal took my mind off the debate with the Lonestars. I will go to Vienna as agreed, a congenial group.

Arguments will be healthy, around Austrian coffee.

Hummingbird. Friday 23 August 2013.

A lot of work this weekend: The Ingenuity conference of *BoingBoing* (presided by David Pescovitz) was at the Regency, a gorgeous ballroom designed for Masonic initiation with a fantastically weird backdrop of dinosaurs and devils.

I also met Garry Nolan. Our main topic was isotopic analysis of UFO samples. He'd read my paper on physical samples (17) and knew of a lab, so I gave him the Bogota and the Council Bluffs materials.

Hummingbird. Saturday 24 August 2013.

Jack and I returned to the Thai restaurant on Solano Avenue tonight. He drew pictures on napkins for the kids at the next table. Jack loves children and is always making kind comments to the parents around us or engaging the kids themselves in conversation.

Back at his place, he let me review the entire set of original pages for *Destiny* while we played his favorite music from neglected composers. Then, Jack 'revealed' to me some of the conversations he held with a Vice Admiral during the Korean War.

He served in a unit involved in naval architecture (he helped design a new type of fast torpedo boat) at the time of the Injun Landing. When Jack asked the captain what he thought flying saucers were, the man took him to his cabin, gave orders not to be interrupted, and showed Jack copies of unpublished logs describing flying objects. He claims to be privy to examples of UFOs kept quiet among the Navy brass.

I presented Jack with DVDs of some of his favorite composers, and as a joke, some special new Number Two pencils and a futuristic pencil sharpener. Our friendship deepens through these conversations.

Kit has brought up the topic of alien bodies again, with references to Leonard Stringfield, little bodies at Wright-Patterson AFB, and early DNA work by a French physician, Dr. Leon Visse, all of which still raises questions. He also mentioned a Harvard geneticist named Matthew Meselson, known for showing in 1957 (with Franklin Stahl) that DNA replicates 'semi-conservatively.' In the 1980s he led an investigation contradicting a CIA statement attributing 'Yellow Rain' to a Soviet biological weapon program. Instead, Meselson concluded Yellow Rain was caused by bee droppings. (The now famous 'buzzy bee.') Later, in the Russian anthrax case of 1979, with the 'accidental' death of 64 people in Sverdlovsk, Meselson had to concede the Soviets did violate the Biological Weapons Convention.

Mabillon. Monday 2 September 2013.

Over lunch with Claudine Brelet yesterday (at Pavillon du Lac in Parc Montsouris) she joked about the current international situation. "François Hollande plays checkers, Obama plays American billiard, and Vladimir Putin wins over both with clever chess moves," she said. "Obama's hollow strategy of declaring a 'red line' for chemical weapons, then doing nothing when it was trampled, represents a major failure, and another loss of trust in the American word."

Claudine has seen UFOs twice—once while staying in a Montréal hotel, she received an urgent mental message to rush to the window where she saw an oval of lights hovering at eye level. She grabbed two people passing in the hall to confirm what she saw.

The second time took place in Africa. She was in a parked car with local officials. A ballet of lights in the sky, obviously not a rocket reentry or an aircraft accident, amazed her but left her African colleagues unfazed.

"The phenomenon doesn't pose the same problems outside our rational West," she said. "Many populations already live among a thick crowd of ghosts, gods, and dead ancestors, so UFOs aren't an urgent issue, they fit easily into their culture."

I haven't accomplished very much since my arrival on Thursday, beyond setting up of a new system for Internet and TV service. Time to prepare for the fifth LoneStars meeting.

Vienna. Thursday 5 September 2013. Steingenberger Herrenhof.

The hotel is only a short walk away from our host's apartments with a fine view over Rathauspark and the busy center of the city. The session began with a phone link, with John speaking of the UFOs' apparent interest in human technology in many overflights by unknown objects at space events, and even in a B-29 case in Korea. He described the case of a pilot who had to maneuver around an airport in South Texas where the runway was blocked by an intruder. Local workers at a nuclear power plant under construction "were seeing them all the time," and the seventh Chinese space launch has been buzzed as well.

The first session dealt, in general, with the characteristics of various types of 'cryptids,' after which we tackled the process of 'confirmation' and its implications, while I reported on recent investigation trips. The later discussion had to do with future formats for research. A new plan, that had been 'approved' under the aegis of the Navy (by a full Congressional subcommittee, and two Senators) got stalled again because of administrative errors and interference again, arguing in the name of "not tempting the Devil."

That reminded me, weirdly, of what Jack Katz had told me about his very private meeting with his Vice Admiral in WW2 days.

Vienna holds great monuments that impress the visitor with their solid magnificence, while horse-drawn carriages gaily carry tourists in style from the opera house to the various parks. As for me, I found a nearby biergarten where I could dine on simple fish and potatoes, alone, feeling a little sad about the wasted opportunities.

Vienna. Friday 6 September 2013.

The day began with a review of the text of a 1961 CIA Special National Intelligence Estimate, or SNIE. Apparently, as an authentic document, it confirmed there were crash retrievals resulting in six recoveries of debris, *including nuclear material*. The document says nothing about fuel or propulsion or specific crash sites, but it does mention structural components and isotopes "not manufactured on earth" used in special alloys. The apparent purpose of the SNIE was to avoid a crisis with the USSR if they, too, detected such objects and assumed we were attacking them. At least, that's the story given, and I reserve judgment: I have no expertise in verifying documents twisted around by deception experts.

We went on to discuss Garry Nolan's research when he says: "*Every stressor is recorded by the immune system within the human body.*" Billions of interactions can be saved: A cell surface marker on a T or B cell is there forever, once created by an actual experience. The immune system elaborates proteins corresponding to radiation, RF, visual stress, etc. and the methylation process is affected. The non-coding (epigenetic) DNA contains information about damages. Bob Bigelow calls this a Cognitive Tattoo, a wonderful name.

The group feels an increasing level of concern about interference, originating not only from the bureaucracy (which refuses to consider any unknown menace) but from religious groups that misunderstand the research and find it "unpatriotic to reverse engineer the demon's technology." As for the phenomenon itself, we've only made small progress, even if hardware is now under study, as well as the evidence for neuro-physiological programming; per our rules, I will not say more.

Vienna. Saturday 7 September 2013.

Not a cloud in sight. The weather remains bright; the red and white flags of Austria fly high above the major landmarks of the city.

Over breakfast, Kit said he had no recollection of the interactions we had in the 1970s when I brought him fresh samples from mutilation sites in Oklahoma and Arkansas, documented at the time in my published Journals (*Forbidden Science 2*). One thing is clear, however: more and more official documents are allowed to come out now, occasionally redacted, supporting the reality of the mutilations, then and now.

Once we reconvened, I gave a presentation on abductions, where I was finally able to show both the positive and negative aspects of regression hypnosis in the hands of competent practitioners. Finally, we tackled the issues of quantum entanglement and nonlocality. The experiments of professor Alain Aspect in the early 1980s have been extended over kilometer scales. Quantum nonlocality is not just happening at the micro scale, it can be maintained at large distances and even at high temperature. It is not eliminated by gravity but stabilized by it, and *nobody knows why*.

The main point is that *there must exist a reality outside of spacetime*: there is no time order (that is, no causality) in the quantum entanglement of particles across spacetime, so causal order is not a fundamental property of nature. Why do we observe causality? It is restored when experimental parameters disentangle particles.

In remote viewing, the subjects enter a less causal, more entangled state. Colbeck and Renner have argued there is no missing information, no hidden influence, and no solid cause-and-effect relationship. *Spacetime is not fundamental: it is quantum entanglement that gives spacetime its structure.*

"Underlying the universe is a sub-quantum domain, and the vacuum is teeming with information. It's information that determines what manifests physically. All of reality is on a 2D membrane located at the event horizon, where galaxies fly away faster than the speed of light."

Doesn't that violate some physical principles, I wondered?

"No: if you conserve energy, you also conserve information. Nonlocal entanglement is common in nature. UFOs use an advanced process, manipulating the information domain; they manipulate the quantum fluctuations, so they look to us like wormholes."

Vienna airport. Sunday 8 September 2013.

Time to kill and to review my notes. Before the drive to the airport, I had breakfast with Eric, discussing the 100-year Starship Project and information physics. He explained to me the flaws in current theories of the holographic universe while I got ready to fly back to Paris.

Eric has the unique ability to focus exclusively on the details of a subject without ever losing the big picture or evading the obvious. The effects can be unpredictable and gently funny. Yesterday, as the group admired a magnificent painting during a break, following learned explanations by an art expert, Eric turned to him and blurted out, "Why does it have so many cracks in it?" and we all laughed, but that led into a fascinating lecture in art history, varnishes, and environmental factors. I've already forgotten these technical answers, but I am certain Eric will explain them to me precisely again in ten years.

Mabillon. Monday 9 September 2013.

Today I had lunch with Dominique Weinstein and Xavier Passot at La Gauloise, one of Mitterrand's favorite restaurants (close to Le Père Paul, where Chirac used to eat). Passot is the energetic new chief of GEIPAN at CNES: a direct, personable type who takes the job seriously. Surprisingly, he'd never heard of Haravilliers, which says a lot about the structure of the French research.

I brought up the issue of cryptids, describing some of the events in America with the tracking of Sasquatch and Skinwalkers by real scientists. That, too, seemed foreign to CNES interests, yet I know that such creatures have been seen in Brittany, and of course around Roussillon-sur-Tinée, from my latest field trip.

In addition to Vaillant at half-time and Passot full-time, GEIPAN employs a full-time assistant/screener and a documentalist, all in Toulouse. Passot invited me to run a session at a two-day closed workshop on databases to be held in July next year.

Today, Rospars joined me for dinner at La Consigne (also at Montparnasse) and I gave him *Forbidden Science 3*. We spoke of abductions and of our recollections of Aimé Michel. Jean-Pierre is a fine researcher, kind and smart.

Hummingbird. Saturday 14 September 2013.

This time I find it harder to resettle in San Francisco, bothered by a lingering cold. The weather turns gray and cool with only filtered sun. At lunch yesterday with our friend Pascal Bouillon who runs Société Générale's leasing operations in New York, we had a pleasant time but our talks about France' economy were pessimistic.

In a recent interview with the *Financial Times* (8 September 2013) John LeCarré comments, "I do think we live in a most extraordinary period of history. The fact that we feel becalmed is the element that is most terrifying: the second-rate quality of leadership, the third-rate quality of parliamentary behavior."

Flamine has resumed her psychology courses, working with intense focus. But she cried in my arms the other night, bothered by a shocking experience with an acupuncture practitioner that left her distraught. At work, I'm closing two small investments in a medical startup called Materna and in Taulia, a clever financial software company.

Hummingbird. Thursday 19 September 2013.

Peter Sturrock told me this morning he was looking forward to a trip to France with Tita and some friends, drifting aboard a barge in Burgundy, with a side visit to Paris. I had brought photographs of the samples he'd given me to secure, and I asked about their provenance. The Ubatuba samples came from APRO through Irene Granchi in Rio, who got them from Dr. Olavo Fontes. The 'Sierra' samples are unique: one composed of iron and one of titanium, found in layers of peat. I will be on my own with this analysis unless I can get Garry's attention.

Hummingbird. Friday 20 September 2013.

Flamine and I just attended a lunchtime seminar by three NASA engineers speaking about the 'asteroid grand challenge.' They remembered Red Planet Capital and told me how upset NASA was when it died: "Your political connections were weaker than the other guys' connections. They just grabbed the money! Yes, a missed opportunity for everyone, and a setback for the US space program."

Hummingbird. Saturday 21 September 2013.

In Vienna, as we discussed blue orbs and their effects, I recalled that in my novel *Le Sub-Espace* (written in 1959, although not published until 1961) I had posited *blue orbs* as the primary system used by alien beings for communication with us. At one point the main character is engulfed in an orb and finds himself back in his study, but everything is reversed: book titles are written right to left and his watch moves counter-clockwise. Two hours have elapsed although he was only inside for a few minutes. The aliens are thoughtforms trying to warn us about cosmic changes about to impact the planet.

Two friends observe another orb in the countryside and enter it, finding themselves inside a large spacecraft *much larger than the orb itself.* The universe rests on the infrastructure of an unknown 'sub-space' through which one can travel at arbitrary speed. Now, in the words of a theoretical physicist from Texas who studies blue orbs, "underlying the universe is a sub-quantum information domain…"

Hummingbird. Tuesday 24 September 2013.

Garry and I met at the trendy Café Flore in the Castro today. As the tall waitress in a high hairdo and very short red dress circulated around the tables, we spent much of the time discussing plans for isotope tests of my first two samples, and the process for his own analysis of the immune response of close encounter witnesses. I've started work on a new prototype for Capella-2, since Washington and Vegas have dropped the ball.

With the powerful machines we now have, I can do the work of a whole team in hours, not days or months.

Hummingbird. Sunday 29 September 2013.

Last night we attended a concert of San Francisco Renaissance, with the music of the Final Judgment and "Rejoicing in the Torah." Good music was a welcome relief amidst Flamine's slow recovery from her traumatic experience, and uncertainty about my own work.

As smart as the LoneStars are, they still pursue a first-order model of the UFO phenomenon, assuming (perhaps to please their sponsor) that we are dealing with cosmic aliens. Even Hal, with whom I have been arguing about this for years, clings to a straightforward engineering interpretation of the sightings (*cum* relativity) without taking into account the symbolic angles. It's as if someone observed Catholic parishioners eating small round hosts and drinking wine in church and interpreting the place as a restaurant.

Two breakthroughs in space today: Orbital Sciences launched a resupply ship to the ISS, and the nine-engine rocket built by Space-X reached orbit. Alain Dupas tells me the French, who once snickered at Space-X, are beginning to respect private US space companies.

Hummingbird. Thursday 3 October 2013.

Alain has set up a financial meeting for Maurice Gunderson and me at EBRD (European Bank for Reconstruction and Development) on October 14th in London, so our project may get restarted—without Russia.

We're sad today: Flamine survives on little food (tea, salad, and cottage cheese) and fights the stress of panic attacks, which puts me on the edge of a depression I haven't felt for years. Her courageous attitude is intact, so I must face my inadequacy and deal with it.

Hummingbird. Sunday 6 October 2013.

Going back over the 1961 SNIE, I do find it important, *if authentic*. The title is "Critical aspects of unidentified flying objects and the nuclear threat to the defense of the United States and Allies" and it is dated 5 November 1961. Most of the text is clear in language and syntax, but with some odd wording: "in pursuant," "at parody." The point is that the Soviet Union could misunderstand UFO flights as an attack and trigger massive retaliation—an argument I'd made in discussions in France and later with Hynek.

The authors go on to reveal that (1) there are six cases in which nuclear materials were involved in the retrieval of unidentified space vehicles, and (2) "past incidents in the US were detected through air sampling and ground monitoring stations." These statements are said to be based on intelligence "provided by MJTWELVE."

The references to nuclear material are at variance with what we know about actual UFO events: no radioactivity was detected at Roswell, and those few recovered materials we know aren't radioactive. So, is it a fake text, just a trick? It looks like there was once (i) an authentic report about the danger of simulated missile strikes (which the USSR could indeed assume to be real), and (ii) a report mentioning that six cases of "unidentified space vehicles" were found with nuclear material. Was this simply designed to catch spies?

Hummingbird. Saturday 12 October 2013.

As I pack my suitcase for another trip overseas (to London with Alain Dupas and Maurice, then Paris for the Genopole medical meeting), I carry some serious concerns: Flamine is still fighting to recover, and my partners, too, have suffered setbacks. Maurice is stepping down for health reasons while Graham tells me he's diagnosed with a nasty tumor, fortunately caught early.

San Francisco is moody and gray. We have lost the Lumière Theater and Fields Bookstore, two points of culture in a fog of neglect, financial turmoil and often violent street life.

Dr. Diana Pasulka, who teaches religious history at the University of North Carolina in Wilmington, agrees when I deplore that ufologists have moved away from science to adhere to embrace shaky new belief systems, but she thinks that may be part of the phenomenon. This resonates with Jeff Kripal's analysis: It doesn't really matter if hypnotic regression of abductees is absurd, or if the fascination with alien cadavers at Roswell is obsessive, that is the way a new system of belief evolves.

London. Monday 14 October 2013.

It was a cool day in the City as Alain Dupas and I traveled on Eurostar for a meeting at EBRD, where Maurice Gunderson joined us to present Runway Partners. We had a good contact with their international investment team and their Russian specialists, who made a point of reminding us of the "complexities" of Moscow.

Mabillon. Tuesday 15 October 2013.

A day for little things, welcome chores. Many people are quietly going out of business in France these days, or applying for unemployment subsidies, while economists draw up new plans. My impression, unchallenged by friends here, is that France is simply becoming slower and heavier, not like a battleship hit by torpedoes, but like those beached fishing boats, intact in shape but filled up by water and mud.

After the shock of the seventh round of layoffs at Alcatel-Lucent (1,000 technical jobs lost in France, 14,000 in other countries) we learn that 12,800 companies have gone bankrupt this summer.

Mabillon. Saturday 19 October 2013.

Flamine arrived at noon, happy in Paris again but somewhat weary, tired of too-intense studies. She'll see three patients tomorrow but feels, as I do, that our priorities are wrong: too much hard work. Fortunately, we leave for Nice on Tuesday. I look forward to meeting investigator Jean-Claude Dufour, one of the last people to speak to physicist René Hardy, who was to meet with him in Draguignan to disclose the UFO secrets he'd discovered. He died the day before the meeting, on the morning of Monday 12 June 1972; his notes vanished.

As I continue to read *Soviet Space Culture: Cosmic Enthusiasm in Socialist Countries*, edited by Eva Maurer et al., I find stories that run from the ridiculous ("Thank you, Comrade Stalin, for our happy childhood!") to the unrealistic, like the project that installed 11,000 planetariums inside the conveniently shaped spherical white domes of Orthodox churches as a futile effort to eradicate religiosity in the wake of Sputnik!

One of the contributors, Michael Hagemeister, notes that Tsiokowsky, grandfather of Russian space developments, described the universe as a "living organism" whose rationality and "absolute will" also defined the actions of mankind and its quest for reason-driven perfection. He believed in immortal beings much more developed than humans and almost incorporeal, "etherous" and therefore hardly visible to us. These alien beings, like angels or ghosts—constructively intervene in the lives of humans, read their thoughts, and send them messages through "heavenly signs." He had seen such signs himself. Indeed, Viktor Schklovski reports about a conversation with Tsiokowsky in the 1930s: he frequently talked to angels.

Hagemeister further writes: "The influence of Gnostic, Theosophical and spiritualist teachings on the philosophical work of Tsiokowsky has been hardly researched up to now because it was to a large extent taboo in the Soviet Union. The provincial town of Kaluga, in which Tsiokowsky lived, was... the most important center (after St. Petersburg) of the Russian Theosophical movement at the beginning of the 20th century... [Thus] Tsiokowsky's notion that part of humanity would become highly developed and ultimately turn into luminous rays is a central motive in the Gnostic myth as it was popularized in Russia through the 'secret doctrines' of the Theosophists and later the Anthroposophists." Bavarian scholar Carl du Prel, (author of *Philosophy of Mysticism*) believed, like Tsiokowsky, "that the initiative to start cosmic history was triggered by inhabitants of another star."

Another author, Slava Gerovitch, writes: "By shifting the focus from debunking the myths to examining their origins and their constructive role in culture, we can understand memory as a dynamic cultural force, not a static snapshot of the past."

The reference to 'Spaces of Memory' is linked to the notion of *espaces-autres* in the sense of Foucault. An *espace-autre* is a *heterotopia*, a place where normality is suspended, neutralized and reversed; also, a place where time stands still. I note all this because the same happens in UFO close encounters.

Mabillon. Sunday 20 October 2013.

We woke up to the soft sound of rain on the roofs of Paris, but a pleasant sun broke through clouds throughout the day. I had a delightful lunch with my son and grandson: boys' talk.

Colm has issued an interesting memo building on his notion of 'Rosetta Stones,' which are similar to my own 'Golden Keys,' instances when the phenomenon reveals its inner mechanism. He gives as an example the moment when Mr. and Mrs. Sherman looked up, saw a silver disc, and felt overwhelming joy for no apparent reason. "A central working hypothesis," writes Colm, "is that an *ancient parasitic intelligence* has been on Earth for thousands of years...A related hypothesis is that there is a small minority of humans endowed with above average *discernment*. This cadre of humans belongs to a particular gene pool considered dangerous by the phenomenon because they alone can create countermeasures to the control system...The phenomenon is likely machine intelligence and frequently can, and does, make errors."

This memo raises serious questions. For me, the phenomenon isn't really a *system* at all, but *a meta-system*, able to generate many sets of situations under diverse logics. That's what my scientific friends at BAASS have been missing.

Nice. Wednesday 23 October 2013.

We've met Jean-Claude Dufour at last. He is about my age, living in the quiet eastern section of Nice, at the foot of the hills. He received us warmly, in a room that held a bed in a corner, next to bookshelves and a shortwave radio set. He told us he had no prior interest in UFOs until the day (2 February 1966) when he saw a very bright, "medusa"-shaped object over the beach and took three photographs of it. *The pictures didn't show the object* but only a downward light cone where he'd seen a "rain of light beams," others watching the scene. He took the pictures to Guy Tarade, who introduced him to Aimé Michel, who advised consulting Pierre Guérin. "That's how it all started," he said.

Dufour's father was a French policeman in North Africa, so he spent his childhood there. He worked in commerce (import-export) and later in domestic Intelligence.

We discussed the events at Col de Vence. He knows a man who observed the orbs and a family who saw what looked like the cockpit of a large airliner, minus the fuselage and wings. He also gathered the testimony of a Mr. Cartelier who owns a house in Audon, where he's seen low altitude lights and stones falling inexplicably. As in most poltergeist cases, the stones occasionally 'fall' along horizontal trajectories and don't hurt the people they hit.

Dufour then brought up a topic about a woman named Wolf and her daughter, in 1969, before Easter: "We used to have local groups, 'Adepts' and 'Cerec,' which met in Nice. There was a small organizing committee that met at the home of an optician. One day, a 50-year-old woman named Wolf came to the meeting with her daughter who was very strange. She had slanted black eyes, dressed like a squaw, and never spoke..."

Later the same day.

One of the topics I was eager to discuss with Mr. Dufour was the strange death of Dr. René Hardy. "I saw him last at the Saint Raphaël train station," he told us. "We used to meet in the square. He was ill and had regular dialysis sessions. There had been some curious incidents during a meeting with some ufologists in Draguignan, organized by Jean Chasseigne, in June 1972. René took me aside and said he'd "found the defect in the armor" of the phenomenon, and he'd tell me about it the next time we met. He was dead a few days later."

"What was discussed at the Draguignan meetings?" I asked.

"We usually had talks by Guy Tarade, Hardy, and others, before about 200 people. I sat with them on stage. That time, we noticed a man in the front row, about five feet tall, impeccable in black clothes that looked brand new. He stared at me with a smile and approached me at intermission. 'Some people search too much,' he told me."

"Did he tell you anything more?" I asked, now very interested.

"Not at the time. But he was close to me when we filed out of the room, down a narrow stairway. He said the Socorro symbol that was shown on screen was the wrong design. He also said 'some UFOs *communicate with humans through light flashes directly into the brain,*' an advanced notion for 1972! I asked for his card but he said: 'We'll contact you when the time is right.'"

"Did you investigate Hardy's supposed suicide?"

"I tried. In Corsica I had a colleague from SDECE (French Intelligence) who warned me not to get involved. There were several investigations but no conclusion other than suicide."

The house on the hill is haunted: a locked window in the attic is regularly found open, there's never any dust accumulation in the rooms, and one hears heavy footsteps along with a sound "like that of a diver breathing through a tube." Lights have been observed at the windows when no one was in the house. There are reports dating back to World War I, of a French partisan disappearing there. The owner of the place was a *collaborateur*, a German sympathizer.

I wish I could spend more time here with Dufour and the local witnesses but I must fly back to California in a few days.

Hummingbird. Tuesday 29 October 2013.

At 8 o'clock this morning Garry Nolan picked me up in his gray BMW and we drove up to Petaluma, where I'd arranged to introduce him to Dean Radin and Marilyn Schlitz. Along the way, he gave me the sad news of the death of Dr. Len Herzenberg, the Stanford genetics pioneer. He'd recently suffered a series of strokes.

In Petaluma, we sat comfortably in a small room next to the big Faraday cage and spoke for two hours with Marilyn and Dean before driving back to the Bay area, which gave us time to review some projects and the skin mark case in Camarillo.

I am reading Winston Graham's wonderful historical novel, *The Grove of Eagles*. Flamine joins me tomorrow. I worry, wondering how I will find her. In our last call her voice was under great stress.

Hummingbird. Friday 15 November 2013.

The drive to Silicon Valley, too, was glorious this morning, amidst the changing colors of the trees, as Flamine and I met Garry Nolan for lunch at Quadrus, a restaurant atop the hill. We introduced him to Peter Sturrock and sat back to listen to a sparkling conversation that ranged from neutrinos and nuclear decay theory to the challenges of evolution. Garry quoted research that suggested that life originated 12-billion years ago (which means it couldn't have started on Earth).

Hummingbird. Thursday 21 November 2013.

Everything is speeding up: I consider a few startups, and I try to study the replications of Poher's experiments at NASA Marshall. As for the Camarillo mark, it only led to argument when I came to the skeptics' conclusion that "the hair dryer did it!" Many more cases are being privately reported, but I need to get ready to fly to Russia for the meeting in Korolev in a couple of weeks.

Hummingbird. Wednesday 27 November 2013.

Yesterday I signed the promissory note for the gravity company, and I gave them my check for the new lab under construction near San Jose. Mainstream physicists still regard the theory as an aberration, which encourages us. They haven't seen what I saw.

In Silicon Valley, the dark days of the financial crisis are over. Money from recent IPOs (notably Twitter) and the high stock price of Facebook, LinkedIn, and other companies fuels a new technical generation. Things are not so simple, however. As was the case during previous bubbles, the new wealth will drive many long-term residents away from the city, including artists and families with modest income.

The Dow Jones average has topped 16000, the S&P500 exceeds 1800 and NASDAQ is above 4000 (thus climbing back to its level of 13 years ago) but this does nothing to alleviate the misery of many Americans crushed by economic adversity and political corruption.

Hummingbird. Saturday 30 November 2013.

Dr. Green called me today and we had a constructive conversation about my letter of late June. "Your letter puzzled us," Kit told me frankly. "I drafted a five-page response and sent it to Hal for his thoughts, but he never did send you the final version. I wanted the three of us to get together, but it never happened."

"The way I see my duty when I'm part of any project is to stay keenly aware of what it does," I replied, "even if some of it is over my head; I should be able to recognize things that land on my desk, even if I'm not an expert. That's how venture capital works."

"Well," Kit replied, "the lead people thought those papers we all wrote for BAASS were a waste of time. There's only one organization that cares, but in the end, Hal had no need-to-know."

"I can see why he's worried," I said with a sigh.

"Not just worried," Kit remarked. "In my case, I don't want any more government affiliation, too many soap operas. Unless they tell me I can have access somewhere, or read actual autopsy reports, I'm no longer interested in chasing this stuff."

A wise man, yet I disagree. I will go on "chasing this stuff."

Hummingbird. Tuesday 3 December 2013.

My son turns 50 today. Our conversations are warm, reflective, vulnerable, and often profound. He leaves his finance career behind but will always be a keen analyst. His life takes a new direction, with two sons and a smart wife.

Something fun: I just read about a stunning exchange between mathematician Ramanujan, on his deathbed, and Professor Hardy, who came to pay a last visit.

- "How did you get here?" asked Ramanujan.
- "I took a taxi," answered Hardy.
- "What was the number?"

Hardy looked at the receipt and said: "1729, not a very interesting number, I'm afraid."

- "You're wrong," said the dying man: "It's the smallest integer that is the sum of two cubes in two different ways."

Hummingbird. Thursday 5 December 2013.

Up at 5:30 this morning for the long drive to Fremont, where I met Garry at the Balazs laboratories, a division of Air Liquide. We entrusted them with samples of the Council Bluffs and Bogota material, which should provide a good test of their ability to discover trace elements (I already know the main composition) and to do isotope analysis. As for me, in a couple of days, a hard trip into the cold and high-stakes negotiations in Russia.

I went to Palo Alto; the leaves from our golden tree covered the pathways and the yard as if gilded by a mysterious artist. The Sun cast a precious glow over the area. I gathered a handful of the golden leaves I'll take to Normandy.

Hummingbird. Friday 6 December 2013.

Dinner with Maurice Gunderson and Alain Dupas. San Francisco has been invaded again, as in the heady days of 1999, by a population of wealthy smart young workers with BMWs, iMacs, and iPads, marching briskly down the avenues with white EarPods, listening to hard rock, ignoring the population. Older San Franciscans watch this with concern as they cling to the few apartments whose rent has not exploded. Everybody at the trendy restaurant seems to be 28, very smart, and equipped with a gold credit card from some startup.

My colleague Maurice is walking with uneasy step and health concerns, but it's too early to give up, as my conversation with Dupas convinced me. In one short week Alain had visited SpaceX, the Tesla factory, and Planet Labs, so he was feeling a little dizzy: In the meantime, the most notable recent decision by the French Parliament was to decrease the added-value tax on condoms...

Korolëv, Russia. Wednesday 11 December 2013.

My colleague Christofer Dittmar and I were the only non-Russian speakers invited to address this conference on aerospace. I was tired, upset, and confused. The conference was poorly prepared, a recitation of old ideas, with no window on innovation. Yet such trips are useful: they shake the mind, forcing new perspectives. There's snow on the streets but no wind. The temperature is ten degrees below zero, which helps us stay focused.

I'll be back in France in two days. We've given up on Russia.

Fig. 11: *The Korolyëv (Russia) Aerospace Venture Forum, Dec. 2013. Cristofer Dittmar and J. Vallée with Russian colleagues.*

Fig. 12: *Flamine at the Rosicrucian (Thomas Lake Harris) 'Round Barn' Santa Rosa, February 2014.*

Bellecour. Friday 13 December 2013 (Saint Hubert Day)

Splendid evening with my son and Maxim near Le Mans. To my surprise, but with my support, Olivier has decided he was fed up with Paris and the machinations of banking. So, he launches a new career after buying this historic hotel with a reputed restaurant. He has a real passion invested here because the area is rich in ancient tales and marvelous food, lush prairies, ancient traditions, and old castles.

Mabillon. Monday 16 December 2013.

Fine winter day, crisp and bright. Our windows are open to the Sun as I work on the Archives, recalling the time when Hynek started asking too many questions in Dayton.

An ESA scientist named Philippe Ailleris is compiling a series of analyses about UFO history. He came from Amsterdam for lunch with me. He told me that the Hessdalen studies of CNES-GEIPAN were not going anywhere, so he asked if I'd join their advisory board. I had to decline again: Optical stations will remain impractical until fast, automated pattern recognition is refined.

Boëdec has sent a progress report on the Brittany UFO of 7 January 1974. The breakthrough came when he was approached at a book signing by a lady who said, "someone has wanted to talk to you for the last 40 years..." and gave him the phone number of a Gendarme who had seen a large orange object *come out of the sea* and hover.

Back in my quiet apartment, I listen to the sublime *Fourth string quatuor* by Shostakovich, a symbol of the best music and most tragic moments of the twentieth century. He composed it in the depths of Stalin's horror, in the winter of 1948.

In the train returning from Caen. Thursday 19 December 2013.

The train and its pleasures, quiet time to read. I wanted to leave business aside and visit Annick, alone in Normandy. We shared a simple lunch, as always, in her sunny kitchen. Afterwards, a drive to the Atlantic shore and a walk along the beach. Now I'm back in cold Paris, reflecting on my son's valiant new business; about the life that comes, and the life that went away.

Mabillon. Sunday 22 December 2013.

This evening, we gathered half a dozen of our friends interested in parapsychology and esoteric matters: Claudine Brelet, who gave me her book of interviews with Pierre Mac Orlan; high Mason Gérard Forgeron, who brought a book about the Hermetic Brotherhood of Luxor; Mario Varvoglis, who runs the Institut Métapsychique; Jean-Pierre Rospars, my best researcher friend from the CNES group; and a woman friend of Flamine's. It was a bright, warm reunion of kindred minds. Claudine expanded on her theory that Egyptian civilization, that cradle of human intelligence and mystical inspiration, borrowed its beliefs from even older African cultures from the Sahara.

She thinks there once was a magnificent prehistoric society when the Sahara was still fertile and that it only survives today through the Tuaregs (whom the French never took seriously, although they used them in local wars, as President Hollande still does in Mali). When the desert supplanted the fertile lands, they migrated to the edges.

Based on her observations as a UN researcher in Mali and Nigeria, Claudine now wonders if Touareg culture may have been responsible for the megaliths in Europe and Ireland. Such monuments litter the landscape of the Tassili, poorly studied because of the extreme heat, which only allows a few hours of useful work or travel per day.

Earlier, I'd read in *New Scientist* (5 Oct. 2013 issue, p.33) that 12,000-year-old ruins of an amphitheater had been found at Wadi Faynan, hinting at the erection of cultural monuments even before agriculture. A new theory suggests that ritual came first and that agricultural societies arose out of a nomadic past, as a byproduct of people gathering for ritual feasts at communal centers. I must ask Claudine if there's any dating of the structures in the Tassili.

Mabillon. Wednesday 25 December 2013.

An investigative website, *Reality Uncovered*, now claims that Hal and Kit were involved in the SERPO saga to a much greater degree than anyone had known. Even if they only joined the hoax in order to expose it, as they implied, such open involvement is problematic. In the Bennewitz case, a good man was driven to extreme stresses and died insane.

Last night the city was whipped hard by the edge of a big storm that rages over Brittany and the Loire. The wind picked up debris in the courtyard and pulverized it against the walls. We played our favorite music: Marin Marais (*Tombeau pour Monsieur de Sainte-Colombe*), Couperin (*Leçons de Ténèbres*), a few anonymous pieces. We spoke of resuming investigations in Provence.

Then, a wonderful surprise at midnight: the bells of Saint Sulpice woke up everybody and ignited a volley of notes that shook the night, flew across space, and resonated throughout Paris. How nice of the Church, I thought, to help celebrate the Solstice with such panache!

Mabillon. Friday 27 December 2013.

The storm that swept through western France has abated. Annick's house is safe. Olivier takes care of Maxim, sick with the flu. The rest of the family celebrates the holidays in Chantilly where Claire's parents live. Flamine and I spent a quiet evening at a sacred concert on Ile Saint Louis, *Solomon's Dream*. The show involves Catherine Braslavsky (whom we'd met at IMI) and her companion, American musician Joseph Rowe, for an evening of ancient Greek, Jewish, Arabic sacred songs. Afterwards, we shared a cup of chocolate with them at the café where I used to meet my grandson after school.

Mabillon. Saturday 28 December 2013.

It's time to fly away from the pretenses that plague France. When one reads an article in left-leaning, 'Bohemian-bourgeois' daily *Libération* about *the places where one must be seen in Silicon Valley*; when the word "innovation" in on the lips of every corrupt politician…Yes, it's time to fly away. I need to catch up with Materna Medical, meet Maurice Gunderson at a restaurant where we don't *have to be seen* by French journalists, and visit the lab that analyzes our samples.

Hummingbird. Tuesday 31 December 2013.

Flamine will stay in Paris for three weeks to make sure her 95-year-old father can return home safely after a disastrous stay in a rest home filled with mindless automata. He's hard at work on a novel.

At the Balboa Café, I showed Maurice my pictures from Korolëv and told him there was no progress. We'll forget about any new trip: there's nothing we can do there. I walked all the way back, feeling light, with strength to cross the boundary into another year of exploration and the mental stamina to face the consequences.

10

Hummingbird. Wednesday 1 January 2014.

A welcome pause, time for recovery. Contemplating the coming year, I trust there will be time to assess what we've learned and to reconnect with close friends. Greeting cards and good wishes help improve the mood; they gush out of our mailbox, reminding us that the world is full of valuable souls, people of good will, and perennial hope.

Hummingbird. Friday 3 January 2014.

Over lunch with Peter Sturrock today we pursued our discussion of isotopic ratios in materials recovered from UFO sites. Most have not been analyzed, as opposed to the Brazilian magnesium. I asked Peter why he thought the Ubatuba sample tests should be reconsidered. His answer was that he hadn't been present during the analysis, which therefore opened the way to manipulation. He's concerned about this because of the long history of samples being lost. "Somebody is willing to go to extreme lengths to get those samples," Peter said. "They must be very interesting."

He went on to say that he'd become convinced there was a high-level secret study, similar to the Manhattan Project. This is based on the strange behavior of several top scientists with whom he tried to introduce the subject—men like Gell-Mann and Wheeler. One time, at a physics conference, he was having a cafeteria lunch with one of the other speakers. They discussed their work, and eventually Peter mentioned his interest in UFOs. Hearing this, his colleague simply picked up his tray and walked away! Some of the cleared physicists in the country may also be privy about decoy projects that mimic UFOs (as Donald Menzel may have been) and won't touch the subject in public.

Hummingbird. Sunday 5 January 2014.

My mind is settling into a quieter pattern after several days of mad sorting through papers, figures, memories, emotions. I spent a quiet day, content to reconnect on the phone with my son and grandson, and with Flamine for a long conversation. The weather is clear, cold, and too dry for the season, but there's a jolly fire in the bedroom fireplace and more piles of documents to sort in the study. I attack them one by one. (But I must confess I also occasionally keep an eye on the 49ers game—San Francisco beat Green Bay in the playoffs, Hurrah!)

Some research issues have been suspended. On our next trip to Nice I must research the crash of an Air-Inter plane near Noirétable on 27 October 1972. One of the victims was Marie-Rose Baleron de Brauwer, a Commissaire with DST in Nice, and a high-level member of the AMORC group.

It turns out an old friend of mine, Rosicrucian scholar Serge Hutin, was her close companion. She was carrying a sensitive report on extreme-right groups like SAC (the Gaullist 'action' goons) as well as on AMORC itself and the ORT, the *Ordre Rénové du Temple*. The report was in a briefcase attached to her wrist; her body was found with her hand cut off and the briefcase gone. (The journalist mistakenly links the case with Indian atomic scientist Professor Bhabha, who was working on nuclear cooperation with France. See: *Le Mercure de Gaillon* no.3.)

Serge Hutin, in a tearful tribute, has told me that Marie-Rose Baleron was on the way to Clermont-Ferrand to investigate neo-Nazi organizations in France and Italy, including the P2 lodge and the ORT created by some occult adepts in 1968 with Julien Origas, an extreme-rightist and former member of the French Gestapo who had served a four-year prison sentence at the end of the war (1945).

There was a high-level AMORC meeting in Clermont-Ferrand on 28 and 29 October 1972; she may have been planning to expose the ORT there. I must also ask about the statements I heard from Patrick Corsi in June 1995 about President Mitterrand's request for a discrete translation of US documents regarding UFOs. His contact was a man who worked in Nice, near Place Garibaldi.

Another track to follow: British military intelligence expert Reginald Jones mentions in his memoirs that Sweden reported picking up debris from 'ghost rockets' at the end of World War Two. Upon analysis, they turned out to be made of slag (*mâchefer* in French) as in the case of my Council Bluffs sample, or the Puget Sound case. All this points to tantalizing developments that Peter has initiated, entirely with private sponsorship.

Hummingbird. Tuesday 7 January 2014.

Lunch in Palo Alto today (at Osteria) with Bob Johansen. He now advises high-level corporate groups around the country and beyond. We reminisced about the early days of social networks when we pioneered the concept of computer conferencing in the hard engineering culture of DARPA.

"Do you remember the presentation of our first software, Forum, before the luminaries of the Arpanet assembled at Xerox Parc?" I asked him.

Bob laughed: "I was so scared...They were reclining on bean bags around the big room, and they looked at us skeptically, wondering why anybody would want to use a computer network to let ordinary people communicate. They were sarcastic, even angry; we were going to "steal valuable machine cycles" they could have used!"

"All those guys have received medals now," I said, "and human communication among people is the primary use of the networks...whether they like it or not!" But it's all forgotten history.

Sadly, Bob told me that our common friend Arthur Hastings, was in the hospital and not expected to survive. When we lose Arthur, one of the most reasonable, insightful minds of this generation of parapsychologists will have disappeared.

Hummingbird. Friday 10 January 2014.

Wonderful visit yesterday evening with Jack Katz, who brought a pad to our dinner at Salute at the edge of the Bay. Over the next hour, while we spoke of his memories, the people he'd met—a violinist in New York who'd worked with Prokofiev, the psychiatrists around Melanie Klein and the brain research at Bellevue—he produced the delightful sketch I had requested: "An Assembly of confused scientists!"

Unfortunately, communication among that elite group is nearly broken now, the physicists whispering on one side and the biologists scheming on another. I have no opportunity to share what I hear from friends in DC. Will that improve in time? I no longer care enough.

The City woke up swaddled in fresh white fog, and a fine rain made the streets glisten when I walked out for lunch. A quiet weekend, tidying up old papers, writing checks, playing some music I just received, the Quartets of Shostakovich.

Hummingbird. Thursday 16 January 2014.

A day for reading, reflection, catching up. There's an irreparable loss, of course, and a poignant time of year, yet San Francisco is sunny, insolently spring-like. I try to unravel the continuing mysteries people report. We're inundated with new data, rich in information.

I keep to my own knitting now, trying to find new paths. So, I opened the sunroof of the truck and played Shostakovich, relishing this freedom, my mind clear. Fear means nothing. Death means nothing. There's only love, mystics tell us. Yet others will be passing on, dear friends like Arthur Hastings. My foolish mood faded.

Reading Manly Hall's *Secret Teachings of All Ages,* I see: "Is the gratification of curiosity a motive sufficient to warrant the devotion of an entire lifetime to a dangerous and unprofitable pursuit?"

A very young Allen Hynek had written in the margin, in pencil, "in science, yes."

Hummingbird. Saturday 25 January 2014.

Loyd Auerbach, a leading researcher on haunting and poltergeist, spoke last night before some forty Bay Area members of the Palo Alto parapsychology group. I had wanted to hear him for years and was delighted to find Arthur Hastings in the audience, weak and emaciated (the doctors give him only a few months to live) but alert and friendly to Ed May and me as we sat with him.

At my suggestion, Garry Nolan came over this morning, and I reviewed for him all the presentations I'd made at the first five LoneStar meetings, including my field work at Col de Vence and Sonoma County, the various proposals for Capella, and a comprehensive review of the abduction field. We covered all that in two sessions, with a happy break for lunch with Flamine, and I gave him a memory chip with all the data.

Garry frankly confided to me that at our first meeting he'd been very annoyed by my position when I suggested that close encounters might be only a *display* presented for our benefit by a very superior form of consciousness, rather than genuine intervention by aliens. He's moving closer to my view, so it will be interesting to see how this will play in Austin.

He's also strongly of the opinion that someone like Paul Allen is more likely to be interested in funding an effort like Capella than a series of medical studies or theoretical physics hypotheses, yet the first draft of the group proposal doesn't include any serious, multi-year effort to detect hidden patterns in the historical data record.

Hummingbird. Monday 27 January 2014.

This afternoon, Garry and I visited the Air Liquide calibration labs at a nano-analysis company in Fremont, meeting with their director of advanced materials and films. We discussed the study of the two test samples I'd given Garry, which have a high probability of coming from real cases. I told our host that our main interest was the isotopic ratio in the most-abundant elements, to which Garry added magnesium as control. The technique will be single-cluster ICPMS with sequential measurement but without sequential time-of-flight.

It was interesting to visit the labs and to peer into their clean rooms. Created by a gifted woman scientist in 1975 to serve Intel and Fairchild Semiconductor, the company remains the standard in the Valley for its extraordinary precision in thin film and materials and to ascertain the purity of the water used in semiconductors.

Hummingbird. Sunday 9 February 2014.

It was raining hard in Santa Rosa today when Flamine and I drove up to meet Oberon Ravenheart and explored the story of the Fountain Grove community, the Rosicrucian group formed by Thomas Lake Harris at the end of the nineteenth century.

Born in England to Calvinist Baptist parents who settled in Utica, New York, Harris became a Swedenborgian, joined in the spiritualist movement of Andrew Jackson Davis, and eventually established a congregation in New York. In 1850 he claimed to receive new inspirations and wrote long poems. When he preached in London in 1859, he was described as "a man with low, black eyebrows, black beard and sallow countenance," according to an authoritative biography cited in *The Encyclopaedia Britannica*.

After creating a winemaking business on the shores of Lake Erie, he took some of his flock to Santa Rosa, creating the Fountain Grove community in about 1875. To his secret circle he taught that God was bisexual, that the rule of society should be "married celibacy," and that he'd discovered the secret of resuscitation.

The sect has left a visible impact on the landscape with its magnificent round barn on top of a hill and the winery of Harris' protégé, Kanaye Nagasawa, who succeeded him at his death in 1906 and led the brotherhood until 1934. Oberon and I sneaked into the barn through a hole between some old boards and marveled at its three-level wooden structure, with its enigmatic wooden dome.

This was an emotional trip because Morning Glory, recuperating from heavy dialysis treatment, had just come home from the hospital. She bravely sat up in bed and signed their latest book, a history of the Church of All Worlds entitled *The Wizard and the Witch*.

Conversation with Oberon is always a colorful and dynamic affair. Whether in the old armchairs of one of the many book-filled rooms at the home in Cotati, or over lunch in the noisy atmosphere of the Redwood Café, we discussed the Harris Community and its possible remnants. Its poetry qualifies as science-fiction. *Wikipedia* states that "He depicted interplanetary empires, major cities entirely covering entire planets, and the 'ancient astronaut' myth, in which space travelers help early humans with agriculture, technology and spirit development."

Somehow, after talking about this year's drought (which endangered frogs and newts, to the distress of Pagans), we came to discuss consciousness and survival, a subject we'd just debated last night at Sturrock's residence in Palo Alto with Adam Curry and especially Federico Faggin. When asked where we go after we die, he stated that death *doesn't mean going anywhere*. "We're already there," he told us, "but we don't realize it because of this body, the instrument through which we experience the world. When we leave it behind, I think we keep our identity, but we're just one more source for the higher spirits. They experience the world through our individual lives."

These ideas dovetail with those of Philippe Guillemant on causality, and mine on information physics; we seem to agree that time, in the common sense, doesn't exist.

Federico thinks that we have free will but only when we apply our consciousness to a hard decision. Oberon said the same thing this afternoon.

The most important goddess in his view is Hecate, who stands at the fork in the road holding two torches, leaving us the choice of the path to take.

Hummingbird. Wednesday 12 February 2014.

Thirty years after Mitterrand, President Hollande came to San Francisco today with an entourage that included foreign minister Fabius, Anne Lauvergeon (former CEO of Areva), and the new CNES director Jean-Yves Le Gall. We were invited to the assembly at the Hilton, where we heard Hollande give a relaxed, mildly amusing talk with all the usual clichés. When it came to discussing expatriate entrepreneurs, however, he was careful to avoid references to 'brain drain.' Instead, he reminded us that he'd just inaugurated an incubator.

Flamine pointed out to me that, once again, France was exporting its bureaucracy instead of trying to learn from the unique culture that allows Silicon Valley to thrive.

Hummingbird. Friday 14 February 2014.

I worked hard today, getting documents and slides together in anticipation of the LoneStars meeting in Austin. We have some serious things to share. I continue to correspond with Robert B, the retired counter-intelligence specialist captivated by the subject. Commenting on the manipulations of data and opinion that plague the field, he writes that "in the last ten years, I have seen countless occasions in which things have gotten twisted around, often deliberately, or just by human fallibility. I've become much more skeptical about much of what I've seen and read on any topic of a historical nature; this is true of the UFO issue... *I'm flabbergasted by how much has been put out there by individuals deliberately wanting to mislead...*"

Austin. Monday 17 February 2014. Hampton Inn Arboretum.

We arrived ahead of our colleagues and seized the opportunity for a leisurely dinner with Hal and Adrienne. Commenting on the alien cadavers supposedly retrieved from Roswell, Hal told us that electronic circuits had been found in their brains, according to unpublished sections of Corso's memoirs. Is that verified?

This morning, we visited the campus, where I'd only returned once since the days when I worked for Gérard de Vaucouleurs. The old astronomy building still stands with its weathered dome. No sign of neglect, in contrast, in the new 'Lyndon Johnson' library, the halls are filled with students carrying late-model laptops. Fleeing away from modernity, Flamine and I had lunch in nineteenth century Texas glory at the Driskoll hotel.

Austin. Tuesday 18 February 2014.

The full record of our sessions is held under Chatham House rules, so I will only mention some highlights. George Hathaway introduced the agenda, stressing that Garry would speak about the role of the immune system. George's own work had begun in 1988 when he and his sponsor decided to "sort out the UFO situation" from a technical viewpoint.

John Schuessler noted again the sighting by General James McDivitt, reported to NASA from orbit. He added that Jan Harzan (nuclear engineer, ex-IBM) was now running MUFON, welcome news after years of sub-standard management. I made some comments as part of a discussion of Professor Harrison's paper about terror, and Colm gave a progress report about the Ranch. Hal and Garry spoke in the afternoon. Dinner was at a fancy place with waterfalls, ferns, and, appropriately enough, ludicrous blue orbs in front of the lobby.

Austin. Wednesday 19 February 2014.

This morning, the schedule called for Kit to present medical cases and *triage*. Summarizing our 'interference syndrome' concept for Garry's benefit, he explained it seemed to use the *dermis* (the deeper layer of the skin) as the place that receives the supposed signal from orbs—or UFOs. Garry followed with an update on his own research.

In the afternoon, to remind them of the long (pre-1945) record, I made a short presentation about Jack Katz, *The First Kingdom,* and his extraordinary experiences of Contact in childhood, passing around enlargements of some of the most striking and evocative scenes in the *Space Explorers' Club*. I left my six panels with Eric, himself an enthusiastic science fiction buff.

Later, I showed the group the new computer tools of historical vocabulary investigation from *Google*, called *n-grams*. Chris Aubeck and I now use those to determine how the meaning of certain expressions changed over the last 200 years; the use of words like *meteors, bolides,* and *aerolites* has evolved, and that must also be taken into account to avoid mistakes in searches of online libraries.

Hummingbird. Friday 21 February 2014.

We've returned from Austin exhausted by the intense arguments. The outcome? No hopes for a newly funded project. Our group now leans towards the view that aliens are physical, potentially very hostile, and exert active influence, a conclusion I only accept in suspended attention, heavily biased as it is by narrow American cultural filters.

Tonight, in proper academic decorum, the Stanford Medical School held a memorial in honor of our friend, Professor Len Herzenberg. Given Leonard's long-lasting love of music, and his daughter's gift for modern jazz, this was a glorious evening, a classic Stanford celebration with some 200 academics from the medical community and luminaries in genetics research from as far away as Washington.

Hummingbird. Sunday 23 February 2014.

This morning, Flamine and I took a brisk walk up to a service at Grace Cathedral, and this afternoon, to a concert of the American Bach soloists: a Lutheran Mass and the Orchestral suite in C Major, but the second part was a Viennese 'pastry': *Hercules at the Crossroads*, a mythological piece where the hero must choose between vice and virtue. (How boring!)

Bach had only composed it to put food on the family's table. The year was 1733 and the Elector of Saxony in Dresden had a son who celebrated his 11th birthday…It's amusing music, sung with gusto. For us, it marked a perfect end to an intense, tiring week.

Hummingbird. Friday 28 February 2014.

Yesterday afternoon, on a visit to Peaxy, the 'big data access' startup sponsored by Federico Faggin, Garry came along to see the advanced software in action. Then, over a snack at Denny's, he gave me his impressions of the LoneStars meeting. Garry is not particularly impressed by the occasional theatrics among the group, but he did agree we needed to define the underlying 'syndrome' by formal steps.

Hummingbird. Sunday 2 March 2014.

On our latest visit to our Santa Cruz friends, Oberon handed to me *The Wizard and the Witch* by John Sulak, a well-documented history of the Church of All Worlds. It made me realize, and cherish, my long association with that clever band. It also made me aware of their travails and peregrinations, and the uphill battle against traditional society upset with any creative lifestyle.

Returning to the business of Capella, I wrote to Robert Bigelow, inquiring about the status of our computer files. I suggested an agreement that would enhance the integrity and security of the data while preserving access for any *bona fides* stakeholder in government, yet preventing trickery until completion of the full AI implementation.

Since our project has been dumped in an unfinished state, I argued, the work should be preserved and tested on those catalogues I continue to maintain privately, untainted, ready for expansion and discovery.

Hummingbird. Wednesday 5 March 2014.

"To those who dare to muse." This dedication by Jack Katz to his readers captures the spirit of his art. I picked him up at the Cove today and we drove over to Solano for lunch. He showed me the first pages of *Beyond the Beyond* and gave me romantic *esquisses* for Flamine. Now that he knows me better, Jack wants to reveal the childhood encounters that inspired him. "I didn't see any of their bodies," he told me, "not even a translucent form, but I felt their hands. They put something into my brain, through my eye."

Whatever his source of inspiration, he's kept a lot of energy (climbing stairs without difficulty, walking as fast as I do) and he works non-stop, classical music playing all the time. Chausson is his favorite today, especially *Chanson Perpétuelle*.

Back in San Francisco, I tried to clear up some confusion. My friends get upset when we review statements that humans are being conditioned. For one thing, that first 'breakthrough idea' has been known for a long time, not from some anonymous 'Source S.' The conditioning has been suspected and described since the days of the *Invisible College*, with Fred Beckman and Douglass Price-Williams, and *we verified that the pattern of UFO waves functioned as a schedule of reinforcement.*

"We've known this," I said. "Why can't we talk about this on an even field and restart on a stronger basis?"

We all feel strong pushback from the stovepipes of classified insiders. In various recent meetings, some heads of Agencies were very upset at people from the Pentagon, reportedly. It seems people "in the know" are preserving their turf, twisting the problem to fit worthless intrigues.

Hummingbird. Saturday 8 March 2014.

Bill McGarity is now on a temporary job with EarthTech. His remarkable career is mentioned on the web. I met him with Hal once, and John Alexander introduced him to Eric and Colm in 1997 when Bill retired from LANL, where he was employed by EG&G Special Projects. I found another contact who verified his experiences: "Bill once encountered a huge UFO over Mt. Archuleta, in Dulce; he knew the Edmund Gomez family; he knew Gabe Valdez, Bob Lazar, and his supervisor; he knew John Lear and Pharis Williams; he and his wife socialized with John and Victoria Alexander in the Los Alamos employee circuit; he knew about John's Army UFO workshops, as well as Col. Tom Bearden; he worked at TA-33 and Area-51; he got originals of the USAF *Project Grudge Report* that were to be destroyed and kept them in his office safe. It turns out that the copies BAASS acquired are not the same..."

Beyond the help McGarrity may provide, EarthTech-IAS/A has also recruited a very experienced bio-computational physicist, Dr. Jim Segala, to assist on reverse engineering EM, nuclear radiation injury, "and other biological effects." There are red flags. Conflicts within the Intelligence community are always brewing; hidden forces haven't revealed themselves yet. Even Kit has become a target over the Internet, all unfair attacks.

There's urgent, constructive work I can do on my own in the real world, with eleven health and technology companies I've helped support since my partners and I exited from previous funds.

Today Colm Kelleher called me at the request of Bob Bigelow, reacting to my proposal. I had little hope that Capella would be released, so I wasn't surprised when Colm explained there were "too many issues of privacy and ownership" for the files to be opened, adding: "Bob hopes there may be a future program, as you do."

They are technically and operationally correct in the current context of grave neglect from the DoD, yet in a joint venture we could have negotiated a structure that would have preserved *both* the rights of the sponsors and the privacy of witnesses.

Hummingbird. Wednesday 12 March 2014.

As I should have predicted, the LoneStars are fragmenting. This is not so much an effect of egos as the diversity of interests, but it's sad. George and Hal have their own project in the testing of advanced physics claims, separately driven by their other sponsor. Kit manages medical research into neurological injuries and skin marks. As a global attempt to understand the global phenomenon, however, this only satisfies special interests, with sparse review or coordination of results into research. There are lots of questions we're pushing aside.

The latest effort to develop a proposal to Paul Allen comes from Garry. It's initially restricted to biological cases, with modest development in information structures. Garry calls on me because he agrees the database aspects are critical, but today there's no funding.

In independent reports I've heard, there were new exchanges *reported* with entities: Was that true? One colleague was once introduced into a formal setting where an FBI officer brought him to an office where there was a human, presented to him as an alien. Was he simply being tested? That's my main question: Who's testing whom?

Hummingbird. Friday 14 March 2014.

Over lunch with Garry Nolan yesterday (at the wonderful Rotunda of the Neiman-Marcus store, an ornate structure filled with natural light, high above the palm trees of Union Square), I was happy to go over the presentations in Austin. We both see the fragmentation of the group as a danger sign.

"If I had my choice," I told Garry, "We would start a very simple LLC, get it funded with no fanfare, stay away from all classified stuff or people, and we'd simply hire good scientists we know, to run the investigations that are needed. All under US law, no tricks.

"Just look at what the government did with Capella," I added. "We hire 20 good investigators, translators and coders. They recompile the entire history of the phenomenon. Then at the end of two years DoD throws them back into the street, and the DoD ends up with a bucket of bits with all the real 'intelligence' stripped off."

"They still have the cold data," he said. "They could turn it over to the Aerospace Corp., or Lockheed."

"No, all they got is a biased bucket of bits," I insisted, "with an amputated internal history they can never put together again. The treasure was in the head of the people we trained and scattered away. It's a dead puppy, poison to a new team. We've lost twenty years."

Hummingbird. Saturday 14 March 2014.

The labyrinth at Grace Cathedral represents a spiritual journey to the soul and a model of the eternal quest. When one gets closer to the center, the illusion suddenly gets broken by long excursions, only to return when least expected, with new insights. I love its well-crafted path of disappointment, patience, and reward, back again to the external world. *"Visita Interiora Terrae,"* the Rosicrucians say.

Now, Alain Dupas has just received encouragements from Airbus about our plans for the Runway Fund from Jean Botti, who heads up Innovation Works, so I am reassembling our team to prepare a meeting in Paris next month.

Hummingbird. Friday 21 March 2014.

Spring Equinox: the clear, warm days and cool nights continue, and flowers have appeared everywhere, trees are in bloom, the cherry blossoms in full regalia. On the open web, Kit explores the theology angle, based on his training over six years in three front-line seminaries: Garrett, Sewanee, and Seabury-Western. He says the meme of the Devil "who will get you, dirty sinners!" was just an allegory designed by clerics for simple people. Perhaps, but it's taken very seriously in American society, especially among the military.*

Flamine and I spent the evening with Diana over a simple, tasteful feast. Much of the discussion had to do with skin marks, affecting up to 5% of people, and her insights about them.

---

* The late Professor René Girard (at Stanford) explored this theme in several books I hadn't discovered when I wrote this.

We were smart (or lucky) to give up on a Russian fund. Putin has just invaded Crimea, the Ukraine is in turmoil. Where would we be if our joint proposals had succeeded? (18) The Kremlin's misunderstanding of what makes a successful modern economy is tragic, for everybody.

Hummingbird. Friday 28 March 2014.

Arthur Hastings, old and very trusted friend, called me this afternoon in a frail voice to say *Adieu,* and to express his love and pleasure at the research we shared over the years. I told him I would go on with that quest as long as my own health allowed. He may be gone within weeks as the blood disease that weakens him finally ends his life. We agreed that time and space were illusions and that in the universe of information we don't need to go anywhere after death because, as Federico said the other day, "we're already there."

Hummingbird. Sunday 30 March 2014.

Calling Garry to discuss a new medical technology, I caught him in Arizona, about to join Kit and Colm for the return trip. "We've had an interesting time," he said, "but we haven't found any evidence of dead aliens. The story has been *considerably* embroidered!"

I had to laugh: "Come on, either those aliens are dead, or they aren't! That's not something you 'embroider'…" He saw the point and laughed as well. In a case that involves Navaho shamans on top of an invisible extraterrestrial invasion, one must be prepared for anything.

Now Douglas Trumbull sends me a description of his current project: "We're just finishing a twelve-minute experiment and demonstration of a new media technology, far superior to IMAX. I am eager to get your opinion." He goes on with an invitation to help present the project in Seattle in May. There will be screenings of a new print of *2001* personally commissioned by Paul Allen. Douglas adds: "I am writing up my thoughts, and just came to the issue of how Spielberg drew from your works in developing the screenplay."

His letter goes on: "The power of this new medium is extraordinary, the ideal way to tell a profound non-fiction story, since it's like a live event. I am eager to hear of your progress and hope we can find a way to collaborate. We are working on remote sensing autonomous platforms for UFO activity research in the sky and underwater...spacecraft cameras such as those on Clementine."

Douglas and I discussed all this at length today. Networked sensors are costly, but technology has moved very fast, even including dust-like self-organizing networks with tiny sensors that don't need external power. I described my own experiments of the late 1970s with the Copper Medic camera, which Douglas asked me to send.

We probably wouldn't get any quick data, unless the phenomenon decided to play with us, as it did with Douglas' camera we locked in that cabin at Marley Woods, forcibly twisted around by forces unknown.

Hummingbird. Wednesday 1 April 2014.

Lunch with Garry Nolan at Le Zinc, nestled in Noe Valley among cute antique shops and trendy bars, flowers and ferns, and elegant fashion displays. He was still puzzled by the scene he'd witnessed at the mysterious Ranch: "a smelly, dreary place with a layer of dirt on the furniture, seven dogs inside the house and a big bird, a parrot, yelling incessantly in his cage near the front door." Alas, no alien.

I wonder at the pattern of such stories where every step, pro or con, ends up shoring up a wild tale. That was the case with the Philadelphia Experiment, the Budd Hopkins 'Case of the Century,' and the Gulf Breeze fiasco. Now Jeff Kripal reminds me that's how every major religion builds up its mythology. Did Jesus bring back Lazarus from the dead? Did the fish multiply? *What difference does it make?*

I told Garry about my paper for the CNES meeting. We discussed the next step in the isotopic analysis of UFO samples.

Dmitry Manakov was in California with Svetlana, so they came here for a relaxed dinner, bringing a bottle of red from Paso Robles and new ideas for our presentation to Airbus next week. Many bright Russian entrepreneurs are coming to the West now, permanently.

Mabillon. Saturday 5 April 2014.

We flew across the Atlantic in just ten hours on the new Airbus 380, a magnificent plane. Paris is luminous, newly dressed in the soft green tones of young leaves. Flamine went to visit her father at the hospital and found him in recovery, but too weak to move around.

A new socialist cabinet has just been installed by Hollande. It drags along the same unsolvable contradictions as the previous one. The French stock market is at an all-time high, most of the profits made overseas, and I see none of the cautionary measures I had expected.

Mabillon. Tuesday 8 April 2014.

The analysis of our samples is on the agenda. Garry just sent me an initial schedule of tests for my materials at the Earth Sciences lab at Stanford: Air Liquide didn't help us. In quiet times I savor Barbey d'Aurevilly (*Le Chevalier des Touches*, which reminds me of Valognes) and *Book from the Ground* by Xu Bing.

Mabillon. Thursday 10 April 2014.

The meeting at Airbus in Suresnes was hopeful. Alain Dupas drove us there, wearing a black shirt, a red tie, and a red scarf. Denis Gardin received us in a small conference room, next to the place where Louis Blériot's legendary office is preserved. He said that Airbus was looking for more than a venture fund, a "global network of industrial leaders." This we can build for them, but the decision process inside their company will be contentious.

Today Xavier Delamarre invited me for lunch at Balzar. He works on the staff of the Prime Minister as an international economic advisor, having travelled in multiple diplomatic posts in Japan, Africa, and Slovenia. His passion is parapsychology; he's even published a translation of Ingo Swann's *Penetration*. He painted a bleak picture of research in France. "The letter that Admiral Pinon wrote to President Sarkozy never got to his desk," he said. "It went into the garbage; nobody takes that stuff seriously. The COMETA Report was treated the same way."

At Stanford, Michael Angelo has started the isotope testing. We'll be able to dispense with the dilution technique in those cases when we only have tiny portions of material.

Mabillon. Sunday 13 April 2014.

We went to Pontoise today and spent the day there under a cool, gray sky. This was Palm Sunday. The church of Saint Maclou, now a cathedral, was filled with a devout crowd, people of all races. I remember Palm Sunday from my boyhood as one of the more pleasant and dynamic ceremonies of the Catholic liturgy. Several of the young priests were of African origin, while the parishioners showed a mix of blacks, whites, and Asians. The Public gardens were as peaceful as ever in their springtime dress, my old street silent.

It was a bit of a pilgrimage. Flamine was curious to know more about my early life and to share the emotions attached to it. We found Cézanne's *Moulin* well preserved and the Sente des *Poulies* as enigmatic as ever, a peaceful promenade along the hidden cliff.

On top of the Château that commands the view over the Oise River, I rediscovered my earliest memories when I was 3 years old—brief snapshots of my parents and my brother, and then the war, destruction, dust and smoke, a broken world precipitated into an abyss of hatred and greed from which it would never fully recover. As I write this, and watch TV, Russian tanks are massing on the borders of Ukraine and conflicts erupt out of simmering passions, from Syria to Mali.

Rochecour. Thursday 17 April 2014.

My son now lives in this peaceful town, a medieval stronghold with stories of sieges and Queen Blanche de Castille, at the foot of its old castle. The moat has been turned into a vast pool where swans are sunning themselves. Cars drive by, on the way to Le Mans or Alençon. I'm reading *L'Ancre de Miséricorde* by Pierre Mac Orlan.

Last night I slept more deeply than I had in a very long time. I review our efforts studying an increasingly urgent alien presence, Kit engaged in theories of brain disturbance by 'syndromes,' and George and his sponsor speculating about subtle energy sources tapped by mediums—all of it confidential as a prized collection of Japanese prints, yet occasionally shared *in camera* with powerful Senators and their staff.

Having lived through the demise of NIDS, and later BAASS, the prospect of another classified project at the mercy of intrigues doesn't appeal to me. The bigger the project, the bigger the politics, the bureaucracy, the side deals—and the stealthy deceit.

Rochecour. Sunday 20 April 2014 (Easter).

This is a rich area of farms and silence, rolling fields along the gentle slopes of hills only broken by hedges and gurgling rivulets. When we open our window, we see the wide pond and the road, the white gate and the fat expanse of green, climbing to the horizon—home to milk cows and horses. English families love the area. They buy second homes here, or move permanently, unbothered by the wetness.

Mabillon. Monday 21 April 2014.

Jean-François Boëdec has completed a survey of 167 sightings in the single area of Finistère between 1950 and 1981, trying to determine the chances for a surveying station to observe a UFO event. The total time of observation was only 2 hours per year, which wouldn't justify an expensive effort. As for the French Air Force survey and tracking project, still unpublished, it hasn't gotten off the ground.

Rospars and Boëdec have found a very interesting, independent observation of the object seen near Pontoise in December 1979, namely the ball of light that is supposed to have abducted Franck Fontaine in a case widely regarded as a hoax (but still open, as far as I'm concerned). The control tower at Cormeilles received two reports of lights at the time, near the transformers. Here again, there was no report to CNES because the witnesses were from Gendarmerie de l'Air, a different branch from public law enforcement…and not subject to the same rules. Bureaucracy wins again.

A researcher also tells me that a certain 'Agent Dumontheil' had indeed confirmed that the affair at Pontoise was a psychological operation of the special service, using the UFO theme as a decoy.

Later in the day.

Coughing and sneezing, nursing a dull headache, I went across the street to my French doctor who diagnosed a new virus raging through Paris. Good news: it clears itself in five days, she joked, adding that my lungs were fine and I had "the heart of a turtle," which I took to be more good news. This encouraged me to pursue my trip to Orange and then to Nice, where a new contact, a researcher named Baker, knows the local stories very well.

The latest LoneStar meetings have convinced me that our sense of awe towards government secrets was exaggerated. Classified Intelligence is precious, but that doesn't mean the analyses are valid. The reverse has been true in dozens of historical decisions since World War Two (Indochina, Iraq, India's A-bomb, and the unforeseen collapse of the USSR, among the ones known).

When it comes to subtle assessments of the nature of UFOs, the secret folks get more easily fooled than ordinary researchers because the classified secrets are so narrow: they fool themselves, their judgment altered by secrecy or tainted by misplaced religiosity. Why would we trust them? Where's their research? Ah! Yes, it was secret.

Nice. Friday 25 April 2014.

On the way to Nice, I stopped in Orange for a long talk with erudite Yannis Deliyannis. We discussed historical records, references to old German and French sightings, and then pursued our discussion over books of exquisite rarity, from Boaistuau to Lycosthenes.

Flamine and I drove up to Col de Vence again yesterday afternoon with Alain Bauquet in great form. We hiked up again to the site where the brilliant white orb had been photographed. At the spot, there's a clearing, short trees, green grass.

The next day, we paid a visit to Mr. Baker. On two topics of inquiry, I thought he might help, namely the case of 'American secret documents' allegedly leaked to President Mitterrand, and the assassination of Marie-Rose Baleron I've mentioned before.

The first case has bothered me ever since a former colleague told me he'd been asked, out of the blue, to translate supposedly classified US documents, evidently fake, having to do with captured UFOs, bases on the Moon, and a grounded craft in Arizona. Allegedly, the Elysée had been working through a cutout, a ufologist named Serge Bernard who ran a small enterprise in Nice. Baker knew the man, better known as Jack Carter. His business was exactly as our contact described it, an *atelier* of women's clothes on Cours Saleya, near Place Garibaldi. He'd worked in Africa in the fifties, as a trucker between Dahomey and Niger. "He's written a book about his sightings and moved to Thorenc, in the back country 'to facilitate his contacts with the space brothers,'" Baker said.

"Did he have information—or disinformation—about American

ufology?" I asked.

"Carter was mixed up with several groups. He was close to Janine Lagneau alias 'Jala,' an artist who was part of a vast esoteric network including L'Ordre Vert, an international, self-described Luciferian organization based in Brussels, thought to be manipulated by neo-Nazis. Carter's girlfriend was Stéphanie Marcow, close to the Raélians."

"But what about the documents supposedly supplied by Mitterrand, who wanted a translation?"

"Carter used to claim he knew a special agent of the Elysée who could have supplied the documents, but they didn't necessarily come from the Palace." He mentioned a name. I burst out laughing: a well-known ufologist, well-informed, a bit scattered, who keeps bragging about his 'secret' contacts.

"It could be the very same man. They met several times on the Côte d'Azur. Carter also had contacts with an American woman who lived in Cannes, Jane G...but he didn't trust her. She closely tracked all the ufology groups in the region."

"Where's Jack Carter now?"

"You won't find him," Baker said drily. "He died on the 12th of January last year, in the hospital in Grasse."

As if to be sure I understood that he mastered the facts and the details, unlike most ufologists, he added coldly: "At 3 am."

At least we've closed the loop on a very old story that has bothered me since I first heard it, twenty years ago in Brussels.

The second case is more tragic. It concerns the crash of an Air-Inter airliner, as I've noted before. Here again, Baker had additional information, as I had hoped. "The Vickers wasn't the only plane that crashed there on 27 October 1972..." he said in a flat voice. I tensed up listening to him. "An American military aircraft on a flight from Frankfurt to Athens deviated from its plan and crashed in those same mountains. CIA managers and other Intelligence personnel were on board. That information never reached the French press."

"A lot of people seem to have been mixed up in esoteric and paranormal research in that period," I wondered aloud, thinking about what had been revealed about the SRI project rooted in MK-Ultra days.

"You're right, and that included people who came from the communist block with their own agenda," Baker said. "There's a fellow named Ivo Brodsky who came to the West from Czechoslovakia in 1968, in search of UFO information and contacts. He was an agent of Statny Bezpesnost, a Czech intelligence service. He was part of a network called Tranzistor, receiving instructions through 'number stations' on the radio. He was very well informed about ufological research; evidently his bosses were under the control of the KGB, eager to get new information about the phenomenon—or military things related to it. He lived in Nice during the summer of 1968 and once drove up to Col de Vence on a borrowed motorcycle... From 1968 to 1972, he collected much information on French close encounters and paranormal phenomena research.

"The CIA caught up with him in 1971 and he was forced to turn double agent, setting up home in Dortmund in 1980. He may still be there! My correspondent in Czechoslovakia tells me it would be dangerous to contact him. He also writes that his father once took part in secret paranormal experiments of the Soviet military near Kalaï Mor, in the Bachkirian savanna." More points to connect.

Mabillon. Saturday 26 April 2014.

This morning, I finished reading *The Memoirs of Catherine the Great* of Russia, whose intelligence and wisdom I find astounding: At a moment when she feels rejected by the Court of Elizabeth, to whose nephew she's married, she knows she's likely to be stripped of her rank and sent into exile: "My decision was made, and I regarded my dismissal or lack of it with a very philosophical eye. I would find myself in such a situation as it would have pleased Providence to place me, never without the resource that spirit and talent provided to individuals according to their natural faculties, and felt I had the courage to go up, or down, without my heart or my soul feeling elevation or pride or, in the opposite case, neither debasing nor humiliation." Quite a girl...Her French writing is enviably clear and direct, without pretense, but her humor and the clever description of the characters around her always shine through the most stressful moments.

The narrative, which was secret and only intended for the education of her children, unfortunately stops as she is about to rise to the throne of the Tsars amidst dark schemes worthy of Vladimir Putin.

Mabillon. Sunday 27 April 2014.

Faced with the complexities of the obese French bureaucracy, some of France's largest companies now leave or pass into foreign hands. After Peugeot, largely controlled now by Dongfeng of China, and Lafarge, a huge cement company smartly merged with Holcim so it could relocate in tax-clever Switzerland, Publicis has allied itself with Omnicom in London and Holland. The latest is Alstom, the big engineering company, approached by General Electric this week.

Economist Elie Cohen warns that France is fast becoming irrelevant, losing its old industries and incapable of spurring new ones, other than luxury goods and weapons, both of which are unproductive dead-ends. It's a sad story I've seen grow over sixty years, but there was always some hope of reform. Now, many of the lies have been exposed.

Still nursing my stubborn cold, I'm reading *Pnin* by Nabokov and short stories by H.G. Wells: "The Crystal Egg," "The Door in the Wall," "The Story of the Late Mr. Elvesham," and "A Strange Phenomenon." The latter is a striking description of an extreme case of remote viewing.

In that story, Wells observes that a possible explanation "...would involve a fourth dimension and an adventurous theory about the various kinds of cases. To say that there was *a knot in space* [HGW emphasis] seems perfectly absurd to me, but perhaps that is because I am not a mathematician. When I objected that nothing could change the fact that the two places are separated by over 10,000 kilometers, he answered that two points can be a meter away on a piece of paper, yet they can be brought together simply by folding the sheet...Wade believes it is possible to live visually in one part of the world, while we live corporeally in another."

As for us, time to board an ordinary plane, back to SFO.

Hummingbird. Friday 2 May 2014.

An old idea—the possibility that some paranormal projects, including UFO events and memes, have been engineered and manipulated since the 1950s by the sinister minds of MK-Ultra—has re-emerged in our Internet discussions. I'd noted that one of my readers (Robert B.) thought that mind control activities were key to many paranormal events such as the Sutton family case in 1955 in Hopkinsville, Kentucky, a town visited by John Mulholland, and that Puharich was working at Fort Detrick Maryland in

1952 around the time when he began the 'Spectra' or the 'Nine' communications. Mulholland and Fort Detrick were both related to MK-Ultra. I also mentioned the curious title of Richard Helms' book of memoirs, *Looking Over My Shoulder*.

At the time, there was a '5412 committee' that derived its name from a directive of the National Security Council approved on 15 March 1954, authorizing clandestine operations of the CIA on a wide-ranging basis. NSC 5412 was declassified in '77. It only deepens the mystery...

Hummingbird. Saturday 10 May 2014.

Yesterday morning I met Dr. Angelo at the Nanofabrication Facility at Stanford and had the pleasure to see him operating that remarkable French machine made by Syseca, a $5 million piece of very complex equipment where our samples can be scanned and studied to an extreme degree of precision. Stanford geologists use it to detect traces of extra-solar dust in Earth rocks, and Garry Nolan uses it to study tumor cells where proteins have been tagged with metals.

Michael Angelo said he was reading *Passport to Magonia* at Garry's urging, so we engaged in a lively discussion of the blinders of academic science as he carefully calibrated the machine's beam.

Hummingbird. Sunday 11 May 2014.

To gain a healthier perspective on local history, Flamine and I took the ferry to Angel Island yesterday and spent the day in wonder, walking ten kilometers around the unspoiled beauty of the park. At every turn, the view changes, from the yacht harbor of Tiburon to the expansive horizon of the Sacramento River delta, to the Bay Bridge emerging from Berkeley, the glorious cityscape of San Francisco, and finally the Golden Gate. Walking before such marvels, I felt I was absorbing the landscape, the landscape absorbing us.

Reading *The Trickster and the Paranormal* by George P. Hansen, I felt a chill when I came to the extensive fieldwork among the Navajos done by Barre Töelken, head of the folklore program at Utah State. He had a particular interest in Coyote stories. In the course of an interview with a native medicine man, he was asked if he was prepared to lose a member of his family. The shaman explained that although some tales could be used for healing, the deeper reality of the Trickster transcended Western science; losing a family member was the price paid for knowledge because understanding beyond the threshold was guarded by Trickster as Coyote, Mercury, Thoth, and Hermes, his five names, part of the trick.

As Hansen writes, "Probing the religious and supernatural aspects of other cultures may unleash forces foreign to Western understanding, even when the information is packaged as fun stories for children."

I intend to verify this, going back to fundamentals, and into the field to talk to critical witnesses in the few very special places I've identified. I won't publish anything unless I'm sure. The field doesn't need one more speculative theory.

Hummingbird. Wednesday 14 May 2014.

Yesterday evening Kit sent word from Washington: "I just finished the day at CIA, with a Deputy Director. He begged me to figure a way to meet you. He's an *experiencer*."

This is new: the man is Jim Semivan. I called him today and we had a long conversation. The plan will be for me to stop over on the way to France to see him and his wife; both were injured in encounters.

Hummingbird. Saturday 17 May 2014.

It all began with an event in 1991, a double abduction from a bedroom. It was a three-part event, over an hour, with injuries to the wife, and medical sequelae that remain unexplained; the man himself became ill, and over several years was near-incapacitated. He left his job for several years to recuperate and is now back at the office, but that isn't the only issue. "There's an offer for additional connections to study large questions, best understood and exploited by Jacques as the lead, and others by George… Garry and I will meet to collect blood for the project…" writes Dr. Green. The witness' wife is an expert in neurotherapy.

Hummingbird. Monday 25 May 2014.

Too many moving parts, and the prospects of another trip to Europe so close after our visit to Paris and Rochecour at Easter. But things are moving within several projects.

The Russian River, an hour or so north of San Francisco, flows through the quirky little town of Forestville and on to the Pacific. Early this morning Flamine and I rented a canoe and wandered down the river for the five-hour trip to Jenner, halting a few times for snacks on the grassy shore and a swim in the swift current. We were alone on the water, marveling at the dark thoughtful redwoods, the silence only broken by the calls of birds, the quiet scene animated by the sudden flight of a blue jay or a proud heron.

I felt welcome again into the redwood cathedrals of the northern forest as I once did at Spring Hill. I am enormously grateful, at nearly 75, to be able to row for twelve miles without pain. I take no medicine and I'm able to work for a full day without much strain. Yet I recognize signs of age creeping up: degraded ability to hear and increasing lag time when I try to recall peoples' names, or to translate words. It only means I need to work harder at the things I love.

Hummingbird. Wednesday 28 May 2014.

An important conversation took place today. It turns out, if we trust 'unnamed sources,' that Jim Semivan *was meant to believe he'd been abducted*. When he tried to explore the nature of his own case at the Agency, that attempt terminated when officers he trusted told him, "We know what it is, you'll get a visit from Z-Division." There are several units by that name, one of them in the Department of Energy.

Very upset, Jim was told, "You can get another clearance for an SAP that will enable Z-Division to brief you in two weeks." He decided that was a bad idea; he only wanted to get cured and resume life as a professional. His wife treats people who've had extreme experiences, whose MRIs have shown neurological impact. So, he took early retirement at 53. He was treated over two and a half years and got better. The Agency called him back. He's convinced what happened to him relates to deception. He hadn't had anything to do with UFO research, especially the one SAP that has 'the hardware.'

They want me to follow up, and various sources of research funding were mentioned, but I want nothing to do with that money either, mindful of what Mr. Frank Pace told me many years ago in New York: "Only do research with financial resources that are impeccable, *never touch anything else*," a principle I have followed religiously.

Then a friend told me: "Well, this doesn't tell us much about the UFO phenomenon. No security protects anyone from an SAP even at his level, because the security of an SAP is inside the SAP!"

I had to let that one sink in. If that's true, it means things are inscrutably murky. Semivan declined access to the UFO secrets, and others did too, afraid that once cleared at that level, they would lose control.

Often, if a lie is big enough, it's more convenient than the truth.

Hummingbird. Saturday 31 May 2014.

In response to Garry Nolan's plans for a proposal to Paul Allen, Colm reminds us that "BAASS was a $22 million project over its 24 months during which time I managed a staff of 40 BAASS employees on a daily basis (not including interacting with the consultants) with field investigators, physical security, document custodians, scientists, technicians, database people, Russian, French and Portuguese translators, and a large admin effort. We investigated about 40 cases per month, including site visits."

Yet our team never did set up a debriefing session. That's the nature of classified projects; zero opportunity to accelerate good things or to stop errors. Colm answered in his clear manner with the lesson learned: "It is a huge mistake to structure a project the way the government established and controlled our study; the phenomenon is too vast and complex to work on it part time. Unless there's a sizeable, well trained, knowledgeable team working seven days a week, we're dabblers with a high risk of failure."

He adds that even for the medical project, a serious effort means a sizeable network where dozens of cases are screened, triaged, and investigated via site visits, per month. So that's a numbers game. At any one time, there needs to be a very large number of cases from independent sources moving through the system at degrees of investigative maturity. But given the layers of deception, onesies and twosies from a single source as we did only wastes time.

He concludes: "It is critical to have top-of-the-line database support. The Capella database was brilliantly conceived ... Databases are the lifeblood of the project. *On a personal note, it was only through poring over the BAASS database that it first dawned on me that this phenomenon was a threat.*"

The only worthwhile information in the USG is SAP'd. When Colm and two others sat in a conference at Bolling AFB with five AFOSI-PJ members, the first words were: "We never talk to people who are not SAP'd." Minutes later, the meeting was over. Colm concludes: "There is a tendency when immersed in these cases to oscillate *between extreme skepticism and extreme gullibility*. It is very important to chart a middle course, and that means constantly refining one's internal bullshit detector and gullibility index."

Kit reacted with some urgency: "Those of us who've seen some of it have had our worldviews shifted. Those of us who haven't cannot seem to grasp that what we say in precise form…could in principle even be true. We've been through three, and now a fourth Agency…in each we are bled and sucked dry of information, especially physics theory, and medical. No one cares about anything else, much.

"We're always told that 'the contract' will be available…'just around the corner…certainly in 30 days.' We've been going on like this for six years. It's embarrassing how gullible we are. As soon as we are SAP'd [ahem] my data…should I give it up…will become ethically unprotected, yet no longer available for me to share. This logical inconsistency led me to submit my clearances for debriefing. The request was rejected. *This means it's pretty late.*"

Hummingbird. Sunday 1 June 2014.

At the Band of Angels' annual Texas-style barbecue, at a hillside home and garage in Portola Valley, Flamine and I were able to admire a nice Model-T Ford reconditioned for racing, and on the way back we stopped at the beautiful hillside home of Federico Faggin. For months I'd been thinking of taking Federico into my confidence about my stealthy personal experiments, but I had to wait until I had a working demonstration. With picture in hand, and an abstract of our patents, I got his interest.

Some neurologists now affirm that the brains of humans have become lighter over the millennia. Supposedly, we have 'outsourced' much of our intelligence to our social environment. Yet rationalism and individual value have declined as well. We now rely on our peers and the web as a social fabric, to know what to think and how to behave, rather than using our individual logic and argument. Does that inhibit our ability to address unexplained phenomena? Of course it does!

Hummingbird. Friday 6 June 2014.

John Alexander, in town for an SSE meeting, paid us a visit this afternoon. I was glad Flamine was able to meet this man who played such an important role in Bob Bigelow's research and has long worked to advance recognition of the paranormal within the military. He's America's leading expert on non-lethal weapons.

He did ask about my current interests, so I stressed the ancient sightings and our current field investigations, but the dialogue took turns that left me puzzled. His background includes extensive NLP training in the Special Forces, author of a paper on clandestine elicitation, and teaching at White Oak. John told us about his travels with Victoria to study shamanism in New Guinea and in the Arctic. They were stunning, a genuine effort at multi-cultural documentation.

Hummingbird. Wednesday 11 June 2014.

Troubled by his experience and that of his wife, Jim went to a colleague, a historian known as 'Duke,' who was surprised at how upset Jim was. "Do you have any marks on your body?" he asked. Indeed, Jim had a scoop mark in the flesh of his leg and a puncture at the base of the neck, one cm in diameter, round and clotted over.

"Well, you've been abducted," was Duke's conclusion. The people from Z-Division will be coming over, they can clear you. In the meantime, read these." Duke went over to a shelf, took down three books, and gave them to Jim. They included *Messengers of Deception* and *Invisible College*. (The third one may have been *Passport to Magonia*.) He went out and bought my other books, and then declined the clearance for Z-Division.

Kit is puzzled at the amount of disinformation and misinformation currently being built up around ufology. "The government keeps misinterpreting and distorting the field. This doesn't mean I believe in your cosmic information theory," he tells me with a wink, "but I'm amazed they don't seek explanations, they just dissemble the facts."

The plan now is to focus the analysis. Garry will draw blood and get the medical records; Colm will follow-up in September. So, what do we have to do with aliens' faulty biology, as some ufologists are claiming? Kit and I disagree here. I told him there would be better ways to run a hybridization program than victimizing humans in the middle of the night. Kit replied that we just couldn't understand what's going on from the limited data. It could even come from us, he said, in an intersection of weird, classified advanced technologies, either in a criminal setting or an accidental connection. My point is: We don't have "limited data." We have truckfulls of very good data.

Arlington, Virginia. Tuesday 17 June 2014. Iwo Jima Hotel.

As my taxi was pulling out of Dulles airport this evening, I received a call from the friendly physics team. They asked me to postpone our visit, on the advice of attorney. The process is producing a surprising quantity of a very scarce and expensive substance, and this requires further tests. He recommends keeping everything under wraps.

On the plane I read *The Mantle of Command*, Nigel Hamilton's account of Franklin Roosevelt's leadership in 1941 and 1942. Everybody should read this. I'm still surprised at how little the world remembers.

Mabillon. Thursday 24 June 2014.

We landed at Charles De Gaulle in such great weather, with only a few puffy clouds in an idyllic sky, that everything looked fine despite the ongoing labor delays. On such days, even Saint Sulpice, with its towering neo-classical curiosities, looks attractive and sweet.

After restructuring computers, making plans to see a witness in Franconville, and planning lunch with Alain Dupas to discuss Airbus, I was so tired that I struggled to read the last few pages of an amusing novel, *Un Mensonge Explosif* by Christophe Reydi-Gramond. The best part is a post-scriptum where he shows a link between occultism and the philosophy of Cosmism that inspired the pioneers of space.

He explains that in 1870 a man named Max Théon resurrected the Hermetic Brotherhood of Luxor, "founded in 2350 BC, recognizing the eternal existence of the Great Light, harmonious power of the universe." With his wife Alma, Théon went on to create the Cosmist Movement in 1900, meeting with an enthusiastic reception in Tzarist Russia.

Mabillon. Tuesday 1 July 2014.

The young witness we met yesterday in Franconville, northwest of Paris, told Flamine and me that the skin mark she'd received in late 2012 wasn't painful and only lasted a few days. She's only seen UFOs once, three years before that incident, while on a beach. She's given me two photographs of that triangle of orange lights.

Now Hal and Eric assure me that MJ-12 really existed and dealt with real UFO crashes: "Doty said the MJ-12 papers were essentially true." But at this point in my work, it's irrelevant whether MJ-12 existed and dealt with UFOs rather than nukes or whether Stalin preferred one brand of Vodka over another.

Mabillon. Tuesday 8 July 2014.

Historic meetings at the workshop organized by Xavier Passot, so I made my way under the rain to CNES headquarters. During the introductions I discovered that Pierre Bescond, who used to be executive director of the Ariane project in Kourou in 1977-82 and director of the SPOT Image project at CNES in Toulouse (1982-86) in the days of Velasco, had used his influence to push for official acknowledgment of the phenomenon.

Bescond (Polytechnique, SupAéro) is a French general, armaments engineer, and former Auditor for IHEDN, who participated in the COMETA Report. He's worked with American space projects over the years. My friends Dick Haines and Jean-Pierre Rospars were among the first presenters, along with CNES investigator Gilles Munsch, lightning expert Philippe Ollier, Frederic Thomas representing the Gendarmerie, and Jacques Py of Toulouse University.

In the afternoon, François Louange presented photographs, Gaelle Fedore of INRIA spoke about sound analysis, Edoardo Russo and Michaël Vaillant addressed databases, followed by Xavier Passot and my presentation on research strategies. The impact was clear: The agency takes the observations seriously; many cases have no explanation, and good research techniques are available.

Mabillon. Wednesday 9 July 2014.

It was a studious audience that assembled at CNES for the second day, sensing a historic turning point. French Air Force officer Christophe Colliard and Jérémie Vaubaillon described techniques for sky observation, followed by Raymond Piccoli's discussion of globular lightning and 'orbs.' Russian researcher Sergey Chernouss spoke about observations beyond the polar circle, and our friend Erling Strand gave an update on Hessdalen data after 30 years, followed by Massimo Teodorani and Nico Conti.

In the afternoon, four presentations on psychological aspects by Thomas Rabeyron, Pierre Lagrange, Jean-Michel Abrassart, and Romain Bouvet who spoke of the biases in reporting. A roundtable concluded the conference, with brief presentations by Ron Westrum, Jacques Arnoult, Bertrand Méheust, and me. (Figure 13, page 262)

Mabillon. Thursday 10 July 2014.

At the occasion of the CNES workshop, Dick Haines had proposed to hold a meeting of the NARCAP organization in Paris, so our apartment hosted nine members including two airline pilots, plus Erling Strand, Dominique Weinstein, and François Louange. Much of the discussion had to do with airline industry policies, hiding data.

Mabillon. Friday 11 July 2014.

Information structures are the key to the system's hidden levels. There's a way to design new software to facilitate the process, but amassing a lot of data is useless if one doesn't secure the trust and intuition of the men and women who understand the information in context, texture, and unspoken extensions. Therein lies the fallacy of the US approach.

Mabillon. Monday 14 July 2014.

On Saturday we took the train to Quimper, where we spent a quiet night. Yesterday afternoon Jean-François Boëdec picked us up for the short trip to delightful Leuhan. Jean-Pierre Rospars had already arrived with the witness we had wanted to interview for years, namely Gendarme Michel LeStunff who'd made an observation in Douarnenez Bay on 7 January 1974.

LeStunff is a big man, jovial and outgoing, who studied to work in law enforcement but later found a more profitable career managing airports in Africa. That fateful day in 1974, he and a colleague were directed to drive to the town of Ted Bruc at 8 am to replace a team of Gendarmes on another mission.

Charles de Gaulle airport. Wednesday 16 July 2014.

Waiting to board the flight, I'm distressed as we linger in a world of mysterious sources with bizarre compartments, and supposed revelations, always denied or distorted. The theory about aliens in search of genetic repair is shaky. Most importantly, Texas friends shouldn't exploit databases when they don't have access to the people who know what the data means. It's a terrible mistake.

Hummingbird. Thursday 17 July 2014.

There's an alignment of articles in *The Economist* and the *Wall Street Journal* announcing, "the death of UFOs," clearly manipulated. This was followed by similar statements in the Dassault magazine *Valeurs Actuelles*. The LoneStars could easily become a ship of fools, since the so-called hardware (or even recovered cadavers) may not have much to do with the ultimate reality.

It's interesting that Colm, at the same time, is inspired to write a note recalling (Nobel Prize winner economist) John Nash's impression of receiving ideas from 'supernaturals.' The early July 2009 events on the Utah Ranch when three military intelligence members encountered a mysterious presence "appear to have triggered a series of cascading events, coincidences, anomalies, and cryptids that have rippled through colleagues, family members, and communities, a multi-year bonanza of the impossible."

He adds: "At every turn, the same song has played, involving the appearance of a supposed 'success only weeks away,' which vanishes like a Cheshire cat, only the smile left. Today in July 2014, the same song is still playing..."

Hummingbird. Friday 18 July 2014.

The mail brings a sad letter from the daughter of my old friend Bill Powers from the sixties in Chicago days: "You may vaguely remember me as a young girl hanging around underfoot at Northwestern University. I was too young to fully appreciate what was going on at the time. Now I find myself tasked with organizing my dad's papers to deliver them to the Northwestern archives. It is as if a special window into the past has opened..."

Bill died over a year ago, on 24 May 2013. They have a reception in Evanston in two weeks. She hopes I can go there with Janine. Our common friend, Harry Rymer, will give a lecture about Bill's work in control theory. But I can't go, and Janine has died. Bill was one of the few solid minds in that fundamental period, and all that makes me very sad.

Hummingbird. Sunday 20 July 2014.

San Francisco has been bathed in cold white fog, but the hills to the south were radiant yesterday when I drove over to meet Garry at his Redwood City home, an impressive and elegant wooden structure that overlooks the coastal hills and the green ranges along the San Andreas fault. I'd brought him a book about Project Azorian (with its hard-edged, often painful descriptions of what real secrets do to people) and my latest CNES presentations.

Garry was just returning from Washington, where he'd collected blood samples from Jim and Debbie. He shared my impression of them as sensitive, intelligent, highly educated, and responsible. He'd also interviewed Axelrod, his wife, and kids, and found the cryptid reports were real. But beyond this, much of the LoneStars' belief about alien hybridization, Indian ancestry of victims, and other themes driving the group's research needs to be challenged.

The best result from all this may come if Jim Semivan is willing to throw light into the dark corners of the manipulation. Driving back on the winding roads towards the Crystal Springs, however, I kept thinking about the old adage, "Be careful what you wish for..." Wasn't it John Alexander who told me that?

News reporters continue to wallow in the unleashing of hatred across the Middle East: Gaza and Israel, Egypt and Libya, Syria and Iraq, all over again. Pro-Russian thugs in Ukraine, a ragged band with missiles, have downed an airliner full of women and doctors.

Hummingbird. Tuesday 22 July 2014.

The English novelist A.S. Byatt writes about the Arab traditions of non-human creatures in her beautiful tale of *The Djinn in the Nightingale's Eye*: "Djinns, as you may or may not know, are one of three orders of created intelligences under Allah—the angels, formed of light, the djinns, formed of subtle fire, and man, created from the dust of the earth. There are three orders of djinns—flyers, walkers and divers; they are shapeshifters, and like human beings, divided into servants of God and servants of Iblis, the demon lord. The Koran often exhorts djinns and men equally to repentance and belief, and there do exist structures governing marriage and sexual relations of humans and djinns."

Such traditional notions may seem fanciful, but a clever reader of *New Scientist* (7 June 2014) remarks that modern science is not much clearer: "We apparently live in a universe that spontaneously exploded out of nothing, which exists in 11 dimensions (most of which we can't see), and that is mostly composed of dark substances we can't detect, where even some of the stuff we can detect appears to be in two places at once, and that all these speculations are an emergent property of electrical activity inside what is essentially a piece of meat!"

Hummingbird. Thursday 24 July 2014.

Only today did I have the courage to lookup our Pagan friends on the Internet, and to read what I'd expected, that Morning Glory had died shortly after our visit in Santa Rosa. Another beautiful soul has been removed from our circle of friends and searchers. I have been trying to comfort Oberon, but I can only tell him that it will be futile to try to turn the page, as it was futile for me four years ago. That sorrow, that pain is to be cherished, not overcome—it's all we have left.

After a successful presentation at the CNES workshop earlier this month, a two-hour live discussion on *Coast-to-Coast AM* with George Knapp was picked up by some 600 radio stations.

Hummingbird. Saturday 2 August 2014.

Linda Howe, who spent the day here, came over with a package of new reports, ranging from the intriguing and the mysterious to the unbelievable, from skin marks to the likelihood of hybrid aliens loose in America. Linda has extensively researched skin marks, so I showed her the Atlantis structure in Plato, which matches them. As we sat down at the table where she'd set up her computer, we had a cup of coffee and spoke about Doty: he'd secretly videotaped her in his office, she said, demanding that she sit in a particular chair in the middle of the room, away from his desk. She'd gone there because of some reports mentioning him in connection with aliens getting shot at Ellsworth Air Force Base in 1978. Doty supposedly had the names of the shooters: "That's not important," he said, brushing off her request. "I have something else for you…" He handed her some papers: "My superiors have asked me to show you this."

They were the papers about MJ-12, now cited as 'evidence.'

We walked to Polk Street for lunch, and Linda gave me a file that had to do with the claims of a young woman who studied at American River College in Sacramento in 1972. Supposedly, she was an alien hybrid, as described in the book *Raechel's Eyes* by Helen Littrell and Jean Bilodeaux, 2005. Many people are convinced that such hybrids are among us, that the human race was seeded and genetically manipulated by aliens. "Was Abraham a Reptilian, with actual scales?" Linda wondered.

"Didn't Abraham have several wives and numerous concubines?" I responded. "Wouldn't they know whether or not he had scales on his body?"

We both laughed, knowing that such rational arguments are irrelevant to the current ufology mindset in America. The legends ooze freely from DC now. I even have friends in Vegas and Austin who buy them. The problem is that such intrigue also oozes from people within the government, with bizarre impunity. Why?

Stanford University. Thursday 21 August 2014.

This morning, I took the stack of plastic boxes that contain our seven most relevant UFO samples (the slag from Council Bluffs, the aluminum from Bogota, the three magnesium samples from Brazil that Peter Sturrock had given me, and some others) and took them to the NanoSIMS facility where I met Mike Angelo. We spent the day running the samples through the

Fig. 13: *Historic UFO Conference at CNES Headquarters, 9 July 2014, Bertrand Méheust, J. Vallée.*

Fig. 14: *Dr. Garry Nolan, at home, with recovered material samples.*

CAMECA after coating them with 10 nanometers of gold, thanks to Dr. Hitzman, director of the lab.

Yesterday, a message from a colleague in the material study cast a little more light on the hardware tested by Lockheed: "The team was able to observe and measure the highly unique nano-structures inside 'recovered' samples. It was the observed characteristics and properties of that nanostructure that convinced them that the samples were…not from here. Their sample collection was bits, pieces, and chunks of recovered items that a 'client' had sent to them for analysis and exploitation, which didn't succeed, due to inability to reproduce the nano-structured materials via human technology about 1980."

Valognes. Thursday 28 August 2014.

I will pass lightly from this life when the time comes. I thought of it as I walked the four kilometers from the station to the cemetery. Two Jeeps passed me along the road, antennas waving in the wind; this is the 70th anniversary of D-Day, its sacrifices have not been forgotten.

I love the hospitable, juicy landscape of Normandy, and I'm reconciled with Janine's last wish to rest here. She once related to me that as a child, her mother had taken her to a medium at a country fair. The clairvoyant told her she would marry a writer and would take many trips with him. Does that mean that the future already exists? Or simply that there is no such thing because Time itself doesn't exist?

Bellecour. Saturday 30 August 2014.

Tall and strong, just back from a holiday in England, Maxim was waiting for me. We spent a lovely evening catching up, while the TV news brought us ugly political debates amidst the debacle of President Hollande. We celebrated Maxim's 17th birthday with Olivier, pleased with new plans to expand his business.

This is the end of summer. My son works very hard against bad economic odds, while Maxim returns to school with new energy. Driving through Normandy and the countryside of Perche and Sarthe, one gets an impression of affluence at variance with the dismal statistics of stagnation. Even in small villages and farm communities, there's none of the squalor and neglect of modern America. True, this relative affluence, these fine roads and the dancing in the streets rely on borrowed money, but it wouldn't take much to put this country back on a successful course.

Mabillon. Sunday 31 August 2014.

About 3 am I turned on the light and read for an hour. There must be someone in the building who reviews new books for the media because piles of novels appear periodically at the foot of our stairs, free for the taking.

By the time I showered and walked out, the Sun was already beaming over the graceful façades. This is Paris as I love it: elegant, uncomplicated, intelligent. The little shops of Rue St-André des Arts were still closed but *Le Départ* was open and gave me some strong coffee as I read a message from Colm, picking up on earlier comments by George Hathaway: "By and large humans have lost discernment, so we are cannon fodder to a phenomenon which over millennia, has perfected the art of subtly invading and manipulating our mental space."

Toulouse. Wednesday 3 September 2014.

Hotel Pullman, close to the Toulouse-Blagnac Airport. Tomorrow my partners and I will meet with the chief of staff of the CEO of Airbus. From my room, I can see the Pyrenees, the most beautiful mountain chain in Europe.

Chris Dittmar and Dima (Dmitry Manakov) met me last night, and Alain Dupas joined us for a two-and-a-half hour meeting with the top echelon of the company. I don't have much hope.

Mabillon. Thursday 4 September 2014.

When I worked with Gérard de Vaucouleurs at McDonald Observatory, back in 1963, I recall our conversation as we were leaving the dome of the 82-inch telescope, discussing galaxies. "They're not out there at random," he said. "They're grouped together in clusters and meta-clusters. That will be the next level of discoveries about the structure of the universe."

At the time, I was editing the catalogue holding the new data he and his wife had patiently measured at Mount Stromlo in Australia, years of the best data about the Southern half of the sky.

His insight has now been fulfilled thanks to a list that gives the velocity of those close galaxies, defining the super-cluster that contains our

Milky Way. A recent paper in *Nature* names it *Laniakea*, "the immense paradise." Galaxies flow along gravitational wells along its flanks, making the notion of a Great Attractor unnecessary. Laniakea contains 100,000 large galaxies each with 100 billion stars. What forms of life, of mind, of spirit does Laniakea holds? And how could we dare to impose limits on its denizens?

Mabillon. Friday 5 September 2014.

"Are you ready to discuss State secrets?" I asked Dominique Weinstein, a bit shocked, when he greeted me in front of restaurant Suffren, our traditional lunch place. I handed him a copy of the latest *Nexus* Magazine, which used this catchy title to introduce a series of articles about our recent meetings at CNES. He laughed.

Joking aside, there were two people from DGSE at the meeting, including a young blonde woman whose job is to track down researchers 'of interest.' We spent most of the lunch discussing his latest catalogue of pilot sightings.

Mabillon. Sunday 7 September 2014.

In a few days I fly back to California, but this afternoon I spent time with Flamine's father and his wife at their apartment on Rue d'Alésia. Mr. Jean-Claude de Bonvoisin, at 95, struck me, as before, with his patient intelligence, his very good memory, and the command of language evident in his manuscripts.

We spoke of the Middle East, of his consideration for Islam despite the atrocities committed by its extremists, and of the impressions of a man confined to four walls. But those four walls can be wide indeed…

Hummingbird. Friday 12 September 2014.

Back in California, walking the labyrinth at Grace Cathedral, I seem to have lost my way, circling and backtracking. An old musician sits cross-legged on a carpet by the side, blowing into a long Tibetan trumpet, sounding bells, hitting precisely tuned gongs. A red-purple light falls on a cluster of people at the center, praying in their own ways. I turn again, circle once more, and suddenly I stand among them.

Daylight is gone now; the tall stained-glass windows glisten in the darkness while hundreds of colored ribbons hang down from the ceiling of the church, swaying. The way to detachment begins here, and it continues for me with an awareness of how unlikely reality can be: this planet, this city, this church, and me.

There was a fascinating discussion over the web today about Robert A. Monroe's *Journeys Out of the Body*. In that book the final chapters on the psychological exercises that relate to the establishment of an OBE are shocking. They were removed after the first edition, considered "too dangerous" by the CIA Project Manager of the Monroe Contract, namely Kit again, who writes "they resulted in my separation from the StarGate program, and in part, from the CIA."

In his book, Monroe begins by stating that initially the experiences of OBE typically generate fear because of the innate struggle to survive (which the out-of-body experience seems to challenge experientially). This must be confronted and overcome in our own psychology.

He then gives a series of exercises along the lines he personally experienced as being effective to generate the OBE. He follows this by discussion of what he labels as *Locale 1* and *Locale 2*, the former being our physical reality, the latter a vast other-dimensional universe that is structured, populated, and real.

The 'other' (ET) contact describes what he labels as a 'hypothesis,' but it is so detailed as to give the reader the impression that it is what he experienced as contact with 'other' intelligences who probe our planet.

Kit recalls: "What he told me personally is not in there, namely the recipe to *get back* after the generation of the OBE experience has occurred... and one is pending return. The issue is that one gets too frightened with what one is confronting to get back.* He had a simple recipe. I tried it, had an OBE that was unambiguous, had a confrontation with an entity, and couldn't get back. Then I remembered his 'recipe,' and it worked like a charm."

Hummingbird. Thursday 25 September 2014.

Yesterday was my 75th birthday. I've lived three years longer than my father, with the feeling of a new plateau, in that zone of indeterminate nature

---

*   I had such an experience once, and it left me terrified.

known as 'maturity' whose inhabitants are charmingly called Senior Citizens. This is a serious time, with bad news from friends: John Schuessler, recovering badly from an operation, won't be able to attend our upcoming meeting. Oberon has lost his long-time companion. Graham Burnette has recovered well after his throat cancer ordeal, but Arthur Hastings and Len Herzenberg are gone. Bob Johansen survived cancer as surgeons removed his prostate.

One source of inspiration is Jack Katz, 87 years old this week, charging ahead. He tells me about another one of his dreams, where he stands at the South Pole (but not of this Earth) in terribly cold weather, snow hiding the face of a stranger who points in the direction of a faraway mountain: "That is *beyond the Beyond*," he says. So, Jack is going there, now on page 25 after 45 hours of drawing without sleep, listening to his favorite composers. He tells me Man is paralyzed by fear; our race is insignificant, it has no purpose, and most of our literature is little more than journalism, and gossip with very few exceptions, like Blake.

Hummingbird. Friday 26 September 2014.

Flamine and I drove to Palo Alto tonight to hear Michael Cremo, author of *Forbidden Archaeology*, who was going to discuss Vedic traditions. He never did, but he spoke of ancient artefacts hinting of civilizations in the distant past and of the existence of "an ineffable entity outside our current scientific understanding," a reference to Rodney Brooks of the MIT Artificial Intelligence Lab.

Much of the discussion was about Alfred Russell Wallace, co-author with Darwin of the Theory of Evolution by Natural Selection, and whose interest in psychical research led academia to drop his name from later recognition! Cremo reminded us of Crookes' research with the medium Home, of Pierre and Marie Curie working with Eusapia Paladino, and Sir Oliver Lodge with Gladys Leonard, William James and the Watseka Wonder, Charles Tweedale and his glowing bedroom apparition…all of it later swept under the rug by 'rationalists.'

Hummingbird. Monday 29 September 2014.

The Gravity Company (and its equipment) has moved to the back of another shop. This time it's an art supplies studio rather than a yarn and crochet shop, but the atmosphere is the same colorful, easygoing jumble. "A great cover!" I told Mike by way of congratulations.

Another man was there, the former chief of staff for the House Committee on Science, Space, and Technology, obviously very well connected. Our main topic was Helium 3 production, now confirmed. They believe they can produce 2 to 5 liters a day, with a maximum of 20, and sell the gas at $5,000 a liter, which would pay for future developments. The current apparatus, enhanced with a 'metal on metal' tube, can produce about 15 kilowatts.

Hummingbird. Tuesday 30 September 2014.

Tonight, we'd arranged for Garry Nolan to meet Ed May over an informal dinner at our home. It was an opportunity to discuss remote viewing, Ed's recent publications (including *ESP Wars: East and West*) and the history of our various relationships with Washington.

I am introducing Garry to as many luminaries as I can to give him a perspective on the LoneStars, and perhaps save him time in the complexities of psychical research and the physics behind it. Given their personalities, the discussion quickly took a very technical turn (statistics, experimental settings, selection of subjects, brain structures, and instrumentation). Ed was moved to reconnect with us, recalling SRI, his project at SAIC, and all the people we've known, and lost.

Hummingbird. Saturday 11 October 2014.

A memorable dinner took place tonight in Solano in honor of Jack Katz and the publication of the last (sixth) volume of *The First Kingdom* ("Destiny"). There were seven of us, including Jack's students and Peter Beren, whose mind was as sharp as ever.

Jack was delighted to be amid this group. Back home, he showed us his latest boards in pencil, in preparation for sending page 27 of *Beyond the Beyond* to his publishers, who were told there would be over 800 such pages. They had asked Peter: "He's sending one per week? And how old is he again? Eighty-seven?"

Later the same day.

In connection with a position paper about the role of imagination in creating reality (for Jeff Kripal's upcoming Esalen symposium), I had a serious discussion with an expert about the current state of physics. My related TEDx presentation in Brussels is popular on YouTube, so perhaps a radical change is coming after all.

Further research has clarified the nature of most, if not all, MJ-12 documents. The leaked papers were never hoaxed; they were official active measures documents produced by CIA counterintelligence Chief J. J. Angleton in the 1960s to track the flow of stolen classified USG documents through Soviet espionage. There were also *non-MJ-12 documents* that were AFOSI disinformation targeting Bennewitz. Counterintelligence tradecraft involves deep secrets... I'm told the 'leak' of the false MJ-12 documents through noted ufologists may have been a botched, ill-conceived attempt by DoD.

Austin. Monday 13 October 2014.

Kit arrived early for seventh meeting of the LoneStars, so we had a friendly conversation at a noisy sports bar. There are problems in Vegas. The Senator inquired about an update on the funding of the BAASS, and he found out that somehow the appropriated money was 'redirected' as a result either of interference or bureaucratic incompetence. Wasted effort.

Next, we discussed the status of Jim and Deb. The object at the base of Jim's neck may not be an implant, but it does mark a site that bled after the encounter, leaving a round mark. Kit surprised me by saying again he wanted to stay out of any government involvement at that level until we had assurances of a *real* project with *actual* material to study, rather than another Washington hall of mirrors.

I do feel the same way, but I'm not an insider with the vast connections he commands.

We discussed Bill Moore's stunning comment to me some years back, that his handlers were "disinforming their own people" in the process of spreading false stories through naïve ufologists. That troubled him because of the implication that some of the higher government staff may have been fed false data as well.

Austin. Tuesday 14 October 2014.

The seventh meeting of the LoneStars began again with key members missing: Bob Bigelow, Colm Kelleher, and John Schuessler. The first order of business was a group blood draw to serve as control of the abduction/interference study Garry has undertaken.

He gave a masterful overview of his work. "The stress-related process at cellular level is a fingerprint of events to which the immune system is exposed," he said. "The body is an antenna for the internal and external environment, and all insults or perturbations lead to the immune system. We try to identify the nature of the signal, in part by looking at mRNA, as opposed or in addition to DNA analysis, which could help us look for alien hybrids, if there is such an animal. Are the stressors acute, chronic, and permanent? Heat, viruses, bacteria, radiation events—all can be discriminated by the immune system."

He went on: "Disease is a communication problem; the brain is picking up on signals from the immune system." (So, why doesn't the brain fix these problems if it's able to detect them?" I wondered.) The new instrument Garry and Michael Angelo are designing will scan eight billion pixels per cell, at 5 nanometers.

Kit shifted to a lengthy discussion of skin marks, followed by George who's developed a taxonomy. Someone asked: "If the marks are made by aliens, why are they not reported by abductees?"

Austin. Wednesday 15 October 2014.

We began the morning session by discussing the 'Vienna Hypothesis,' which led to a discussion of hypothetical biological robots. I felt obligated to point out that the entities involved in a particular example of a case we know in Virginia were not Grays but tall, dark-skinned, non-Negroid humans. There was agreement that Disclosure could create mass confusion, that it might 'hotwire the Apocalypse.' Nobody would be able to halt the trauma.

One member then made a series of personal remarks, obviously deeply felt. He reiterated that abductions were a central factor in the future of civilization. He mentioned that my *Messengers of Deception* was a key book, opening the mind to the complexity of the problem. How can we proceed if humanity faces competition from advanced space-faring empires? Are some of them dependent on us for evolution to continue? What about reincarnation (ours and theirs)? Do they have souls like humans? And, more specifically, can we contact a higher-ranking alien through a medium?

This led us back to speak of the Collins Elite. Father Ray Boeche, an Anglican Catholic priest, was once visited by two Intelligence agents who spoke of their contact with aliens, "devils in disguise."

Next, the agenda called for a review of Rendlesham Forest events.

During all these discussions we had subjected ourselves to blood draws, Kit doing the needlework, and Garry securing, labeling, and treating the tubes with appropriate chemicals.

I gave an update about the death of Marie-Rose Baleron, the French counterintelligence officer and high executive of AMORC killed in the crash of that regional airliner in 1972. At the scene, someone cut her wrist and took away her government briefcase…with the extensive report it contained, alleging supposed links between the Imperator's office, neo-Nazi groups in France, and Lodge P2 in Italy.*

The splinter groups they created (ORT and the Order of the Solar Temple, of nefarious memory) allegedly introduced their members to "actual extraterrestrials" (suspected to be animated statues seen at a distance) in their initiations.

Austin. Thursday 16 October 2014.

I slept poorly and was up at dawn. The country wakes up to the sad spectacle of a world in peril, shaken by the realization of the poorly controlled Ebola epidemic in Africa and of spreading financial dysfunction in Europe, an ominous combination. My assessment of US industry is that it is well-managed and modern, but its recovery from the near collapse of Wall Street six years ago has been too slow and too strained for the US to pull up the rest of the world.

Every morning brings new worries. Stock prices continue to fall, a 10% drop in two weeks, but it's the uncertainty that keeps me awake. What contribution can I still hope to make?

---

*   None of these allegations should be meant for AMORC as a whole, since it keeps a number of branches, and its Headquarters are in San Jose, California.

Hummingbird. Saturday 18 October 2014.

Reviewing my notes from Austin, I emerge again with contradictory impressions. Sometimes I feel like I've been railroaded into sloppy conclusions and a faulty model. I am tired of extracts of catechism that are as unsound theologically as they are inappropriate in this study. My colleagues' bizarre 'findings' about syndromes and abductions of *Scandinavian women with Cherokee ancestors* are not even funny…

Flamine and I worked hard all day yesterday. She is preparing for a psychology exam; I assembled my notes from Austin, without a break to recover. Then we set that aside and happily walked over to a baroque concert from the time of Versailles: Delalande, Charpentier...

Esalen. Monday 20 October 2014.

The Symposium began in the main room of the Big House with an introduction by Jeff, always in control of his material and his audience.

Greg Shaw spoke of Iamblicus and Plato, with the intriguing concept that *we are being written* by divine inspiration. He spoke of 'authorization' in the 'authoring' of our own paranormal experience. The finer spiritual body comes down to animate the physical body, and the role of imagination is to glue us into the world. There isn't a separate world of *ideas* in Plato, contrary to common opinion, since the divine world is on the same plane as we are.

Later, Wouter Hanegraff, co-author with Jeff of *Hidden Intercourse*, spoke about modern esoteric movements. I had great expectations of learning more about the state of magical scholarship, but his talk appeared closed, even oddly skeptical (when it came to synchronicity, for instance). His presentation became interesting when he spoke about Jung's *Eranos* meetings that had put Ficino and other 'irrational' authors back on the map. Diderot and the Encyclopedists had discarded such esoteric writings, following Booker who wanted to "dump them into the sea of oblivion."

After Jung, there was Scholem, Corbin, and Mircea Eliade, and later Antoine Faivre who continued Eranos, speaking in terms of "The Alchemy of History at the End of the Future."

In the evening, Tania Lurhman discussed her experience of religion as an anthropologist who had joined fundamentalist groups that spoke and walked with God in everyday life.

Esalen. Tuesday 21 October 2014.

This morning it was my turn. I summarized the CNES meetings, the changes in perception of the phenomena, and the evolution of physical theories about material reality, entanglement, and the role of consciousness; modern science is revisiting ancient esoteric ideas about the role of the imaginal, I said. The latest entanglement experiments indicate that time and space are 'emergent properties' of the universe, not fundamental entities. Reality is negotiable, and contemporary researchers like Guillemant argue that "our intentions in the future (i.e., imagination) *create the past causes of the present.*"

In anticipation of this presentation, I'd also asked Eric Davis about his latest insight. At Esalen I read out his response, framing it around two 'hand grenades' thrown into the debate about reality.

Eric had responded with a historical summary: "Spacetime was invented as a classical, simple model of the dimensions that were joined together because of the mandates of Maxwell's electrodynamics theory, Special Relativity theory, plus Minkowski's theory, plus Lorentz's theory plus General Relativity theory... Spacetime became a geometrical 'fabric' upon which all of physics could be expressed in terms of quantities that are dependent on spacetime coordinates.

"Ten years after General Relativity was published by Einstein (ca. 1915), along came quantum mechanics (ca. 1925), which was highly unusual. Its mathematical structure was built on a flat, Newtonian-Cartesian coordinate system of space and time, *not spacetime, as in relativity theory*; furthermore, it required the co-existence of an infinite dimensional Hilbert space of complex vectors and scalars, and other concepts in the formulation of wavefunctions and state vectors that totally supersede the simple Newtonian-Cartesian model ('spooky action at a distance' plus nonlocality, entanglement, tunneling, etc)."

Eric went on: "The foundation of quantum mechanics is counterintuitive to that of classical mechanics as originally formulated by Newton, Lagrange, Laplace, Euler, d'Alembert, Poisson, Hamilton, and others. Three years later (ca. 1928), it was Dirac who successfully reformulated quantum mechanics in order to merge it with relativity, thus deriving a fully relativistic quantum theory in spacetime (abandoning the Newtonian-Cartesian coordinate system). He's recognized as the predecessor of modern quantum field theory.

"The old Newtonian-Cartesian quantum mechanics was all about individual particles and waves (via the wave-particle duality of de Broglie), but Dirac's theory was all about quantum fields: the field replaces individual particles and waves, and it exists in spacetime."

Eric throws the first grenade by writing: "The phenomenon of quantum entanglement and its associated nonlocality has now proven that spacetime (and the Newtonian-Cartesian space and time) is NOT a fundamental property of our universe. Instead, spacetime is *an emergent property*. It emerges as soon as entanglement takes place. Before an entanglement takes place, *there is no spacetime!*

"Many academics understand this, but undergraduate lectures haven't been updated. They remain stuck in the old paradigm. It's in the PhD stage when graduate students see the research papers, specialized monographs, colloquia and physics conferences, and become exposed to the quantum entanglement experiments and the spacetime reality paradigm change. There's now a community that accepts that spacetime doesn't fundamentally exist, because it emerges as a byproduct of nonlocal quantum phenomena."

Now comes the second grenade: "Quantum effects are not really 'quantum'! Many lab experiments and astronomical observations have shown that *quantum effects are fully macroscopic* as well. So, it's no longer true that quantum theory is confined to the scale of subatomic elementary particles, atoms, and molecules in carefully controlled conditions. *We now routinely observe quantum phenomena acting on the scale of the universe, in macroscopic matter*, and in extreme, naturally occurring or even laboratory produced environmental conditions."

Eric concludes: "This brings me to my last questions for us 'serious Ufologists': Is there a relationship between information and quantum nonlocality spookiness, and thus reality? *And do the UFOs have a technology that exploits this to manipulate reality?* The fact that spacetime is emergent, and that there's no reality before entanglement, supports your thinking that a physical model of reality should dispense with dimensions: they're artificial constructs, and not fundamental."

Later the same day.

Diana Pasulka spoke next. She's the author of *Heaven Can Wait: Purgatory in Catholic Devotional and Popular Culture*. She had not submitted a position paper because she wanted to speak freely. I perceived her as an extremely astute observer and investigator, aware of the diversity (and possible 'perversity') of the group, mired as we are in the quagmires of 'our' ufology.

She began with a stunning statement of the origins of the modern Purgatory concept in *luminous orbs* that visited French Ursuline nuns in the 1880s. The orbs were thought to be souls of dead people seeking repentance, held waiting between the worlds. (Committing a minor sin, I recalled with pleasure the little trip Flamine and I took in June four years ago, when we slept at an ancient Ursuline convent in Autun... We hadn't noticed any orbs.)

Diana quoted: "Sister Maria, a child novitiate, had nightly visits from a flame that would float above her in her cell. The mother superior sat up with her one night, witnessed the flame and determined it was a soul from purgatory. The next day, she requested that all the members of the convent pray for that soul."

Purgatory is related to virgin births and physical scars (Thomas Aquinas) and "things that harass good Christians at night." The doctrine was problematic for Protestants but a major devotion for Catholics in the 19th and early 20th century. It certainly was taught to me most dogmatically in catechism in the late 1940s.

Amusingly, Purgatory was first thought to be an island called Hibernia, located off Ireland, where a cave existed. People went there on pilgrimages and came out free of sin! The belief ended in the 1400s. In recent years, the contact with flames and orbs has been redacted, and the modern Church has allowed the belief in Purgatory to fade away, although it hasn't been purged from all the books.

After lunch, Robbie Graham, a young Englishman, the author of *Silver Screen Saucers*, spoke of UFO movies and Hollywood's attitude towards the phenomenon. He was followed by Christopher Laursen from Ontario, doing a PhD on the history of poltergeist phenomena. He was one of the most charming and entertaining members of the group. Laursen had hidden various objects in the room, thinking they might move spontaneously as the group discussed poltergeists. Unfortunately, nothing happened.

In the evening, we saw *Jodorowsky's Dune: A documentary about the greatest movie never made*.

Esalen. Thursday 23 October 2014.

Driving back to Esalen by noon, a three-hour journey, we were on time to join the group at lunch and to hear David Hufford, perhaps the most captivating speaker in the room.

He described his pilot study of unusual experiences of combat veterans. He finds they have 50% higher experience of sleep paralysis than the general population and 60% had 'visits' from dead comrades. In his theory of the imagination, he notes that perception is a constructive process and a social process, limited by language. "What was that noise?" is the type of question that already imposes a framework. One always deals with a memory of an experience, not the experience itself. We report it with ambiguation: "I had a dream..." He pointed out that "mystics see what they have been taught to see. If belief produces these experiences, then it's not rational to use the experiences as evidence for that same belief."

There was a celebratory dinner, where we had the opportunity to share Dulce Murphy's table and speak at length about Russia.

Hummingbird. Friday 24 October 2014.

When we woke up this morning at Esalen, the sky and the sea were uniformly whitish gray. Then, as light began seeping into the landscape, bands of faint blue and pink appeared, merging at the horizon. The fog retreated and another day dawned on Esalen.

As we prepared to leave, a small group of us reviewed the highlights of the week. We were all puzzled about some antagonistic arguments we'd heard, esotericists in apparent denial of everything esoteric. "It's like trying to have sex with multiple condoms," said Mike Murphy, "just to make sure the last one remains clean..."

We started the long drive back to the Bay Area with two passengers: Robbie Graham, the young Englishman on the way back to the UK, and Diana Pasulka who expected to visit her brother in Oakland. We were in no hurry, so I stopped in Santa Cruz for lunch at the Crow's Nest and we spoke of all the topics we hadn't been able to address, including the ethnography of abductions, Diana's special interest.

Robbie told us that George Knapp was reviving the Bob Lazar claim of secret work, complete with Element 115, again! We discussed the inconsistencies in the plot made in Vegas, re-heated for Hollywood. I asked him about Rendlesham forest. Robbie pointed out the contradictions, people re-inventing the case as "the next Roswell."

I can consolidate the data threads. My books seem to remain current in the confusion about the "weaponization of religion" that sweeps aside what should be sober research into the destiny of Man. But now I learn of the death of that great researcher, Professor Douglass Price-Williams, in Los Angeles on 20 October.

Hummingbird. Saturday 31 October 2014.

In the streets of San Francisco everybody is dressed in black and orange colors, either for Halloween or to celebrate the victory of the Giants baseball team in the World Series (or both!).

Yesterday I met with Garry over lunch in Woodside, our first opportunity since Austin. We reflected upon Kit's surprising statements about 'experiencers.' Verifiable data or case background is scarce. Together with new media claims about "a thousand abductions a day," and the obsession about Bentwaters, I see people losing their grip on reality in a sad, alarming trend. The phenomenon has been known to derail excellent minds.

On Thursday night we had a very pleasant visit and dinner with Michaël Vaillant and Clarisse. I've introduced Michaël to Taulia and to LinkedIn. They didn't have a ready opening for him, however.

Now the news just broke that SpaceShip Two had crashed in the Mojave Desert. Access to space (or even, in this case, to the top of the atmosphere) remains a very dangerous business. The failure of an Antares rocket earlier in the week had destroyed about 5,000 pounds of supplies for the space station, but in this new accident one pilot has been killed, the second one gravely injured.

Hummingbird. Saturday 8 November 2014.

This morning, the light over Stow Lake was so limpid and the water so subtly blue from reflections of the autumn sky that our boat seemed to glide over the surface of a dream. When we stopped for lunch, the oars tied to the reeds of the shore, fat seagulls came begging, their impeccably white necks puffed up like the fancy shirts of ceremonial waiters, their eyes eager, envious of our food.

It was good to take this break from the arrangements for the visit by Airbus executives. We'll meet with Jean Botti and I'll see the CEO, Tom Enders. We're told to expect a decision by year-end, but the ways of that company are a puzzle, given all the upsets.

In Austin, I'd repeated my presentation to CNES, asking for research into the basic questions, but it was tabled while we discussed the possible outcomes of a supposed alien invasion. In the astute analysis by Dr. Salisbury, it acts through a *display*. In Utah, it played games with the toughest combat veterans the project could find, and won. It mimicked the space program before a confused midwestern family. In Brazil it deployed a tapestry of lights and flying vehicles, mocking a taskforce of Air force intelligence. Everywhere, it confuses both witnesses and investigators. In Weinstein's database the objects behave differently against military planes than commercial ones. They love to play with our gadgets, buzzing rockets in the 1950s, and zipping up into space before helpless F-18s launched by the Nimitz in 2004. We do have pieces of metal, slag, aluminum, magnesium. But when we test their isotope ratios, as we just did with Michael Angelo at Stanford, the results come back (so far) as terrestrial elements.

The phenomenon can use any technology we throw before it and quickly 'own' it, as we found out with Douglas Trumbull at Marley Woods. This implies that scientific expeditions are doomed. It takes control of sensors and resets measurements. Bigelow is right to see this as a game, like trying to get dolphins to solve riddles, while they redefine the rules to engage in new riddles we're not smart enough to solve.

What is needed is a different environment, one that listens to parallel thoughts, tolerates contradiction; that's hard to do in a classified environment. Until that happens, I have to be content with faint signals.

Hummingbird. Friday 14 November 2014.

In response to a question about a documentary showing a supposed alien in a glass coffin, possibly a Progeria victim, Kit writes: "The CNN producer from Chicago … showed me hundreds of photos.

"The key person was photographed with the body at the Walter Reed Army Institute of Research Classified Forensic Museum…which I have visited often, as recently as a couple of years ago: over 1000 photos. I also brokered the team doing the investigation to NIH, and the National Progeria-Genomic Lab to identify the body (12 orthopedic physicians who look at bodies to establish weird body dysmorphia) and 8 fellows of the American Academy of Forensic Sciences who are Forensic Anthropologists. These initiatives terminated their investigation... and seem to have killed the documentary."

Hummingbird. Saturday 15 November 2014.

Diana Pasulka has sent me her book *Heaven Can Wait* with a note that contains important insights: "I made the decision to keep references to UFOs out of the book because I wasn't sure yet what I thought of the connections, and I wanted to keep the book strictly about Catholic history. Now of course I believe Catholic history IS the history of UFOs...only veiled and culturally determined. I have either lost my mind, or found it, but I'm in good company with you and Jeff."

Hummingbird. Thursday 20 November 2014.

Regarding the Rendlesham-Bentwaters case, some statements released by the British Ministry of Defense are relevant, notably this: "It may be postulated that several observers were probably exposed to UAP radiation for longer than normal UAP sighting periods. There may be other cases which remain unreported. The recipients of these effects are not aware that their behavior/perception of what they are observing is being modified."

The MoD adds: "The E-field strengths which are known to affect the brain are of the order of 50 millivolts per centimeter at between 1 and 100Mhz and experiments have shown that effects can be produced at levels of 10^{-7} to 10^{-8} volts per centimeter."

This document, which retraces the research Poher did at CNES in the 1970s, and older tests by Richard Niemtzow, goes on to note that "the reported effect of (presumed) UAP radiation on humans is that it is quick-acting and *remembered*—although, curiously, there is little or no recall of events as a continuum. In short, the witness often reports an apparent 'gap' or 'lost time'—often several hours. It is described as though the exposure causes a temporary memory erasure."

The authors clearly had access to considerable research and laboratory data: "Experiments on animals have shown that low (3mW cm^{-1} at

450MHz) exposures affect brain calcium ions and that these are known to play an important role in the transmission of nerve impulses. Various modulations (e.g. 5Hz, 16Hz) were imposed on the radiation. However, whereas earlier experiments went on for tens of minutes—or even for hours, as such they were not (and not intended to be) representative of the very short UAP exposure times of a few seconds. In 1981 it was discovered that E fields (at ELF) of 10-50 or V.com^-2 at a modulation of 5Hz for only 5 to 10 seconds could increase the excitability of nerves for hours."

The British conclude that if some UAPs produce EM radiation (as evidenced by effects on car ignition) they can affect the human brain when close to an object, "either outdoors or indoors." The report adds: "This causes the brain to interact in an unusual way with the imagination 'library', causing reports of visual activity *which are not in fact a true representation of the facts.*"

Then they add: "It is of particular interest that a perusal of a limited number of old UAP reports (pre-20th century), while often similar or even identical to modern reports, does not produce evidence of 'spacecraft' aliens (gray or green), portholes or searchlight beams!"

That last conclusion is wrong, as Chris Aubeck and I know well from our recompilations of the *Wonders* database. Besides, the phenomenon may be selectively projecting images adapted to a particular era. In the 12th century a spacecraft would have had no meaning other than in a religious context (remember Ezekiel!), and an alien, grey or green, would have been readily mis-classified as a demon.

Hummingbird. Saturday 22 November 2014.

Our visitors from Airbus have gone home; I now have time to work on new research. Lack of a systematic approach to field research is a mistake. One cannot tackle large, messy situations like Bentwaters or the Arizona story without contradictory interviews, formal confrontation of witnesses, and access to site history. This is my father speaking in my ear, as a *Juge d'instruction*: "No substitute for the careful judicial process!" he would say. Confront all players, carefully cross-checking the statements by the cops. This was never done. What we call the Interference Syndrome, while real, may be a side-show of larger phenomena where UFOs are only one parameter, and this may lead to false conclusions.

Adding artificial intelligence indiscriminately will only deepen the errors. Instead, in the last few days, I've restarted Capella-2, building a new distributed superbase as a fresh prototype, from scratch. Now I can forget the demise of BAASS and put the study behind me.

Hummingbird. Sunday 23 November 2014.

We've worked hard all weekend, the City engulfed in wetness that yielded to a pale blue sky brushed with high clouds, the air gentle and fresh. Thanks to Peter Beren, I've completed *Wonders*. This evening Flamine and I walked over to St. Marks to hear Greek tunes from the 14th century.

I'll be happy to fly away, business be damned. America recovers from the subprime crisis, the Mideast crisis, the Tech Bubble crisis, but it drags along two demons: racism and greed.

Hummingbird. Sunday 30 November 2014.

Last night, Jack Katz and I walked again to the old Berkeley restaurant called Spengers, an enormous cavern filled with nautical artifacts and decorated with frighteningly huge fish hanging from the ceiling.

We spoke of his new *magnum opus*. He showed me the first 37 plates. His mastery of the pen is unequalled, and the scenes continue to gain in intensity. Jack's mind is curious, precise, and thirsty as ever.

Hummingbird. Thursday 11 December 2014.

Airbus had some bad news today: a severe 10% drop in their share price as Qatar reneged on a promise to take delivery of the first A350, while trouble continued to haunt the production of the A380, which may replace all the engines. *This may kill our project.*

In a summary of research at the University of Washington in Seattle, I read that a team under Adrian Raftery challenges common wisdom about world population supposedly leveling off at 9 billion after 2050. Their model predicts that population will rise from the current 7.2 billion to between 9.6 and 12.3 billion by 2100, primarily because African birth rates will remain high. This is in line with the models I once built but didn't publish: a very dark path for the world.

Now the storm intensifies, covering the roads in torrents, flooding houses and shops, and causing a five-hour power failure. I took the bus to

Fillmore and found a small shop open for a sandwich of stale bread and my first cup of coffee. The sky was dark all day, the mood a mix of an odd excitement and suspended time. Flamine left me a message in a raspy voice. She was in Paris, waiting for me.

Mabillon. Wednesday 17 December 2014.

France was freezing when I landed but it has warmed up pleasantly under a mild rain. Yesterday was sunny and bright, the news filled with the usual squabbles. The economic pages speak of the vertiginous drop of the ruble. Flamine and I keep watching the rumbling of the world. Over lunch with Dominique Weinstein yesterday I learned that he suspects many pilot sightings of the recent period (post 1980) to be the product of tests. The unknown factors in the Roswell tale should teach us to remain prudent. We spoke of my progress with *Wonders*, and he generously offered to turn over to me his files of pre-1947 cases.

Bellecour. Friday 19 December 2014.

Last night, in a wonderful medieval setting, I had the great pleasure of dining with my son and Maxim. We spoke of the luminous trips we'd taken together. What emerges is an ever-increasing admiration for what he's done for Max, and the way he turned around the business.

On my desk, a presentation by Tim Taylor and two long letters from Diana Pasulka about her research that she can "only send through hard mail": "I could not talk about this in my presentation at Esalen. I used the term 'weaponization of religion' very consciously in my email to you, amidst a lot of other things, to find out if you would pick up on that. You did."

It feels oddly appropriate to read this in room no.5 of a small hotel as this country town wakes up, the lingering darkness only broken by the silver lights on Christmas trees across the street.

She adds: "Based on my work in Catholic studies, I believe, just as you have pointed out in almost all your books, and so eloquently, that this phenomenon has been at the basis of all religions and folklore. It is real. What is different now is that governments can manipulate it better. There's no ontology, as you say, but there are effects."

Tim Taylor's presentation is called "The Search for the Truth." The author of *Launch Fever: An entrepreneur's journey into the secrets of launching rockets, a new business and living a happier life*, he worked on the Shuttle from 1979 to 1991, then founded Eudris in Boston for the spine-ortho market and sold it to Zimmer in 2007. He has 44 electrochemical and surgical device patents and one (1994) on ceramic-metal coatings. He's also a board member of Tutarus (encryption) with Dan Wolfe, the NSA director of the Information Assurance Directorate, and Alan Wade, former chief information officer for the CIA, a serious man I've met in the startup world.

Mabillon. Friday 26 December 2014.

Holidays, stretching ahead. Flamine has filled the living room with brightly wrapped gifts her clever mind has assembled. It's also time to post New Years cards, since we detest automated wishes on the web.

We visit friends in reconnection with a past that had seemed frozen. Yesterday we went to see Simonne Servais. At 92, age and illness made her hardly able to get up to open the door, but her mind was clear. We spoke of Valensole, of course; of Maurice Masse and his 'mission,' and how the two beings spoke to him telepathically. Bitterly, she complained about the corrupt systems of new French politics, the leaders reducing so many people to poverty.

Among my presents was a gift from Flamine entitled *Les Lieux de Pouvoir entre Mythe et Histoire*, by Louis de Maistre, all about hermeticism. I share the author's judgment about Aleister Crowley: "very intelligent and gifted with a fine sense of irony." On the relevance of Gurdjieff, I differ. I see him as a smart, egocentric guru with an easy ability to control people.

Mabillon. Tuesday 30 December 2014.

A "soirée" Sunday brought us Mario Varvoglis, Claudine Brelet, and two journalists from *Le Monde*. Claudine described her experiences in Nigeria and Mali where local tribes (especially the Dogon) have long-standing traditions about gods from the sky. Tonight, an evening concert of Russian orthodox chants at Saint-Louis-en-l'Isle. On the way back, a cup of hot chocolate in a cozy, antique-style *bistrot* before the walk home in the cold.

I spent most of the day computing least-square figures for the 147 ancient cases in *Wonders* that provide sufficient date and location data. The weather puts a sunny, freezing winter touch on the whole city. Peter Banks came over for lunch; I briefed him about Runway Capital.

Every morning, Flamine selects a recording from her large collection of ancient and sacred music, rarely venturing later than the magical year 1700. Reality slowly filters back into our minds from the far reaches of time, replacing the dreams and occasional anxieties of the night. We plan a few more visits with friends we haven't seen for a while. Then it will be time to rejoin the Game in California.

Part Twenty-Three

FIELD WORK

The only new thing in the world is the history you don't know.

— President Harry Truman

11

Hummingbird. Wednesday 7 January 2015.

Another year, another bloodbath. This morning, Islamic terrorists broke into the Paris offices of the satirical magazine *Charlie Hebdo*, in reprisal for cartoons mocking their fanatical version of Islam. Well informed, they surprised the working staff in an editorial meeting and bravely fired at the defenseless employees, military-style, with bursts from Kalashnikovs. Back on the street, they executed two policemen, yelled that Prophet Muhammed had been avenged, and fled.

As I write this, twelve people are dead, including two of the best-loved cartoon artists in France, Cabu and Wolinski, whose work I've admired since my teens. The world reacts with revulsion, aware that another ugly page just turned in the long history of tyranny. This is a war on human conscience, a despicable product of the 21st century that began with the destruction of the Twin Towers—and gets worse.

Hummingbird. Saturday 10 January 2015.

Flamine, still in France, tells me she'll attend the massive demonstration in Paris tomorrow. Several heads of State will march with François Hollande. Emotions are surging in the world, so perhaps the bloodbath will have some long-term positive effects.

Hummingbird. Sunday 11 January 2015.

A small demonstration for 'freedom' in support of *Charlie Hebdo* took place today on the steps of the San Francisco City Hall, so I joined a few hundred people from the French community and their American friends in the afternoon sunshine. I went there in honor of Wolinski and Cabu, not in support of the magazine, which I never admired. Ironically, it was about to go bankrupt.

Dialogue among the LoneStars goes on through the web, but why do I have the impression that it's empty? I no longer find the heady sense of progress that the team exuded. Perhaps the change has come because several members believe they are now engaged in deeper classified work, or soon will. In contrast, editing *Wonders* enchants me. Then, I walk through San Francisco when the fog is burning up, admiring the scene as it lifts its veils from the great bridges.

Hummingbird. Saturday 17 January 2015.

I've reorganized my technical files in anticipation of another year of management. I cleaned up the apartment and then I turned off the lights and listened to an old recording of Victoria's *Officium Defunctorum*. Watching the City through tears, I tried to come to an understanding of this milestone, five years, while the memory of my parents, who also died in mid-January, continues to fade, gently.

Hummingbird. Sunday 25 January 2015.

Flamine is back. We went to a concert last night, Voices of Music, and had the pleasure of meeting friends Béatrice and Paul Gomory.

The LoneStars' messages bring echoes of their theories, although we no longer use the web for anything confidential. Ideas that were mere hypotheses long ago, such as the tale about Aliens at Holloman Air Force Base, have now been accepted as fact, with zero data revealed in support. Yet a new generation of investigators is arising in Europe, serious people like Michel Turco and Chris Aubeck.

This morning, Flamine and I took a trip in time and sunshine, a visit to the old Villa Peralta in San Leandro. It was home to a family of Spanish settlers who came from Mexico with their cattle, their pride, and their noble projects. Local volunteers have turned it into a small museum, preserving a bit of history, pre-Gold Rush. A wonder.

Hummingbird. Monday 26 January 2015.

More arguments around the supposed Holloman footage. In *Forbidden Science,* I expressed doubts about that film, recalling Hynek's concerns at Norton Air Force Base. My friends freeze in sectarian temptation: The film must exist, they "almost" saw it, a friend "was told" it existed (showing an Alien out of a saucer being led away, coming back less than an hour later, and taking off). Eric "knows a DCI who was told it was real but was denied access." We go down every rabbit hole, like the zealots we mocked.

Hummingbird. Wednesday 28 January 2015.

Financial gyrations. The extreme Left in power in Greece, allied with the fascist Right, threatening the Euro zone. Banks are shaken all the way to Paris, while the Swiss are forced to revalue their money. As for France, it seems powerless to reform itself and create real jobs. The financial projections are fairy tales.

Alain Dupas and my contact at Airbus are scheduled to meet separately with Jean-Louis Gergorin, Tom Enders' consultant on innovation. I am both hopeful and leery of this move.

Hummingbird. Friday 30 January 2015.

Breakfast at Bucks with Garry Nolan, just back from Washington where he joined our friends and two patients: one from Naval Intelligence, who knew about the Tic Tac case, and a soldier fresh from Iraq who's had spooky experiences since her return. "I could hardly say a word," he recalled. "They were talking all the time. The patients didn't say much; I just drew their blood for the study."

One physicist and an unnamed physician friend have gone off on Sandia Mountain at the insistence of 'The Ambassador' who predicted sightings and landings. Nothing happened. I said it all reminded me of the converts of the Solar Temple, initiated in contacts with fake Aliens, or my private *Cult of Trolls* (manuscript), an assessment of true believers. I've mentioned my concerns about Jim Semivan's experiences. Jim saw another hooded figure in his bedroom last week.

Hummingbird. Monday 2 February 2015.

I've written to Colm Kelleher that I didn't plan to attend the LoneStars meetings anymore. The group only valued having me around because I could provide an international view, but they don't share my concerns about research misdirection, or the structure of the databases. We simply cater to a few members' personal interest or fancy nowadays: still useful, pleasant enough, but disconnected.

Hummingbird. Sunday 15 February 2015.

A storm has passed, with much-needed rain all over northern California. Early spring has put flowers on every tree, new leaves on every bush, and sunshine on the hills.

My message to Garry about my leaving the LoneStars: "That saddens me because of the long friendship we've enjoyed and the respect I have, but I miss the opportunity for long-term research.

"Also, quite bluntly, I am leery of hearing sermons based on bits of mysticism. It has infected the way we assess important cases like Hopkins' 'Case of the Century,' which I believe never happened."

In truth, I've seen this slow irreversible drift before. One member is convinced the Aliens come here in search of bodily fluids to restore their ability to make babies; another swears by Linda Cortile's stories; some are sure we'll soon build a propulsion system to go to the stars…After a while, such visions obscure the real issues, the hard work.

As someone remarked to me once, "The true nature of the phenomenon only reveals itself once you leave ufology behind." I'd thought the LoneStars would be more than ufologists, and for a while they were, to their credit. There is, of course, some glorious research to be done, but the conditions are not yet right. Most importantly, there is no stable management in place.

A woman named Morgan B. called me two days ago. She had picked up my letter to our friend Diane Hegarty. She is the neighbor who found her unconscious. The firemen had to climb to the balcony to gain access to her condo on December 26. She was close to death, dehydrated, pinned under a heavy bookcase. Arrangements are being made to release her into a friend's care.

Hummingbird. Friday 20 February 2015.

The Saint Francis Nursing home is a sprawling, motel-like facility stretching on the hill near the expressway, a few blocks from the wide cemeteries in Colma. It's a noisy, busy place, but the staff is friendly and the sun shines over everything. I found Diane curled up on her bed, awake and aware; she recognized me right away.

Visiting time wasn't long, but I felt reassured her mind was busy and agile. She spoke a lot, very softly; she enjoyed my photographs.

Now, Denis Gardin calls from Airbus in Paris, saying his strategy memo about our team had been "positively reviewed," that our Runway fund was considered; their CFO Harrad Wilhelm will come see us in late March. Denis also told me there was a new book by top executive Jean-Louis Gergorin, *Rapacités*, which describes an aircraft industry filled with corruption and rampant violence amidst a financial world increasingly dominated by crooks of all stripes (1).

Hummingbird. Sunday 22 February 2015.

Last night, Flamine and I gathered a few of our friends for dinner at the Fairmont. We've been married for three years of gentle care and stimulating travels, of conversations and long walks in the City as the evening crawls up along the hills and the fog drifts in to blur the sounds of the night. Zoila and Marla, Flamine's long-term artist friends, were there as well as John and Michèle Forge, Rod Swigart, and Paul Gomory with Béatrice, just back from France with passionate stories about the magnificence of Chartres. We missed Diane; she had been at our wedding.

Hummingbird. Friday 27 February 2015.

Today brought the kind of break where many things become clear, where simplification takes place. I found the right words to tell my LoneStars friends that I'd re-assessed priorities, eager to go on working on parts of the problem I could study by myself, without support from Washington or others; I have a framework for new data. George Hathaway jumped in right away, probing my motivations. They are simple: the phenomenon used to consist of objects in the sky, lights buzzing around in the woods, and small humanoids that evoked images of space visitors. This is still true, but the newer cases involve bedroom apparitions, monsters in the backyard, poltergeist effects, and impossible influences, none of which fits the hypotheses of 50 years ago. Even the LoneStars are working with obsolete tools.

Add the delusional theories evidenced at our last few meetings and I no longer find a place in the conversations, even if the medical efforts (now with Garry's help) do present a potential breakthrough. The cases are real, but the Intelligence guys' focus on 'threats,' which only feeds their friends in secret places, can't lead to good research.

Hummingbird. Sunday 1 March 2015.

Last night, Flamine and I had dinner with Jim and Deb Semivan and a very urbane couple who live between London and Pacific Heights (he's a former colleague of Mitch Romney at Bain Capital). They suggested that I collaborate with a new database plan, using Mr. Frascella's funds and Jim's management. Since both Flamine and I are busy with new projects, I had to decline politely, my focus on a few things "that just won't ever get done if I don't do them myself."

Jim asked if I thought Bigelow might ever release the databases I designed for AWSWAP (unlikely, I said) and if involvement with the Frascella project would be construed as subtle attempt to control the data (I said yes, of course). For the first time, I have a sense of foreboding, given the growth of disinformation. It may have begun in the days when Sidney Gottlieb, who headed the Central Intelligence Agency's 1950s and 1960s assassination attempts and mind-control program known as Project MK-Ultra, started collecting UFO books, as Hal told me.

Hummingbird. Friday 6 March 2015.

This afternoon I went back to see Diane at the nursing home. She was sitting idle in her wheelchair but recognized me warmly. The nurses were getting her to walk, she said. She recalled falling backward, then nothing more. She was nervous, her hands always in motion, brushing hair from her face, and she had trouble speaking, her mouth very dry, but she knew where she was, and had good contact with the staff.

I left with an intense awareness of our recent losses. The leader of the gravity company also got hurt but should be home this weekend. Against the backdrop of human frailty, including mine, the mysteries of the paranormal look ominous. It is a new feeling for me, one of danger and confusion, so I made a list of problematic areas.

First, the public is exposed to negative articles that paint witnesses as drunks and morons. Nothing new here: "Make your readers feel superior." We've seen such accusations before, but they now appear in *The Economist* and the *Wall Street Journal:* a professional job to discourage testimony.

Second, there's a US blackout about our meeting in France, where *a major space agency gathered 100 experts from seven countries for two days to discuss the reality of UFOs*. Shouldn't that be news?

Third, I fail to get through to the LoneStars about the biased use of bad hypnosis in abductions. They're seduced by the twisted Alien interpretation of Hopkins. As Diana Pasulka remarks, we're dealing with the weaponization of religion here, not with a real study.

Fourth, I'm bothered by the connection with MK-Ultra and Sidney Gottlieb. He was the man to whom Hal referred when he told me, back in 1971, that he'd met a senior Intelligence man who had my books in his briefcase. Gottlieb would not have bought *Anatomy of a Phenomenon* just to entertain himself. How much of the phenomenon, post-1970, points to a deception campaign borrowing or stealing UFO themes?

I stand by the statement that the phenomenon is genuinely real; there are other modes of reality, even more real since they disrupt and infect our own. Attempts to bring them into our scientific templates may be futile. Science has never developed efficient tools to document and classify so-called paranormal effects. Denial is so convenient, and cheaper!

Hummingbird. Saturday 7 March 2015.

Ever since I worked at the Paris Observatory, I have wondered about the bright satellite we tracked on a retrograde orbit. Now I've done more research, using new declassified records to identify it.

The first Keyhole (KH) spy satellite on a retrograde orbit was launched on 29 April 1965, when I was already in Chicago, but earlier programs might account for what we saw back in July 1961. The Pacific Missile Range at Vandenberg was already launching sun-synchronous retrograde payloads. The secret Discoverer program used to recapture film exposed on retrograde orbits aimed *west of south* from Vandenberg.

I suspect there were clandestine test flights before the KH series. The Russians may also have launched into novel orbits from Plesetsk Missile Test Range prior to US flights. Was that what we saw at Paris Observatory?

In the warmth of early spring, Palo Alto and Menlo Park display masses of flowers; the trees are burdened with flowers, and sidewalk terraces are shaded by the wide branches of magnolias. No shortage of bright technical ideas here: The Thiel Foundation invited me to meet an Indian physicist working on antimatter propulsion; he came to lunch with Flamine and me last week. And I was briefed by a team of computer scientists at Ames working on software for quantum systems on an early model of a D-Wave machine of 512 qubits. As for the ponderous wheels of the Airbus corporate machine, they are finally moving; a visit by their CFO should happen in three weeks.

Hummingbird. Sunday 8 March 2015.

Having expressed my new concerns to the LoneStars, some immediately stated they felt as I did. George Hathaway asked, "I presume your increasing unease has more to do with frequency of human-human activities (whether in collusion with the phenomenon or not) rather than phenomenon-human interactions? If there's an iota of validity to the 'Vienna Hypothesis' (2), the last thing the phenomenon wants is to reveal anything substantial about its interactions with humans."

What's changed is the recent evidence that the interference isn't coming from small, deluded groups or isolated bureaucrats, but is in fact massive and well-managed. John Schuessler comments in agreement: "A lot of folks in this field are afraid, and that's why they hedge their bet on everything they do or say."

Later the same day.

Leaving frozen Wisconsin (minus 20 and covered with ice and snow) Bill Calvert and his grandson Cole have come to California for a few days. We're able to discuss plans for a return trip to Brazil in late July. Bill is eager to track down Carmelita, the woman who took over the affairs of the former patients of Dona Cicera, the extraordinary healer. He's tracked down the daughter of their chauffeur, a man named Jahir.

While the four of us were speaking, an urgent message came from Garry Nolan: The 'visitations' to his contact are escalating, several in the last few weeks.

The incidents involve 'presences' in his room at night, and two possible poltergeists. "He's a stereo fanatic," writes Garry. "Came home one day and the settings were changed…"

Why would first-level extraterrestrials do that?

Hummingbird. Friday 13 March 2015.

Bill Calvert came back last night with his friend Beverly and her husband. We ate at a big round table by the window, while all kinds of lights were passing in the night sky over the San Francisco cityscape, eliciting a few jokes.

Bill had brought his photographs of surgical procedures done by Donna Cicera, many of which Beverly had attended in 1977. It's clear that the woman was psychically gifted and followed strict rules of ethics that brought an exceptional acknowledgment of sainthood by the Church. She wasn't interested in that sort of recognition, however. (Neither was the local bishop, who told Bill that "saints were nothing but trouble, with all the tourists!") She even fled her family, moving to faraway towns every time they tried to exploit her ability.

She had little education because she heard so many voices as a child that she couldn't study in a normal school program. In the West, psychiatrists would have drugged her extraordinary gifts out of her.

While the 'operations' were obviously real, as Bill's pictures show in bloody detail, they were sometimes symbolic rather than medically relevant. This doesn't explain anything, however. In one photograph he took, Beverly stands over the open belly of a patient, the gaping wound clearly seen, holding a bloody mass with a pair of forceps (Fig.15). While Donna Cicera was working in the patient's abdomen, Beverley, uncharacteristically, had just picked up the forceps in a kind of trance and reached down to probe the mass. Unable to explain why she did it, she heard Donna Cicera tell her that "she liked people to have their own proof" whenever a medically-trained person was around her—a case of 'professional courtesy,' beyond the will of the visitor.

She once told Carmelita, the assistant of Donna Cicera, "You're a very old soul, and you'll only come back here a couple more times. But I'm getting out this time, I cannot take much more."

She died at age 33.

She claimed total recall of her past lives and visited locations that she recognized. Whenever she did surgery, a 'Lady' was inspiring her, and she often saw two beings watching, floating rather than walking, dressed in shiny, elaborate clothes.

Pope Paul VI, who suffered from a chronic condition, wrote to Donna Cicera three times. She replied politely that 'The Lady' told her that healing the Pope wasn't part of her mission, so she shouldn't travel to Rome. The Church acknowledged that she had 'a charisma' and left her alone, as did the local pharmacies, which routinely supplied the drugs and the injection systems she prescribed.

The photographs show a remarkable, clearly inspired woman, surrounded by calm, intense teenagers with the beautiful, pure faces of innocent helpers. Blood everywhere, and patients undergoing these fantastic operations without sedation, in the arms of some incredible faith...Yes, I long to return to Brazil and learn more.

Hummingbird. Wednesday 25 March 2015.

Spanish researcher Jose Caravaca has engaged me in a discussion about the nature of the UFO phenomenon. I find myself sympathetic to him. He calls his conclusion *Distortion Theory*, proposing an interesting equation: "If we know many details of the witness (personality and knowledge, studies, hobbies), we can guess what kind of UFO experiences he will have."

I responded that indeed everything worked as if the phenomenon displayed itself with elements in the environment or human consciousness—at least in close encounters. This distorts the kind of analysis that can be done. But it doesn't explain everything.

He reminded me of what Jenny Randles has called The Oz Factor. Caravaca went on: "This absurd factor registered in hundreds of UFO incidents is the exclusive result of the oneiric, unconscious creativity on the part of the witness at the moment of 'gestation' of the encounter. What's observed during these anomalous experiences is the result of the interaction and communication of the psyche of the observer with an unknown external agent able to externalize and project a visible and tangible 'scenography.' These incoherent elements within the experiences with UFOs and their occupants don't have any special meaning beyond the value provided by the witness himself in a process similar to a dream: images and feelings are intertwined, and present iconographic forms."

I replied in agreement that we're dealing with a type of display, an advanced form of virtual reality drawing imagery from the collective (or personal) unconscious. The cinema, too, has real physical technology behind it, which the spectator never sees: projectors, lights, strips of film. We need to explain the material traces and the devices; we cannot solely rely on witnesses.

Jose Caravaca argues that the Distortion Theory stipulates that the external agent is able to manufacture *Materia Ephemeral* to give sightings corporeality and support the belief that the phenomena are outside the witness, and physical: "Even if we dissect the type of traces and tracks that Alien visitors often leave, they coincide with the evidence that anyone could assume in advance, such as ordinary burns, oil stain, or radioactivity. Following a UFO landing, one has never found traces unpredictable for the witness. Thus, the traces produced in the landings are also the result of distortion."

Here, he may be wrong. The case of Trans-en-Provence (3) was outside the range of predictability and foreign to known human technology.

Hummingbird. Friday 27 March 2015.

Harald Wilhelm is a dynamic man, close to 50, a product of German executive schools with an impeccable financial career at Airbus that propelled him to the CFO position. He came over to meet our team at the Quadrus Conference Center in Menlo Park, accompanied by his chief of staff, a friendly and equally dynamic Frenchman. This was a key presentation for us, and it covered the critical areas, but the conversation only led to more delays.

Hummingbird. Saturday 28 March 2015.

My son tells me that in November I will be a grandfather for the third time. His business, now in second year, is reaching breakeven despite the sluggish French economy.

Jack Katz has made giant progress with his work, now on page 50 of the enormous volume that will cap his career. I drove him to a big dinner at Spenger's Fresh Fish Grotto in Berkeley, that beautiful and chaotic place full of ship models, huge fish, and nautical memorabilia. He needed someone to listen to his extrapolations on science fiction, his theory of 'memory factors' that he sees as the scaffolding of all life and matter, and the reasons they're so weak and depressed in our miserable corner of the universe. His characters move away from the First Kingdom through inter-dimensional conduits into *Beyond the Beyond*, seeking illumination through their heroic migration.

My friendship with Jack is deepening. He treats me as a young disciple.

He gave me a copy of an unpublished graphic novel, *A Flower in Winter*, in which a young girl grows up in a circus. It's a romantic story; the portraits are powerful, oddly desperate.

After dinner, I pay for his groceries for the week, so we had to rush over to the store before closing. Of course, we didn't remember the street, frantically driving and lost, joking about parallel Berkeleys.

Hummingbird. Thursday 9 April 2015.

A massive cyberattack against TV5Monde today has caught the French government by surprise. It infiltrated the servers, starting with the social network links and eventually turning off all broadcasts, replaced by an Islamist video with threats against French soldiers.

The French government has trotted out three ministers to reassure the public; security experts were interviewed, issuing carefully worded technical statements. The real fact is that, following the execution of journalists at *Charlie Hebdo* in the middle of Paris, this is the second act of the Califate's war against France, and Paris hasn't grasped the implications. Parliament reacts with its own version of a Patriot Act that has no chance of catching the sophisticated criminals. Spying against citizens can only facilitate a future police state. We've witnessed that shameful disaster in the US.

Hummingbird. Friday 10 April 2015.

Airbus has narrowed its options for access to open innovation, so they don't need venture capital, or so they think. Their CFO made it clear there would be no investment in any independent fund. I feel oddly relieved to have that burden lifted.

Garry Nolan, who has just had papers accepted—no mean achievement—in three major journals (*Science, Nature,* and *Cell*) tells me he had dinner with Hal and Eric in Austin on Wednesday and is still trying to get a handle on the ongoing implications of all the withheld data: "Very difficult for me to work under those conditions. In my field I am accustomed to complete access. It is fruitless to do analysis when other data might dispute conclusions."

UCSF Hospital waiting room. Monday 13 April 2015.

I write this in the waiting room of the oncology department, trying to keep my mind distracted from yet another test, a biopsy. I slept badly last night, preoccupied with the venture business, and my recent mistakes when I imagined that a company like Airbus would appreciate what we could contribute in technology.

We're not alone in this situation. The news describes an anemic recovery of the US economy. The crash and Great Recession of 2008 was badly mismanaged by Washington, naively pretending that big banks would resist excesses that might damage the stability of the system. Self-interest and short-term greed soon proved otherwise. Less surprising is the fact that no criminal banker has gone to jail.

Now I review a confusing case, a variation on the usual 'target' design on a woman's patient's body. Dr. Terence G. Banich, a trauma & thoracic surgeon (emergency physician, reportedly tied to Nellis and the Las Vegas police), will perform a mastectomy to eliminate her risk for cancer. He promised to save the tissue, marking it for study at Stanford. There is more, however, another hair-raising chase after supposed Alien creatures.

The story is plain on the web: He describes meeting an Alien…a tall Nordic…the most handsome person he'd ever seen, present in the Emergency Department where he (Banich) was on duty.

Dr. Banich reportedly took the patient and her mother down the hall to his private office. Moving to his display case, he showed them a large anatomical bell jar holding another Alien, a Grey, in preservative liquid. Very long arms and fingers; very long legs; very large head with pitch-black wrap-around eyes.

Another 'planted' tale for our benefit? A Tall Nordic and a Grey? Who is testing whom? And to what end? Aliens, in Vegas?* Or was all that just a big joke?

---

*   A couple of months after a visit to the Utah Ranch in 2018, Dr. Banich died in his garage, where his son found him. The police gathered evidence, inspected the site, but kept the case quiet. Suicide was thought unlikely.

Hummingbird. Sunday 19 April 2015.

Four in the morning. Fully awake. Flamine contemplates her return to France. Here, she misses her patients, the sense of being useful, and the density of life in Paris. Studying for her American degree is intense but redundant, and her heart, understandably, is with her father in the last phase of his life.

Hummingbird. Friday 24 April 2015.

Garry sums up our impressions of the LoneStars, whose separate plans are a cause of dysfunction: "The money sets agendas and limits free thought. It's mixed with lack of expertise on their part around database systems. They're so convinced they know what they're talking about that they dismiss your efforts.

"Everyone is trying to take as much and give as little as they can. I suppose that's humankind's biggest weakness. It's not how I like to operate. Especially in a scientific endeavor. I always feel there is so much to gain from complete sharing of thoughts that whatever comes of things, there's discovery enough for everyone. Time and again, though, I have found that people don't all see it my way. Once they get a morsel of a result, they want to run with it and claim it as a prize.

"To play your psychiatrist for a moment... *You need to let it go*. You're looking for validation from a group of people who are set in their ways. They're not involved in a team effort. It's guys coming together to share common interests. Occasionally, they share data."

Garry's view is precious: "You're willing to let variables stay unbounded. You delight in pointing out how facts don't fit the conclusions, or how alternative hypotheses might apply. It's the rare scientist I've ever met, willing to live for long in limbo. You've hit the nail on the head with the 'hidden agenda' notion. It not only causes people to limit what they say in order to fish for ideas from others—but once an agenda is in place, it starts to define assumptions, which limits hypotheses. That's anathema to you, hence the tension.

"I am willing to live in an unbounded definition of the phenomena because *the data suggests it is many things at once*. Others, after decades investigating it, just want closure. So, I am taking a 'participate, wait and see' approach. I'm willing to support anything reasonable to the extent I feel I am contributing or learning something and it's not distracting to my work. I find it an honor (despite the frustrations) to be involved with such accomplished men, and especially around a subject like this, where I can talk openly."

Hummingbird. Monday 27 April 2015.

The television news shows riots in Baltimore, where a black teenager died a week ago in police custody, without explanation or apology. There's been a series of such crimes on the part of law enforcement in the last few months, exposing the sad reality of enduring racial hatred in the US. Demons are loose in spite of a black President and all the evident goodwill within populations of all races and colors.

The Dow-Jones index has climbed back to 18,000, the NASDAQ above 5,000, and the S&P above 2,000, but it's taken 15 years to erase the Internet bubble burst of 1999, followed by the sub-prime crisis of 2008.

Olivier tells me about the joys and agonies of running a small business, even one as well-located and loved as his establishment. He was giving a bottle of milk to Raphaël and enjoying a glass of red wine, contemplating life in socialist France, and running out of euros.

Hummingbird. Saturday 2 May 2015.

Flamine got up early this morning, and we spoke for a long time, suspended between two continents. Her unease with America is real.

California has hurt her, and she doesn't find rest. She loves me but there's nothing I can do if her soul is not here, with mine.

For me, a return to France would be suicide, intellectually and financially, as it was when I attempted it in 1968. As glorious as it superficially seems, the Paris elite is petty and spiteful. Many of my present and former colleagues are leaving, like Bruno Combe now in Brussels, others in Switzerland or London. The various governments of the last ten years have emptied the country of energies that could have renewed it through inspired vision.

Flamine is still willing to live here but only in short stretches, in anticipation of going back to an older world. I am sad because I know that her father's death is imminent and will tear up her life. What did I miss? I charged ahead, thinking the fragrances of California and the light of the

Pacific had enough force to sweep her into this life. I must acknowledge and respect her choice, and the torments of solitude.

Hummingbird. Sunday 3 May 2015.

We walked down in the cool evening to a Bach concert at St. Marks, conducted by Jeffrey Thomas. They added a Vivaldi piece (*Nisi Dominus*) that was marvelous, but Flamine was very sad all day because her father is now unable to leave his bed. He's written the last line of his novel and now seems detached, angering his tired wife.

We fear what we may face when we fly to Paris in a few days. Our plans include Madrid, a conference about research.

Pontoise. Tuesday 12 May 2015. Auberge du Cheval Blanc.

This is a 'charm' hotel of only three guest rooms in the northern part of town, the only remaining such establishment in Pontoise. All the modern hospitality chains have relocated to modern Cergy. I arrived this morning and found my old town very neat, even smart-looking under the spring sunshine.

At the Guillaume Apollinaire library, chose to the Château, the attendants told me I could freely look up any old records, so I spent a long afternoon going over the local papers dated between 1939 and the Liberation of France in 1944, the first five years of my life. I even found a short notice of my birth and a few references to my father as President of the Criminal Court. I located key dates in my parents' lives, from the first German bombs and their flight to Normandy, to the arrival of the Americans.

Pontoise. Wednesday 13 May 2015.

I've reached a plateau in my understanding of the phenomenon. I now think it represents a global, conscious system outside humanity, sharing the planet. It needs us for its function and works by imitating or anticipating us. The Midwest cases proved this to me but the same can be seen in the ancient cases, the airships of the 19th century, and the structures that emerge when I analyze the files of pilot reports with the new screens (short software self-executing pieces of code) I've built on the ruins of the study in Vegas.

The phenomenon is particularly fond of technology: impossible pacing of new rockets, fake 'attacks' against military jets, interference with missiles, even simulacra mocking human explorations. The result is a display so bizarre it's hard to take seriously. It functions like religious imagery, replete with stunning miracles and its array of fabrication and lies, supporting those lofty dogmas *that may not be criticized or tested*, as Jeff Kripal teaches.

We've come full circle, dear Mr. Manly Hall, dear Dr. Hynek, to an esoteric system with a consistent mechanism outside space and time.

Bellecour. Friday 15 May 2015.

This time I picked up a small car at Chartres and explored a new road. I drove in wonder among the forests and luscious fields of the Perche region. I found my son tense, worried about a car mishap (harmless, fortunately), and I was able to spend time with my grandsons before Philippe Favre joined me for lunch with Kathryn. I had been impressed by the 'new measures' of the Hollande-Valls government to help small companies and facilitate the return of successful expatriates. He told me they were only the usual poppycock.

Mabillon. Monday 18 May 2015.

A pleasant meeting, as always, with Dominique Weinstein who turned over to me his large file of pre-1947 sightings. At home, he updates his pilot sighting compilation, but his time is limited since he was brought back into action after the attack on *Charlie Hebdo*.

Speaking about past German ambitions, he told me of his astonishment when visiting the place simply known as La Coupole, near Saint-Omer, that housed the most advanced Nazi rockets. Hitler was much closer to dominating the world than we've been led to believe by historians; it was only a matter of months before he could have achieved both the atom bomb and the intercontinental missile.

Dominique also said the management of the CNES project was about to change again, with brilliant Xavier Passot moving on.

A new book about François de Grossouvre (*Le Dernier Mort de Mitterrand*) has me fascinated. The author doesn't spell out the fact that he was assassinated, but the weight of evidence supports it, as Thérèse de

Saint-Phalle once assured me (4).

Life in Paris seems easy, so we forget hard realities. In ten years, four giant French firms (Alcatel, Alstom, Arcelor, Areva) have lost their place among the top 500 companies, and two others have moved their headquarters elsewhere. Even economic minister Emmanuel Macron, in a stunning understatement, recognizes that "our major firms are leaving because our fiscal policy is ill-adapted." The CEO of Schneider has moved his office to Hong Kong while Total transferred its treasury management and financial communications to London. As for Lafarge Cements, they've already relocated to Switzerland.

Bellecour. Friday 22 May 2015.

My daughter joined us in Paris yesterday, with Rebecca and her sister. We all went out for a walk through the Luxembourg Gardens to admire the great Medici Fountain in the sweet air of May.

Flamine and I drove to L'Escoublère today to visit her 83-year-old aunt, who whipped up a tray of cookies and strong coffee, and took us on a tour of the beautiful fortress "which was never taken."

L'Escoublère rises in the fields about an hour north of Angers. The Mayenne River flows to the west. It's a place of strength and beauty, well-proportioned behind its moat. There are frogs in the water.

Bellecour. Sunday 24 May 2015.

Annick, my dear sister-in-law, is 80 today, and this is my daughter's birthday, so it was appropriate to celebrate our reunion in Olivier's splendid dining room. Later, we took a historical tour of the old town with a local expert who brought to life centuries of human experience, labor, and local talent. It's the wonderful gift of such small European towns, where the deepest lessons are still found in the fold of ancient walls or simple sculptures, to help us get a true measure of passing time and changing moods.

Madrid. Wednesday 27 May 2015. Hotel Paseo del Arte.

Chris Aubeck, whom I hadn't seen since Porto in 2003, welcomed me warmly as we toured the fine exhibits of the Casa Encendida, now dedicated to art and space exploration. Several contributors to the unique

Magoniax research group on the Internet are here, starting with Theo Paijmans from Amsterdam who spoke eloquently about the development of the imagination in the industrial age. Nigel Watson spoke after Theo of a lively exposé of unexplained objects in World War I that he entitled "Flights of Fancy."

Madrid. Thursday 28 May 2015.

Jesus Gallego believes UFOs feed on our beliefs, borrowing our own imagery in order to manifest. They've had a plan for centuries, he says, and we're puppets. We had a good roundtable discussion before a lively audience: lots of fine imagery, interviews in Spanish, a sense of renewal, and people with fresh ideas, away from American obsessions with worn-out concepts of super craft and rancid religiosities.

Madrid. Friday 29 May 2015.

This morning, I spent a pleasant couple of hours alone with Chris, going over our plans for *Wonders*. We walked through central Madrid in the warm air, halting for a cup of very good coffee, and catching up on topics we hadn't been able to discuss over the net. His girlfriend took pictures of us. He told me about his own sighting in 1995, an orange rectangle in the sky in Talarrubias (Badajoz province) seen with a group of friends, oddly foreshadowed by a passing view of an ordinary orange object on the side of the road—the usual sense of confusion the phenomenon seems to need to manifest.

At the roundtable discussion, Theo said that *Passport to Magonia* "liberated UFO research from the narrow nuts-and-bolts model that had prevented research from growing."

Mabillon. Saturday 30 May 2015.

This afternoon, sunshine lingered over the Latin Quarter in such a supremely quiet mood that I urged Flamine to take a stroll with me and took her to a simple dinner close to the Luxembourg gardens. I felt so happy with her that I regained confidence.

We walked back through Odéon Square where young crowds assembled at the sidewalk cafés. Everything was peaceful, a perfect Saturday evening. That must have been the time when her father took his last breath. We didn't know it as we walked home with a sense of great peace. I now think of that evening of Beltane as a special blessing, his parting gift to a beloved daughter.

Flamine was downstairs, finishing some work on the computer, and I was getting ready for bed when word came of his passing. We took a cab to the hospital, a dreadful place surrounded by half-finished roadwork and dusty palisades, hallways flooded with crude lights.

I felt I might bring some support to Flamine but she said I shouldn't change my travel plans; she'd face the next few weeks alone. Not to share her pain, and support her in her sorrow, saddened me. All I could do was cry with her and hold her as the cab drove back across Paris.

12

Hummingbird. Monday 1 June 2015.

The first day home has its routine. Assembling whatever seemed important or timely among the accumulated mail, I made a bundle of it for Flamine and took it to the Post Office.

There were two new girls at the counter. They fussed with my envelope and mailing sheet, pored over a book of regulations, and looked up directions for Paris under "United Kingdom," the only overseas place of which they seemed aware. I protested, the three of us standing there like silly characters in a TV slapstick comedy.

My daughter and her friends are still in Paris for a few days. Flamine plans to stay there till month end, preparing for the funeral, the legal meetings, and the heartbreaking paperwork.

Hummingbird. Thursday 4 June 2015.

An explosion of pinks and reds on the deck, sunshine everywhere. The plants Flamine had bought still thrive as they await her return. I hope it will be a hopeful renewal for us, even if tinged with lingering grief. She doesn't want me 'to force her to be happy.' How could I?

Later the same day.

An urgent phone call comes from a colleague, mandated by a potential new owner for the Utah Ranch, driven by a great sense of urgency. He asks for my informal appraisal. He's decided the phenomenon "isn't Alien, or UFOs," but then, why are we doing this? I only have three issues: (1) He wants Bob to remain involved in the ranch, but why should he? (2) Why should the new project inherit liability? *Are there potential issues with injuries to personnel?* I ask. A need to transfer records? I recommend prudence. (3) What about the government? "We'll tell them anything they want to know," he said, "but they may not be interested; they already know about cloaking."

Yet, Jim Semivan says that "people working on this stuff" (at CIA) are isolated with no time to develop serious interest. The answer left me hanging; once again, lots of unstated assumptions.

Hummingbird. Friday 5 June 2015.

Monsieur de Bonvoisin was laid to rest this afternoon in Paris, after a religious ceremony attended by his French and Belgian family. Flamine tells me that her sisters and their children were there and helped place the sorrow in perspective, as he would have wanted it.

I only knew Jean-Claude de Bonvoisin for the last three years of his life, as a man with a sharp mind: an aging French novelist reflecting on characters he remembered, an old warrior who had fought and suffered and drawn lessons from various sides of history. At this stage of my life, he inspired me to think more widely, more serenely. He'd been happy to see Flamine and me together.

Hummingbird. Sunday 21 June 2015.

I spoke to Richard Niemtzow today, intending to put Garry in touch with him. Richard has become an expert on 'battlefield acupuncture,' now recognized by the Pentagon and acknowledged in China.

Next, a fruitful conversation with Jeff Kripal about the archives. Since I mentioned that I'd prefer to turn over my records to a local institution like Esalen (where I could retain easier access), he reminded me of the Terence McKenna's archives, which Esalen deposited in a house in Carmel. The house burned down and nothing was saved.

There's an argument, he says rightly, for a secure University with a library built like a fortress, so we agreed I'd visit Rice to give a lecture to his class, meet the special collection staff, and survey their building.

Hummingbird. Thursday 25 June 2015.

Flamine flew back here yesterday, and I briefly pictured a renewed happy life. Today, however, when the San Francisco skyline emerged from the fog, I had to face her new reality: Her need to resume her career in France can't be satisfied with occasional trips, or links to clients over the web as we once planned, she told me. More likely, it will soon be the end. I am sad, disoriented. Afraid.

Hummingbird. Friday 26 June 2015.

The annual dinner of the Band of Angels was held in the Allied Arts Guild gardens, a magnificent property built a century ago in the style of a Mexican hacienda, in a secluded area of Menlo Park. Across the table from me was my old friend Harold Shattuck, an IBM supercomputer engineer who joined Ray Williams at Amdahl in the 1970s. We had the kind of conversation one could best have in California, arguing about quantum memories among tall Redwoods and a rose garden.

Flamine and I continued our own debate, carefully stepping around the brittle shell of broken dreams no supercomputer can fix.

Hummingbird. Sunday 28 June 2015.

We went boating on the lake yesterday. We had our customary picnic in the canoe among the flowers overflowing the shore. But Flamine was silent and I was sad.

The LoneStars, too, follow a dysfunctional path. Some of us now understand that the BAASS Brazil expedition never reported the full picture.

Last night we drove up to San Anselmo for a performance of Purcell's *Dido and Aeneas*, for which Flamine's friend Marla Volovna was stage director. Rob Swigart, who'd met Marla at our 3rd anniversary dinner, attended the stunning event filed with magicians, witches, and a monstrous dragon, amusing on a tiny stage in an open courtyard.

San Francisco airport. Thursday 2 July 2015.

Max's flight must be crossing the Nevada line right now. I wait for him with much love, eager to give him all the support I can, and a taste of Campus life as he works at a Stanford lab for a couple of months.

In the last few days, my exchanges with the project have taken a new tone. In Brazil, the BAASS team was composed of six investigators who travelled across the country, interviewed 17 officials, including General Uchoa, and came back with a report that led to a conference at the Brazilian embassy in DC, attended by Hal, Colm, and Eric, all under Air Force 'sponsorship.' A security officer determined that distribution of the report, after review by a military group, would be closed.

Since I'm the only one of the group who met with Colonel Hollanda Lima and was briefed afterwards by his top security staff inside the Air force base where the project was managed, I stand confused by all the obvious discrepancies. Many unreported details were shared with us, and we saw the original, secret data. I could have flagged errors and misquotes in later years. Now, I can't even read 'our' full report, even though my friends confess "it doesn't reflect reality." Are they blind to the full story?

The second subject dealt with medical effects. It builds on a strong paper written for EarthTech that surveys the domain of radiation impact on human subjects, particularly in the microwave range. It is excellent and also chilling, even though it only addresses the issue in a general manner. Details are in other studies, all of them classified.

Hummingbird. Saturday 4 July 2015.

Several evils crept upon me today, a kind of fear I haven't felt before. I confess lingering concern about the way the Brazil trip was handled, a missed opportunity. Apart from a partial listing of medical cases, what I've seen rehashes Bob Pratt's data, unaware of the details Bill Calvert and I had been shown in 1988. Bureaucracy at work? Or something more nefarious that will demand that we forget what we'd uncovered about the beings? For me anyway, cooperation is over. If we can't even do this, why bother with wider projects hoping to understand the whole phenomenon?

Even more importantly, there are stern warnings from a friend whose messages hint of hidden data and bizarre projects: "About three years ago, as part of a totally different investigation, I was asked to meet a deeply undercover federal officer. One thing led to another, and he informed me he'd been involved in an arrest in which there had been evidence of a post-MK-Ultra program to research *Dissociative Identity Disorder as an intentional outcome.*" (As in Pascagoula, I had to ask?)

My correspondent weakly discounted it, but he gave me a long document about over 100 relevant subjects written by physicians and psychologists and sociologists. Reportedly ICE, Border Patrol, and even the FBI had special instructions regarding future events. He discounted that again as 'scifi.' Then, more serious things happened, so he took the documents to the Department of Justice for validation.

After an intense and courteous hearing, they went away to think, *for a week. Then they said the documents and the charges…were true.* That was three years ago. It's a sickening story.

Later the same day.

Two coincidences have come to awaken my interest. Flamine reads Tobie Nathan's anthropology book *Nous ne sommes pas seuls au monde* where I find a good dissertation about the Djinns and the Djinneyas, including the conjuring one is supposed to do when pouring boiling water: *Destour y a as'hab et arb* ("Let it be according to your law, o my owners…") which confirms the story of my young Saudi friend who became possessed because he'd neglected to say it.

Hummingbird. Tuesday 14 July 2015.

Over a phone call from Italy, where he's vacationing, Garry tells me that the seventh group meeting in Toronto "went fine." George Hathaway demonstrated an experiment, and members spoke about evidence of the interference syndrome in the table of medical codes. No conclusion yet, except that the conditions affecting close encounter witnesses are not seen in other people with autoimmune diseases. In particular, the caudate-putamen connection remains significant. Our subjects exhibit complete connection on both sides.

At the same meeting, Garry tells me he gave a tutorial about epigenetics, a subject he relates to Colm's study of retrotransposons and remote viewing abilities, but the relevant genes haven't been identified.

Hummingbird. Friday 17 July 2015.

With polite notice in advance, a senior analyst with the Institute for Defense Analyses in Washington stopped over in San Francisco yesterday to see me, on the way back from Australia. We spoke at the Marriott *sans téléphone*. He wanted to go over the *Nimitz* (Tic Tac) case again for my benefit. He told me there'd been a similar incident with pilots over Virginia Capes. When they reported their sighting, their superiors confessed they'd never heard of anything like it.

He's discussed such items with Richard Niemtzow, who told him how amazed he'd been when agents went into the hospital where the Cash-Landrum witnesses were treated and carefully preserved the record on their medical charts, which can be compared to Rendlesham.

The main subject my visitor wanted to discuss, once we had disposed of cellphones, was that of coincidences. Recently, as he made a stop at a Sierra tourist place for a day of skiing, he sat at the bar between a man and a couple with an empty chair on either side.

A second man came in, sat between him and the couple, and spoke with them. The man happened to be a former investigator for BAASS, who'd traveled around the world to research UFOs. He said the important sites were: Colares, Mount Hood, and the Northern Tier, which makes little sense. My visitor had another coincidental meeting during a visit to a military base with a man who'd been a mechanic on a KC-135 flight where secret records had been confiscated.

Hummingbird. Saturday 18 July 2015.

In a restricted email circle, we've been discussing the work of an English scholar calling himself Isaac Koi, who's built an extensive database of declassified CIA papers: "Koi's website is the most complete searchable database of government documents on UFOs, remote viewing, and how they relate," I was told: "Over 90,000 pages on just declassified CIA documents, never seen."

Reportedly, the data includes DoD documents on Cameleo-type activities, and primary data on Dr. Jolyn West suggesting he was the CIA physician supporting Sid Gottlieb, instilling dissociation in human subjects. Similarly, our group had never heard of a highly classified Rendlesham remote viewing publicly released ten days ago… Hal was unaware of it, among people we knew well.

I still don't know who Isaac Koi is, but his methods seem impeccable. He's put the final nail in the lie that governments don't study UFOs, remote viewing, and mind control. This showed the extent to which the situation was being manipulated. Some 25% of Interference Syndrome patients die within 7 years of diagnosis, average age of 50.*

Hummingbird. Sunday 19 July 2015.

Through the parapsychology group, I learn of the death of my former SRI colleague and friend Dick Shoup. He'd fought lung cancer for three years. An expert in mathematical logic, he may be testing the Square Roots of Not in another dimension, but it's another big loss for us.

In this sad moment, music comes on the radio: Ketelbey's *In a Monastery Garden*. Jack Katz had told me of this fine composer. Men like Dick Shoup represent the very engine of Silicon Valley, the unknown princes of a calculus that defines our future.

Hummingbird. Wednesday 22 July 2015.

Before dawn tomorrow, I leave for Brazil again. Maxim is still in San Francisco, so last night we took him to Berkeley for dinner with Jack Katz, who was in fine form, warm, deeply human, my only true friend at the moment. Jack is lonely, speaking of the latest woman in his life as a "prisoner of conceit," afraid to think freely beyond conventions.

Yesterday the situation with Airbus resolved itself as I uncovered the fact that the new structure was a closed corporate fund after all. It makes direct investments but won't be a player in venture capital.

Miami. Thursday 23 July 2015.

Our flight landed here after circling a thunderstorm that briefly closed the airport. The view was of extensive swamps, marinas, and brand-new high-rises by the sea, which makes me wonder about the fate of the city

---

* This was true, but the numbers were very small: 40 cases, of which 10 showed the signs. Now 150 cases are known, 75% are relevant and 22% show signs of autoimmune disease.

when the ocean's level rises a few feet more, as it surely will.

For now, Miami is growing and vibrant. It attracts not only Latinos (two thirds of the population speak Spanish at home) but Italians and French as well. "These European transplants praise the beauty of their homeland, but they're quick to bemoan its tangled regulations and inefficiency, dismissing it as an impossible place to do business," writes a local journalist with good judgment.

An article in *New Scientist* reminds us that the Earth circles the Sun at 30 km/second and that the Sun travels around the galaxy at 200 km/sec. Feeling lost under such conditions is a reasonable reaction. Feeling lost in love, however, is far worse.

Later the same day.

Bill Calvert has joined me from Chicago. Over dinner, our conversation went back to reminiscences of Andrija Puharich, whose interest in Brazilian psychics and close relationship to the CIA continue to inform the research. It was General Uchoa who advised him to study Arigo when Andrija came to Brazil to observe mediums 'under the sponsorship of NASA,' as if NASA needed mediums. He'd brought obtrusive scientific gadgets with him, another typical North American mistake.

Fortaleza. Friday 24 July 2015.

On the plane, Bill and I reminisced about our previous trip when Janine and I joined him and his wife Regina in Fortaleza in July 1988. Now it's just the two of us. At the time, we did an analysis of the massive UFO wave that had arisen from the south, moved up to the coast, and west to Colares. *Operacion Prato* was only a point of culmination. We recalled the efficiency of Agobar Peixoto whose maternal uncle, known by the nickname Peixotinho, had been a physical medium.

Fortaleza is "a city of 4.3 million people that nobody has ever heard of." We're staying at a clean but Spartan hotel on rua Desembargador Praxedes, after the seven-hour flight. We're three degrees from the Equator, but the weather is far more pleasant than Florida. It feels like California, with a tepid wind and flowers everywhere, but of course this is winter. Agobar was waiting for us at the airport, video camera riveted to his eye. We made a stop at his house in a funky area of the city where people sit outside for sidewalk dinners, young and old just happy with the cool night.

I must find peace again, heal myself as the pain of losing so many treasured companions (Arthur Hastings, Dick Shoup) lingers on.

Fortaleza. Saturday 25 July 2015.

Breakfast is one of the happy moments of the day in this bright morning sun. They serve tapioca and papaya and other flavors and tastes unusual in my 'other' world. Bill has reminded me of some aspects of the Midwest case that I had forgotten, the witness' near-death as a child, and the way he was revived amidst family prayers.

Brazilian spiritism is evolving here, Bill said, with Kardeckian researchers taking an attitude of extreme academic caution for fear of official doctors, legal obstacles, and skeptics. In the process, they renege on some of their psychic principles and exceptional facts, as US parapsychologists do in their misplaced attempts to look respectable.

Bill says, appropriately, that we should go back to the writings of men like Humbolt, Sir Richard Burton, and Alfred Wallace, the latter eclipsed by Darwin because he dared consider psychic phenomena.

Later in the afternoon.

We sat at the wide table in the shade on Agobar's patio, dominated by a giant mango tree. In that equatorial setting we started opening piles of documents and passing around photographs, maps and artefacts.

We spoke about history, and about Colares, picking up where we left off 28 years ago. He recalled that the founder of Brazilian ufology was João Martins, who worked at *O Cruzeiro* in the early 1950s. A friend of Agobar in Rio named Fernando Cleto was another leader along with Dr. Olavo Fontes. Those pioneers initiated the research.

I reminded him that I'd met Dr. Olavo Fontes in Chicago in 1965 and that I had received his records of early observations, as his death was imminent. Fontes, a medical doctor, was the man who researched the famous abduction case of Antonio Vilas-Boas.

"Do you know that a research institute from Las Vegas has sent a sizeable team from the United States to obtain the data from Colares," I asked: "Did they ever speak with you?"

"Sure, I met three Americans at a conference in Sobral," Agobar replied. "They were taking lots of notes, trying to make contact with the group from Rio, but looking in all the wrong places..."

There are new cases in the area, including an attack by *chupas* just two months ago when a woman, Martir Melo, was lifted and shaken. She hurt her foot when she fell back and had to walk for 20 minutes to get home. She had mixed up speech, couldn't be understood. She also had a puncture in the left arm, had lost blood, and suffered an inflammation that lasted two days.

A new scenario: A light, a beam, the witness lifted, loss of consciousness, and the body dropped like an old rag doll, not at all the same type of incident as in Colares. The conversation came to Bob Pratt, whose cover was the *National Enquirer*. Agobar mentioned he'd known him for 36 years: "He took lots of information from me, but didn't give anything back, the earmark of an *honorable correspondent*."

He also knew Gary Richmond, whom Bill had met as an employee of a wood company who had paranoia after he began taking drugs.

When Agobar was young, his passion was enhanced by reading *Passport to Magonia*. In 2006 he witnessed the landing of a disk in Quixadà among a group of researchers.

Later the same day.

In the evening, as the air cooled, we felt refreshed after some sleep, as we met Jorge Nobriga, a 68-year-old geologist, one of the many local friends of Agobar. As a young child, he saw an oval object, about 9 meters in diameter. It flew over the yard where he was playing. It was red orange "like frozen fire," shaped like a Zeppelin, and made a low humming noise. He believes he was abducted; he remembers his father angry at him for disappearing from the garden. In meditation he recalled more details: the object was partially transparent, it "sucked him inside," then he saw the Moon as a huge object, close to him.

A more specific experience is that of Manoël di Queiros Meneza, who suffered an extraordinary abduction at age 47 (in Mulungu, near Pacoti, on 22 October 1999) and has the sequelae to prove it.

He was at a birthday party for his friends in the countryside. In the middle of the night, he had gone out to relieve himself when a beam of light seemed to explode around him and lifted him. He was found in the morning, missing four hours, muddy and dirty, with sharp incisions (stitched) all over his chest. Agobar was called. He shot a dramatic video and took him to a doctor. We spent time reviewing the case (and the video), then Agobar drove us back. Bill and I went through the new data again once we were at the hotel.

Brazil seems relaxed, urban, and modern when you first arrive. The deep cultural differences only show up later, taking you by surprise. Bill knows the country well, but I'm just beginning to come to grips with its unique culture, and its conflicts.

Fortaleza. Sunday 26 July 2015.

The Manoël case was the topic this morning. Agobar had recorded the DVD at his house the day after the incident, documenting the multiple cuts and other wounds he suffered. He also interviewed the man who had found and helped him. (Fig. 16, page 322)

Following that review, as an experiment, Agobar arranged for me to meet with a medium who channels extraterrestrials. We went to her house where her sons, tall black men with the powerful muscles of athletes, seemed unconcerned about living among such phenomena. I asked her about the differences she felt between contacting ETs and speaking with spirits.

"The spirits are the ones who open the connection," she replied, "I just make myself available. The actual contact is more intense, refined: ETs are superior beings, with the ability to heal." Then she went into brief trances where her voice changed, speaking of beings who came here for reconnaissance, creating bodies like ours with their own mentation. Mindful of the cases we'd studied, I asked why some witnesses had died.

"They're not interested in hurting people," she insisted. "Humans can meet them in the flesh; it depends on each person."

That last remark reminded me of General Uchoa's theories, and his reliance on séances with coordinated energy. But we left without new insight. Her guides have names like Azazi, Codaki, Toati...

Tomorrow we'll meet Manoël in person. Agobar says we can't go to Missi until Thursday, so the schedule will get crowded. Next, Bill and I do want to travel to the place where Donna Cicera had her clinic, which will take a couple of days in this immense country, and we'll still have plenty to do here with Agobar.

Much of Fortaleza is a giant maze of poorly paved roads lined by low buildings. The city does have a good bus network with efficient, clean Mercedes vehicles that would put San Francisco to shame. Once beyond the dirty walls topped with barbed-wire you find houses with high ceilings, well-painted walls, decent shelters, and the occasional elegant villa, like Agobar's home and its fresh garden where the only sound is the occasional burst of a mango dropping from a high branch.

Fortaleza. Monday 27 July 2015.

We just had a long talk with Manoël di Querois Meneza, at his home, about the things he saw near Pacoti 18 years ago. I asked him to describe only what he recalled consciously, in his own words.

He said it was 4:30 in the morning. He was going outside to relieve himself when he saw a light overhead. He only recalls raising a hand, falling back. It was a rainy day. People found him late in the morning. He had a bleeding cut over his right eye. He was surprised to be holding a coin in his left hand and his necklace in the right. When they took off his shirt, they found the cuts. He was driven home but suffered amnesia again during the trip.

We spent another hour going over details. He confessed that he occasionally was going back to that place now, trying to initiate contact again. He's a Catholic, 69 years old, and doesn't feel that his personality has changed. Nothing happens inside the house, but he has dreams of 'them' *taking him through the walls*.

Fortaleza. Tuesday 28 July 2015.

I only had short periods of sleep, bad sleep in the heat of the night. Agobar is with his family today, so Bill and I went looking for maps of the State of Ceará. Brazil has superb bookstores, with new editions on topics from high tech to spiritualism; Alan Kardec is the best-seller.

I got back to take a conference call with my partners from Frankfort and found that both Christofer and Dmitry wanted to continue to try raising a new fund with Russian money. I argued it was a pipe dream. Then Bill and I enjoyed a quiet lunch of fish and rice while looking at maps, trying to figure out whether we should fly, drive, or take a bus to Aracaju, the town where Donna Cicera once had her medical clinic.

When I can't sleep, I read *Le monde enchanté de la Renaissance*, the life of Jerome Cardan l'halluciné by Lucas-Dubreton (5). Cardan says he never had a personal 'daimon': "It is silly to rush towards the knowledge of such things, and even greater folly to attempt the interpretation of what is closed to the human mind." That sentence sounds fake to me: Was he only saying it to escape the persecutions of an ignorant, cruel, and close-minded clergy?

Bill and I had dinner at the Churruscaria Wagner down the street amidst some 30 tables filled with local families and the laughter of lively kids. Along with the crowd, we applauded the Ceará soccer team, beating ABC on a 1-0 score. Since our arrival the weather has been perfect, "but this is the dead of winter!" Bill reminds me.

Fortaleza. Wednesday 29 July 2015.

Agobar now tells us the story of Marilan, a girl (15 years old) who served as a medium in one of the many spiritual centers in the region. Some of the members were in the military; the junta was still influential, although there was a civilian President, Castelo Branco.

Marilan, whose guiding spirits were Umbanda archetypes, was a trance medium who contacted 'the Dead,' but self-described extra-terrestrial entities were said to drop in and give her information. Then, one morning at 7 am, as she was walking to school, she saw what she thought was a star. It turned into a sparkling white crystal that became yellow orange. It lifted her, and later dropped her. More experiences happened, including contacts at home; something pushed her from her bed. She became disoriented; her life took a negative turn.

We had an intense morning today, analyzing a series of videos, case after case, including another incident with a skin mark studied by Agobar, who has preserved the dry crust, shaped like a spiral. After a full round of debates about the possible meaning of the marks, we enjoyed a fine lunch of white fish and drank passion fruit juice on the patio, while the cat chased lizards under the giant mango tree.

I tried to call Flamine when I got back to my room but even with all the modern technology (4G, Skype, WiFi) I failed and found myself in despair. At 4 pm I finally got through. She was fine, having dinner with Marie-Claire on Place de la Sorbonne. We promised each other we would find a solution, "better than before."

Fortaleza. Thursday 30 July 2015.

We left the hotel at 6:30 this morning in Agobar's car to beat the traffic on the road to Sobral. We stopped on the way to pick up Monica, a volunteer who assists him by filming his interviews and keeping notes.

A year ago, Monica (a nurse at a local hospital) saw an intruder inside her house. She had been ready for bed and, as she turned off the light, a short brown being appeared, about 1.3 meters tall. She had just time to see him clearly before he vanished.

The road to Sobral is modern. We drove among rich plantations, many palm trees, and Bill reminded us that the region had been developed (and richly exploited) by the Johnson Wax Company.

None of us had had breakfast, so once out of the city we stopped at a truck stop in Caluena for good coffee and 'tapioca,' the indigenous staple that replaces bread in the Northeast. It is made from manioc, which was brought to Brazil by slaves from West Africa.

Driving on, we passed a dramatic landscape of basaltic mountains, an arid land with low vegetation along the slopes. After the turnoff to Missi we were on a long dirt road, the landscape increasingly dry.

The main case we wanted to review was the death of Antonio Martin, 40 years ago, never reported. Agobar had only heard of it through neighbors. It's a classic 'killer beam' case that took place on 3 November 1975. Martin was a cotton farmer working in a *bueno* (a family farm, a *sitio* in Portuguese). Two children saw that brilliant sphere in the sky above the farm, then two beams (red and white).

When the father didn't come home for lunch the family looked for him and found his body badly burned and his polyester shirt fused. He had a hole on the back of his head, missing flesh on the forehead, and deep sunburn on both cheeks. We tracked down the daughter-in-law of the victim, who gave us more details.

Agobar and Monica had interviewed people in Missi, so when we stopped at a local Internet café, some of the neighbors clustered around and we were able to track down witnesses describing recent blue balls of light, silent, some of them attached to more complex devices. And at one of these interviews luck was with us: a neighbor showed up *who turned out to be one of our missing witnesses,* a nephew of Antonio Martin! As a child at the time, he too had seen the body of his uncle and could recount the facts, independently.

Fortaleza. Friday 31 July 2015.

We began the day watching video recordings of various new cases and in return, telling Agobar about cases we'd studied. I pointed out the proceedings of the CNES-GEIPAN meetings, the conference in Madrid, and the presentations in Riyadh.

Agobar had extracted from his files some data about Fernando Cleto Nunes Pereira, an Air Force consultant who died in Rio two years ago. He'd told Agobar that in the 1970s the military didn't interfere with UFOs, except for a conference in Brasilia in July 1974 to assess the phenomenon. There's a special file of the official Operacion Prato, the copy that belonged to Hollanda Lima. On that subject I had my own counsel: Janine and I had spent the day inside the base with Hollanda, and the whole evening with three officers who told us about the missing data and showed us the evidence.

"What about the medical documentation?" I asked, thinking of the BAASS report, with its summary of physiological effects.

"Dr. Wellaide was the only medical person there," Agobar said. "I think it's a lie that the Air Force had medical personnel at Colares. Back at Headquarters, they feared the phenomenon was a communist plot, a precursor to triggering a revolution!"

He added: "The most important cases happened at Crab Island. There was one death, and injuries. Dr. Silvio Lago did the hypnosis."

I knew that was the case; I'd met Lago in Rio de Janeiro on an earlier trip. "What about the Brazilian Navy?" I asked.

"The Navy is close-mouthed on the subject," Agobar said. "One time, an old friend, an admiral, was over at my house, and I brought up the subject. He got angry. He didn't want to talk about it."

We spent the rest of the day discussing strange animals (the famous Chupacabra) seen in the days when José Sarney was President of Brazil. Then, we reviewed the photographs people had taken near Sobral.

Fortaleza. Saturday 1 August 2015.

We're free until this evening, when we board the flight down the coast to Salvador da Bahia. So, I suggested going to the beach, with time to reflect on everything we'd seen so far. We had some rain last night, but the blue sky has returned, with the fine breeze we'd enjoyed since our arrival. A young cabbie drove us around to the harbor and to a section facing the ocean where Italian investors have erected cliffs of high-rise hotels and condominiums.

Bill wanted to see his old house again, where he used to live with Regina and where Janine and I stayed in 1988, then we drove to the city center near Teatro San Luis where 'Mister Hall' and the founder of Johnson's Wax used to reside during World War Two. They played shadowy roles in active Intelligence for the US.

Our conversation turned to parapsychology. When Bill lived in San Francisco, he knew Arthur Young, Lee Sanella, Puharich, and met with Bob Monroe whenever he came into town. He even experimented with a combination of meditation and biofeedback to reach a state where the ego was set aside and the experience of leaving the body took place.

Salvador da Bahia. Sunday 2 August 2015.

There was no bus to the State of Sergipe until tonight, so we took the opportunity to walk around the Convento, now restored as a fine hotel, and to admire the views of the bay. In Pelourinho, the area of the ancient public punishment, narrow streets lead to the Hotel Villa Bahia where we had lunch on the patio. We visited the church of Sao Francisco, whose chapels and statues are covered in gold—fifteen tons of gold leaf!

That church's special treasure has a unique history: The Portuguese army wasn't allowed to give money to the church in a direct manner, so they elevated one of the statues (Saint Anthony) to the rank of lieutenant colonel, and put him on the payroll!

The bus ride to Umbaùba on the road to Aracaju takes four hours, a luxurious ride in a Mercedes coach, with a very convenient rest stop at Esplanada. Conversations with Bill about the history of Brazil, the early mapping projects of 'the Interior,' and his recollections of life with Frank Lloyd Wright and his wife made the trip enjoyable, a rich source of insights.

Umbaùba. Monday 3 August 2015.

The bus left us at a street corner and local people directed us to the only guest establishment in town, a primitive place calling itself Hotel

California. It's a poor excuse for a motel, a dozen whitewashed rooms and rough beds. But we're blissfully cutoff from the world.

This morning, heavy, tepid rain overwhelmed the hypothetical gutters and drenched the farmer's market. Unperturbed, Bill had the idea of going to the Post Office to try to locate Jair or Carmelita, respectively the driver and the assistant of Donna Cicera. Interested in our story, the manager took us to a shop he knew had been connected to the healing group. Soon, we were being passed from relative to relative until we ended up in the home of Donna Cicera's own 'administrator,' a nice fellow named Raimundo, who dressed in monk's robes. He blessed us, listened to Bill's story and soon treated us to a wide flow of news: Jair had died in a car accident, Carmelita had married and moved to Italy. Raimundo was one of the last links to the legacy of that extraordinary woman.

This evening, the owners of our motel took us to a large modern *churriscaria* that serves the truckers. Both had known Donna Cicera. The wife, who works in police administration, recalls her mother taking her to 'the clinic' for a minor eye operation as a child.

Umbaùba. Tuesday 4 August 2015.

At 76, I'm entitled to well-deserved retirement, yet I feel that I've simply gained a new freedom to explore, create, and help others in a search so long suppressed.

Tomorrow night we fly back to Fortaleza, and on Friday to Miami. We should congratulate ourselves on a job well done, with Bill fulfilling his mission of a return to the site of Donna Cicera's extraordinary ministry. As for the people we met in Umbaùba, they had no real interest or awareness of UFOs, only anecdotes about lights in the sky.

Salvador da Bahia. Wednesday 5 August 2015.

An important historical remark in David Hess' *Spirits and Scientists*: "There is some evidence that some variants of the 'theory of the unconscious' were direct translations from the doctrine of spirit communication. This is most plausible for Myers' concept (1903) of the subliminal self, an idea that James adopted in his later work, and a tradition that Jung continued. Likewise, Pierre Janet was quite familiar with it." And later: "Camille Flammarion established the principle that it is not only Planet

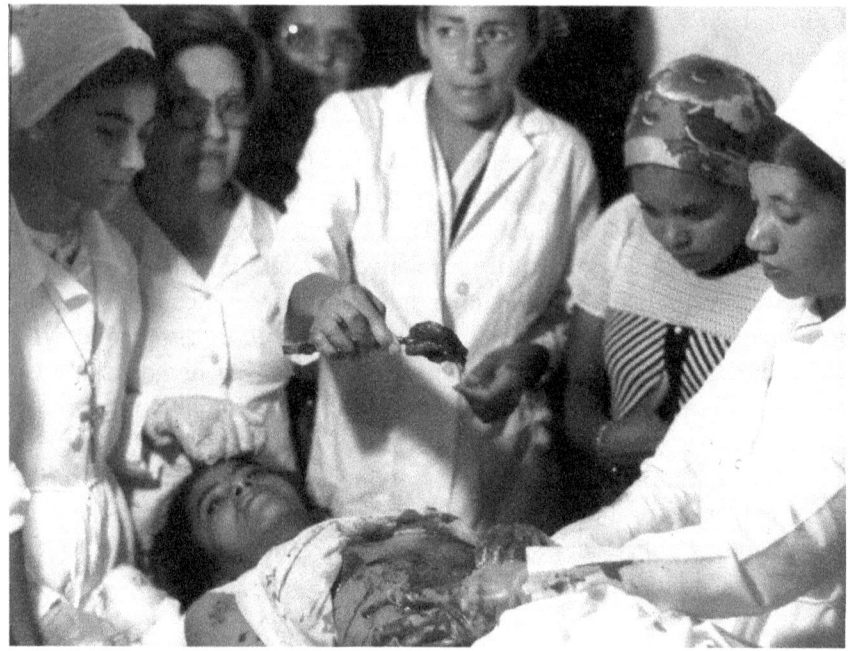

Fig.15: *Brazil 1977. Donna Cicera (left) with Dr. Beverly Morgan (holding forceps) and patient (photo by Bill Calvert): Our reason for going back.*

Fig.16: *Investigation of the abduction and injury case of Manoël di Queiros Meneza. Fortaleza, 27 July 2015. (Photo by Bill Calvert)*

earth that serves as the home of the human species, but all the stars, spread out throughout space without end."

Fortaleza. Thursday 6 August 2015.

We began our work today in Agobar's pleasant patio where Bill gave a summary report of our trip to Umbaùba, stressing the perspective he'd gained on the work of Donna Cicera. Then Agobar pulled out one more file, from which he told us about the abduction of a seven-year-old boy named Janiel in 2011, one of his most important cases.

Janiel lived on an isolated ranch, far in the Interior. On that day at 5:30 pm, he was playing soccer when a star-like light came closer, circled, and seemed to leave, but suddenly it circled back, shining a beam on them. *It became a solid object that hovered.* As both families watched, a door opened and a tall being came out. The being grabbed Janiel and cut him under the left armpit, creating an incision the width of three fingers, before releasing him. The boys ran away while the machine took off, brushing against a cable, creating a discharge.

Janiel was taken to the nearest town for the incision to be sutured (there was no blood, and later no inflammation but a normal, dry scar). The beings had large eyes, no pupils. They were bald and wore suits, white in front, brown black in the back. The family called Agobar. Again, absurd behavior, emphasized by physical scarring.

We had a very nice home-cooked meal tonight, with Agobar's whole family, then we parsed yet another series of files, including the unpublished records of Operacion Prato.

TAM flight to Miami. Friday 7 August 2015.

Lingering impressions of Brazilian cities. Black mold over any surface that hasn't been covered in ceramics or can't be washed every day. A juxtaposition of poorly-built, heavily barbed-wire walls and narrow doors, next to large modern apartment towers that will probably decay in ten years, said Bill the architect, because corrupt contractors have mixed the cement with beach sand. Uneven sidewalks dumping rainwater into clogged gutters. But every morning, the breeze in the palm trees, and the delight of wildflowers over graffiti-covered walls.

Miami. Saturday 8 August 2015.

Florida greeted our return with its heavy soup of damp air, reheated by the hot pavement, enhanced with more than a zest of diesel fumes. But Brazil still resonates in my mind and body with echoes of passionate weirdness. One of the puzzles left unsolved is the role of talented Bob Pratt, who wrote such accurate reports of his UFO investigations in the *National Enquirer*, buried into a tabloid known for its sensationalized speculations—and its remote Intelligence connection.

Hummingbird. Tuesday 11 August 2015.

I met with Garry (at Café Vida, in Menlo Park) and we spoke of his meetings in Salt Lake City. The purchase of the Bigelow ranch proceeds, with some legal issues unresolved: Who carries the liability for illnesses on the property? Who has mineral rights?

There's a belief that we have 'incontrovertible proof' that the Caudate–Putamen sign is not genetic in our data set. The specialists believe there's evidence that it is epigenetic, and not injurious. Yet Garry warns that to prove it, we'd need full family genetic records, which are absent. Saying "it isn't injurious" means that the growth wasn't triggered by an injury or shock. Here, we agree.

Petaluma. Thursday 11 September 2015. Noetics Institute.

A doe and her two fawns strolled below my window this morning when I got up to join Jeff Kripal and Mike Murphy for breakfast at Earthrise. The symposium on cultural, religious, and anthropological aspects of the phenomenon doesn't begin until tomorrow, but the three of us were asked to come in early for a presentation to the sponsors.

I was surprised when Mike Murphy stated that "Obama should reveal the truth," as if there was a single Truth, and as if the Intelligence community had the final data. Mike's reference to the mind-control issues and the growing implications of witnessing a restart of MK-Ultra under new guise discouraged me.

Hummingbird. Saturday 12 September 2015.

David Hufford gave a very cogent account of his involvement with *The Terror That Comes in the Night* that must have resonated with Garry. Putting them in contact was one of my goals. Whitley, emotional from the loss of his smart and generous wife, Anne, of a brain aneurysm, spoke lucidly of his Visitors: "They're shadows, projected from inside us, and the implants are bait. The Intelligence community has long known uncontrollable things were happening."

Whitley doesn't perceive his experiences as evil, but more akin to being in school. Like me, he distrusts the idea of Disclosure. "It would act as a tripwire," he said, "making Them part of reality, and the door could never be closed again."

Wasn't it General Nathan Twining who said, speaking of Blue Book, "The coverup is the Disclosure, and the disclosure is the coverup"?

Flamine enjoyed the meetings at Earthrise. Most of all, I will long recall walking with her through the magical grounds in the crisp morning and the star-studded night.

Hummingbird. Friday 18 September 2015.

Stunning new commentary from a medical expert: "We've had a conversation before about the LIDA-IV that went to a Brain Research institute (UCLA, 1983) where Dr. Russ Agee tested it, but I never considered any applications of interest to us; now it seems that stimulation of near-surface cortical edge—receptors in the visual cortex—may affect our Interference candidates. The device put subjects to sleep and made them 'see certain things.'"

Is that a clue to the unfolding of the more sensational encounters?

The researchers had a few videos and lots of still shots. In general, they didn't believe the results until 1983. They were not published till 2011...28 years later, (as "Diathermy, whether achieved using short-wave radio frequency (range 1–100 MHz) or microwave energy (typically 915 MHz or 2.45 GHz), exerts physical effects and elicits a spectrum of physiological responses, the two methods differing mainly for their penetration capability. Some of which penetrate skull.") (6)

The article shows that the culprit is in fact intracellular heating. John Schuessler has circulated a reminder that much more was done. A Los

Alamos study on these technologies ended in 1985 and has been declassified. But that was 30 years ago!

Diathermy devices had been used to cause interference to German radio beams targeting nighttime bombing raids as early as World War Two, at the so-called Battle of the Beams (7). Suspicion was that the operators became confused, but the link hadn't been reported.

"That was 70 years ago!" continues our expert. "You may recall a certain slide on ultrasonics as an explanation for several of our observations in CE-3 and other cases…not necessarily requiring ET. That said, I fall on the side of Jacques, to quickly point out that specificity is not unique to CE-3 cases and apparitions in history."

Still catching up with past messages on this topic, one correspondent writes there are mechanisms where non-thermal RF can disrupt DNA. Could the process account for some of the injuries of UFO witnesses?

Five years ago, DARPA was briefed on the first 30 UFO patients studied, with signs of the inflammatory reaction in white matter and Virchow-Robin spaces. After the briefing, a very disturbed DARPA scientist, in tears, told Kit: "It's us! Technology from Sandia…meet me at a briefing room…" But when Kit got there the security personnel told him the session had been cancelled.

Hummingbird. Saturday 19 September 2015.

A difficult discussion with Flamine as we argued in front of a very expensive mediator. We finally emerged from the nightmare, yet I was devastated when I drove away to meet Garry for lunch at Café Flore. His friendship helped me regain my composure. He said that, on the drive up to Noetics, "Flamine spoke of how close the two of us were, how it felt as if you'd always known each other. How she loved you."

We spoke of the research. I begged Garry to clarify the medical jargon and to stay in touch with Richard Niemtzow. There must have been a specific French interest in all that, which would explain why they kept seeking out Richard's involvement to explain their cases of witness paralysis, Valensole in particular. But there is more: everybody wants to understand how to make a paralyzing weapon.

Brandon Fugal, the potential buyer of the Ranch, is a commercial real estate leader, but he also was an investor in Motion Sciences, a free-energy

R&D company started by Joe Firmage, based on earlier work with Bernie Haisch and Creon Levit.

The company included several of the Lonestars as well-paid science board members, so that may be where the connection came from. I was not involved, or consulted, and only later did I discover they'd met four times, three days each time. Based on Puthoff and Haish's concept of vacuum energy, the theory was interesting but too speculative, and no product was ever developed.

Hummingbird. Sunday 27 September 2015.

Last night, a pleasant dinner for my birthday (no.76) with my daughter, Rebecca, and Flamine (at Harris' on Van Ness). We spoke sadly of the world around us, so brutal: over 700 people dead at Mecca after sudden, unexplained crowd panic during the Moslem ceremony of the stoning of Satan. Such a crazy idea, to lapidate the Devil!

Flamine and I spent the day in Healdsburg at an emotional reunion of my colleagues from InfoMedia days. It was hosted by Ruthie and her husband, David, proud to take us on a tour of their new facility, which makes specialty drinks for national markets.

This evening, we're treated to an eclipse, a blood moon enlarged by its proximity to Earth. Is it a sign?

Hummingbird. Monday 28 September 2015.

With a heavy heart, early this afternoon, I drove Flamine to the airport once again. I had to tell her I felt a sense of personal failure as we parted, that I hadn't known how to sustain the joy of our times together.

She'd left everything behind to join me, yet the challenge was too great when her father died. She assured me that everything was possible once she recovered her balance.

I had some patching up to do myself, I said.

13

Hummingbird. Thursday 1 October 2015.

A new creative phase has begun in Silicon Valley. Our 'Band of Angels' enjoys nurturing the progress of startup companies in which the group invested bravely (including Taulia in finance, and Materna in medicine). The presentations are filled with new insights, a move towards improved opportunities for health and (we hope) a cleaner environment.

Early this morning I launched a fund-raising campaign for a collectors' edition of *Wonders in the Sky*; we raised $3,200 in the first few hours. Recalling our good talks in Porto, I mailed our early manuscript to Brother Guy Consolmagno, director of Vatican Observatory in Tucson.

Hummingbird. Sunday 11 October 2015.

Over dinner in Berkeley, the 88-year-old Jack Katz repeated that he'd "known" many women during his long life, something I blamed for his current laziness and distraction, but he ignored the jab, laughed good-heartedly, and told me every one of them had attempted to control him. "It was my duty," he said, "not to allow this domestication to happen."

Well, there is only one woman in my life, I said, but she's away on another side of the globe. I'm flying there with hope and a long list of projects with fellow researchers in France.

Mabillon. Wednesday 14 October 2015.

My physicist friend Philippe Guillemant asked me to dinner with his colleagues, so we met over dinner at La Fontaine Gaillon, owned by French film star Gérard Depardieu. François de Witt (*La preuve par l'âme*) was there, along with Henri Kaufman (*La Serendipité*) et Jean-François Vézina (*Danser avec le chaos, Les hasards nécessaires*).

The conversation turned into a debate about the influence of the future on the present, the 'spiritual dimensions' of the web, and the sense of free choice under quantum gravity. The goal was to define a colloquium to be held next year, but plans were vague.

Mabillon. Friday 16 October 2015.

This morning Flamine made an offer on an attractive, cosy apartment, close to the Montorgueil area we both love. I went with her in the rain, happy to see her regain confidence, yet aware I was staring at a future without her. We saw the latest Woody Allen movie, *Irrational Man*, a light, run-of-the-mill comedy with a very thin philosophical theme. Now Bertrand Méheust has kindly sent me his latest book about the history of Christianity, *Jésus thaumaturge*.

Bellecour. Sunday 18 October 2015.

Paris has been clean and dry, the weather affording agreeable periods of autumn sunshine, but France remains poised on a political edge that reveals deep faults of discontent. Last year, when a mob of disgruntled Air France employees tried to lynch two executives, the French were surprised when *The New York Times* put the picture on its front page, his shirt torn.

Now I'm in the country, a guest of Olivier. I read *Ovnis et Conscience* (8), a physics book that buries the old concepts about the phenomenon. In it, Philippe Guillemant offers novel ideas in physics to account for the features left unexplained in the extraterrestrial hypothesis.

Mabillon. Tuesday 20 October 2015.

The classical, first level ET theory of UFOs is dead; how many times do we have to say that? A new book by Daniel Robin about Fatima (9) quotes my TEDx conference of November 2011 about Information Physics, as does Guillemant in his book, but most physicists still expand on quantum physics or relativity in conventional ways.

They validate their speculations with narrowly selected observations. They skip one obligatory step, namely looking hard for unexplained facts beyond their theories. It's that direction that I need to explore, and better articulate what we learned in Fortaleza.

Mabillon. Thursday 22 October 2015.

Alain Dupas waited for me in vain at our favorite restaurant today. I was so distraught by legal divorce maneuvers that I forgot our appointment. I rushed over to apologize and join him. We spoke of the latest space projects, CNES and Thales, Airbus and all its window-dressing venture in Silicon Valley.

I said: "Airbus never understood how venture capital could help, and still doesn't. We were collateral damage of internal feuds."

Alain replied: "It's always the same with big fat firms, and we should know it. They get money for the big things and they ignore too many essential details. No way to get to the Moon, guys! And even less to Mars."

Mabillon. Saturday 24 October 2015.

Age has a unique way of bringing novel sensations and an awareness of human flaws. A cheerful young Chinese doctor did some electronic tests; the lining of the sensitive nerves in my legs was damaged, so he gave me some pills. I read the disclosures. The medicine was a neurological drug, something to do with a neurotransmitter named GABA, and it didn't work very well. Among the side effects, it clobbers your short-term memory, so I stopped taking it, wasting another opportunity to advance medical science as a docile subject.

For a long time Flamine and I had planned quiet visits to hidden passages of Paris, best known to long-time residents. The weather invited us to walk everywhere in search of bookstores and odd boutiques. There was tenderness, and sadness.

I told her: "Last June, when you came back to San Francisco I had a picture in my mind, of a simple, wonderful world with you…"

She remained silent so I added: "Your father had just died. I thought a page would turn. I was with you that evening. I had been with you for the last two years of his life. I thought I could mend the grief and the pain… I would sweep before your door, make everything right."

"I need time," she said softly. "Happiness isn't that simple."

"To me, happiness is very simple, on the contrary. It means being able to help you, to give you a secure place and a loving home."

"Those are things that involve the mind," she responded. "But the heart is important too, to put everything together."

"Then there's going to be a big hole in my heart, and in my mind too," I told her, "A void."

"Sometimes it's good to feel the void," she snapped, and then I didn't know what to say.

Mabillon. Tuesday 27 October 2015.

The air is so still that the drapes in our bedroom didn't shudder when I opened the window. Sidewalk cafés serve pastries and cappuccinos to tourists in ridiculously expensive dark glasses. Parisians in no hurry to return to their offices stand around, smoking one more cigarette.

Yesterday I went to Nanterre to consult Boëdec about Brittany and to show him the mock-up of *Wonders,* where his research is quoted.

The urban landscape could be a backdrop for a sci-fi flick: rectangular lawns in a mile-long perspective to the threatening skyscrapers of La Défense, lined up on both sides by tall steel and glass boxes flaunting the names of insurance companies. A loud helicopter kept circling.

Mabillon. Thursday 29 October 2015.

The Internet crowdfunding campaign for *Wonders* reconnects me with friends and colleagues. One of the recent subscribers is Claude Poher who reports: "As far as propulsion is concerned, my 'emitters' have a ratio (thrust/electrical power) 1000 times more powerful than space thrusters ejecting argon plasma, and there's no matter ejected. They work at ambient temperature, not in liquid nitrogen. I have understood how to generate electrical energy in a massive, static manner, without using fossil resources. A prototype is being built; it should be finished in 2016." Now a German lab is duplicating his experiments.

Mabillon. Friday 30 October 2015.

At noon, standing at Saint Michel fountain, waiting for Bertrand Méheust, I saw a bespectacled fellow with a yellow tie walking up to me: It was Jean-Luc Rivera, waiting for Jean-Pierre Rospars. I hadn't seen him in ten years.

The four of us happily walked over to a Tunisian restaurant near the Sorbonne, where we spent the next three hours discussing current research and case studies, especially Valensole.

The day ended with a long session at Institut Métapsychique, listening to Erlendur Haraldson, a senior psychologist from Iceland who spoke about a materialization medium named Indridi Indridason, active in the 1905-1909 period.

That group left records of studies of levitations (of the medium), and once a column of light that announced a major fire in Copenhagen that only became known in Iceland several days later. No such medium is known to exist today. Like Donna Cicera in Brazil, Indridi had a short career, dying at age 28.

Mabillon. Saturday 31 October 2015.

Jean-Luc Rivera alerted me to a new art exhibit in Paris on the theme of 'The Invisible.' It includes one room dedicated to Commandant Tizané, the gendarme who was the leading investigator assigned to claims of hauntings in France in the 40s and 50s. This morning, I went to see the show and I fell again under the spell of his astounding cases.

The exhibit includes his manuscripts, charts from his investigations, and even his *képi*, the official hat that a mischievous spirit lifted from his head and moved slowly through the air in front of his eyes.

In the afternoon, this being Halloween, we took Flamine's young cousin, Joseph, trick-or-treating in the neighborhood. His grandparents are neighbors of Simonne Servais in Neuilly, so I was thinking strongly of her in the evening when the phone rang and I heard her voice, strong and clear at 93. She has moved to a nursing home where she's alone, miserable with chronic anemia.

I never published what Simonne and I learned about the phenomenon when we visited Maurice Masse in May 1979. Neither did Marie-Thérèse de Brosses, who had his confidence. And so much for the folks with security clearances, when they're not trusted. The witnesses are the ones who know.

The same day, at 6pm.

Another boy has come into the world, my third grandson. His name is Xavier. It was a difficult birth; Claire must be exhausted. I will see all three of them tomorrow, and the next day I fly home. I called my brother to tell him the news, which Gabriel greeted happily. His wife had undergone another hip operation and long hospital stay. Both have slowed down, they said, but I found them clear-headed and warm.

"You're still young!" She chided when I described my current life and growing health concerns. But Montaigne writes, "There is no man so decrepit whilst he has Methuselah before him who does not think he has still twenty years of life in his body…"

Hummingbird. Monday 9 November 2015.

After the long drought, rain has returned to California, bringing unusual lightning strikes in Marin County. Shower storms are looming. I've been paying bills and rescheduling appointments. All in due time, but when I read the galleys of the new book by Jeff Kripal with Whitley, *The Super Natural*, I am eager to resume this unique research. At 76, I may not have a lot of time, energy, and mental capacity left to undertake the work.

The Limited edition of *Wonders* has raised close to $30,000. This is long-term, complex work, the culmination of a decade of study with Chris and his friends.

Hummingbird. Friday 13 November 2015.

Over a hundred dead so far, hundreds more wounded, torn-up bodies all across Paris. Islamists struck again in the heart of the city today, turning a popular concert hall into a slaughterhouse, executing anyone within range and, in separate action, blowing up explosives at the main stadium during a soccer game between France and Germany.

When I reached Flamine, she was safe at Mabillon, but both of us readily understand that this challenges the very concept of Europe and the survival of democracy, already upset by so many technical and bureaucratic encroachments.

Tomorrow, nothing will move: schools and universities closed, public events cancelled. A new season: a time for blood and fear.

Hummingbird. Saturday 14 November 2015.

One of the LoneStars medical research projects calls for MRIs of all our brains, so I had occasion to call friends today, reconnecting after a month-long interlude. Not much has happened at the Utah Ranch, no deal has yet been signed. Bob keeps his own records about NIDS, thousands of pages. Nobody is sure about the results of the electromagnetic readings at the Ranch, or their meaning. There was microwave monitoring through the 3-to-5-megahertz range, based on previous detections (the RB-47 over Oklahoma, and Ron Blackburn's own device tests) but they only got a single signal, never repeated.

Next, we went over recent discussions with of Isaac Koi, with whom my friends spent hours in London. He's taken a skeptical stance, negating many cases, concentrating on abductions, compiling medical databases, and he claims to have collected some 20,000 science fiction books. He has a computing degree, uses Tor extensively.

Isaac spent two full months on the Bentwaters case and concluded it was a waste of time. It seems that an 86-year-old woman who lives in a cottage in Rendelsham forest had shot videos on both nights. Koi went to have tea with her, saw the videos, and then dropped the case.

"I questioned him," my friend said, "that was the only time Isaac seemed troubled. He said he couldn't talk about it, but it had nothing to do with Aliens. The answer would surprise me."

Hummingbird. Tuesday 17 November 2015.

Tonight, Harvard astrophysicist Lisa Randall gave a lecture at the Jewish community center about the structure of the universe.

Dark matter, she said, is now thought to be scaffolding. It forms a spherical halo over the disk of our own galaxy and probably accounts for the lensing effect as we look at distant objects.

I didn't emerge from this with any tools that would enable me to pursue my own understanding of it, while some astrophysicists have come to doubt the very existence of dark matter.

Hummingbird. Friday 20 November 2015.

Federico Faggin, inventor of the microprocessor and pillar of Silicon Valley, unveiled his model of the physics of consciousness tonight at a meeting of the Foundation for Brain-Mind Research in Palo Alto, defining every term, posing boundary conditions, explaining the need for every concept. The goal was nothing less than a framework for the structure of reality and its emergence within consciousness.

He places his newest research under the concept of cognitivism. "Until 25 years ago this topic wasn't considered worthy of scientific study," he said. "When fMRI technology emerged, it became of interest to neuroscientists and a few isolated psychologists, but it wasn't a problem in physics. Move over to 2015, and the prevalent view is that consciousness is an emergent property of matter, a product of quantum fields in the brain."

He laughed: "There are still unexplained anomalies, but the doctors want us to wait until they've mapped the whole brain, and then we're supposed to understand everything!"

Federico's approach, on the contrary, seeks to unify science and spirituality. "Consciousness is a sense of self, a feeling. I know it because *I feel that I know*—an experience that is both inner and outer, and gives *meaning* to life." He went on to define *qualia*, items of information like the smell of a rose, the taste of bacon, *the feeling associated with a thought*. We perceive through qualia; we *know* through qualia.

"What about robots?" someone asked. "AI can recognize images and perceptions, too!"

"A robot can recognize an image; it doesn't associate a feeling with it. Conscious reflection and response aren't possible for computers," Federico said. "*There's nobody home*. They can only imitate."

Hummingbird. Tuesday 24 November 2015.

In Santa Rosa, where I went to consult my friend Ruthie, the Sun came out after the rain. Red and yellow leaves covered the sidewalk. I asked her how I should think about a permanent separation from Flamine. Happily married after a stressful breakup of her first marriage, and experienced as a pastor, I knew she'd understand.

"Are you joking?" She asked, "Many women would be thrilled to live with you." I'm not so sure. I bring multiple layers of personal history and complexities, hardly in synch with the desires of most American women with challenges and aspirations of their own.

I drove back to San Francisco, watching in wonder the gentle landscapes, the pastures brilliant green after the first rains, the forests deep with the quiet bounty of springlike expectations, the wineries growing orange, rust, and purple in autumnal robes—all of it a tapestry celebrating life, mocking my inadequacies and fears, my trepidations.

Hummingbird. Wednesday 25 November 2015.

The latest LoneStars brain project prescribes an MRI for me as "a member of a special cohort for which we are investigating sporadic and etiologically uncertain white matter inflammation..." Where does that fit? Garry clarifies it: "The overall project has to do with brain structures able to process anomalies, through an information transfer into the sensory system: auditory, visual, emotions..."

He adds, helpfully: "When we write 'etiologically uncertain' that's just hand-waving, but it's not wrong. There are several cohorts: remote viewers, high-functioning subjects, psi-capable subjects, interference patients...How they overlap is an interesting question."

The study posits that some individuals have a putamen-caudate connection that density-normal individuals generally lack, so he's pushed hard to find an automated approach to metric the structure.

Hummingbird. Monday 30 November 2015.

On Friday, such a privilege: Thanksgiving dinner at my daughter's home, and a quiet evening. In the guest bedroom are paintings by my mother and a chair from our old library at Spring Hill, memories of past years. In the living room, *The Mill at Pontoise*, from deep time.

Over breakfast we caught up with impressions of life, swapped local news of the Bay area, and we laughed at the games of the two little dogs playfully barking canine insults at the squirrels. Then it was time for me to drive away in another quiet, sunny morning in California.

Hummingbird. Friday 4 December 2015.

At Todd Pratum's place near the lake in Oakland, I met with music and esoterica scholar Joscelyn Godwin this afternoon. I'd heard much about his expertise in hermetic matters and ancient texts. As for Todd Pratum, one of the most knowledgeable rare book dealers on the West coast, I hadn't seen him in years. While we admired the thousands of volumes in his apartment, I wanted to quiz both of them on authentic Rosicrucian history. Joscelyn has written a book on Atlantis and is familiar with the nested symbols. He deals with the works of the alchemists, currently planning a new edition of Robert Fludd.

Both agreed that modern remnants of the mystery schools, including the Freemasons, had degraded into narrow ideologies. Todd said that among his customers, Freemasons were the ones with the most limited range. "They've been told they have the secret knowledge," he said, "so once initiated, their only curiosity is all about freemasonry itself."

Joscelyn isn't part of anything and feels that no Lodge could add to the creative work he's planning in retirement, a lesson I must remember.

He added, "I think there's a plane, or a plan, concerned with the spiritual destiny of humanity. It includes both living and discarnate entities. As it manifests on Earth, it becomes colored by the ambitions, beliefs, and psychologies of the bearers, who are doing their best—as are we! My own contact was via Brunton."

He added with humor: "Other fixed stars in my firmament are Blavatsky, Guénon, Evola, and Charles Fort—none of whom would have cared much for each other!"

Hummingbird. Saturday 5 December 2015.

Rumblings and ramblings from the LoneStars. Several are involved disseminating the profoundly deceptive SERPO story, which saddens me deeply. Don't they understand the danger of inventing fake UFO stories, while trying to study the real ones?

Garry is absorbed with the complex bureaucracy of Stanford, so our group functions with few common objectives.

The ongoing medical correlation studies, including the analysis of cerebral MRIs, discloses that my brain is healthy and 'age appropriate,' with no evidence of demyelinating disease. It "looks like the brain of a much younger man," I'm told. I take this to mean that I would have no excuse for slowing down my activities, intellectual or otherwise. I also have 'the sign,' the unusual nerve bundles connecting the putamen with the caudate, but we still don't know what it means.

I note in *New Scientist* (16 January 2016 p.33) that when researchers activated the orbitofrontal cortex in mice, they became more goal driven. This behavior relied on a connection between the cortex and the medial part of the striatum, while habitual behavior corresponded to its lateral part, analogous to the putamen, but the article didn't mention a connection between them.

Hummingbird. Sunday 6 December 2015.

This was a week of hard work but also new contacts and novel ideas. Once again I marveled at the intellectual density of this region, where one could hear about dark energy on Tuesday, genetic control of the human body on Thursday, and novel mathematical models of consciousness on Friday.

After another round of analysis and corrections on *Wonders*, I suddenly felt a re-awakening in me of impressions that had been dormant—a tangible sense of the future, an awareness of transcendence beyond years, tears, and sorrows. It was heartening to feel that I still had access to the trapdoors of sub-space, in the recesses of the soul.

This morning, over breakfast with Garry at the colorful Bucks' restaurant in Woodside, he made me laugh when he threatened to draft ads for me to post on Internet dating sites. Somehow, in the heady atmosphere of a Sunday morning in Woodside, it all made sense.

When I got home, I read about the results of my brain MRI, "consistent with the state of a younger man," but the caudate-putamen connection showed up prominently. Since I am not a *savant* and have only spontaneous remote viewing abilities, my scan leads to revision of some hypotheses. This didn't surprise me; no one's ever asked what I thought my talents (if any) consisted of, or a list of actual happenings that might have cast light on 'other' abilities.

Mabillon. Sunday 13 December 2015.

One month after the killings at the Bataclan theater here in Paris, one senses the crowds' eagerness to return to the shops and re-occupy the sidewalk cafés, encouraged by the mild weather, filtered winter sun, a feeling of suspended time. The terror is blunted, not erased. Fear hangs in the air, abstracted in the untold pains of a wounded people and the hidden agendas of its politicians.

This is the second day of the regional elections. The Front National of Marine LePen is riding high on ugly hatreds and the nostalgia of simpler times. Also, deep fear of further killings; if they want to renew indiscriminate fanaticism against everything Arab, all the Islamists must do is to keep throwing bombs.

Paris has survived, however. It has kept its place as a diplomatic center, witness the climate treaty signed today among 195 nations supposedly committed to "maintain the planet's temperature growth within 2 degrees." Fine promises, meaningless numbers.

Mabillon. Monday 21 December 2015.

My grandson was full of projects when I met him at Montparnasse station yesterday. He arrived from Le Mans, dragging two suitcases and his roller blades. On the way to holidays in Tunisia, he took time for lunch with me, speaking of his many interests before dashing off.

In the evening, Flamine and I assembled our favorite guests from French parapsychology: Mario Varvoglis from IMI, Gérard Galtier and his wife, Flamine's friends from *Le Monde*, also Bertrand Méheust, and Philippe Guillemant with Laurance. I'm reading Bertrand's *Jesus thaumaturge*. In my phase of life, helping build such networks is the most valuable contribution I can make.

Flamine will fly back with me, but only long enough to wrap up her life in California, then true separation begins.

Rochecour. Saturday 26 December 2015.

Christmas at my son's new home: a garland of blue lights on the tree, happy kids, easy country walks in the mild weather. Flamine and I slept in Maxim's studio, fighting the Norman wetness that seeps through the walls, Atlantic weather rushing inland with the tides.

The year is over. It brought tragedy to the world and distress to our home. The disgrace of extremist Islam, absurdly tolerated by major countries, has deeply wounded the French and devastated the Middle East, setting the stage for unending bloodshed.

Space exploration regains momentum, however. Europe, with its successful mission to comet Tchouri, has shown it still had the ability to excel; the mission detected organic molecules including glycine, a component of proteins such as DNA. The Philae robot also found water with deuterium (hydrogen with two protons instead of one) at triple frequency compared to Earth, confirming that our oceans didn't originate from comets after all.

Hummingbird. Wednesday 13 January 2016.

At the Quetzal Internet Café today, a snack with Ron Brinkley, back from New Mexico for a few days. We spoke of Socorro (he is still searching for a cattle brand that might correspond to the actual design of the symbol seen on the craft) and of Chameleo, the invisibility project. His life in New Mexico brings him close to the Indian tribes, the major ranches, and the strange dealings of the US government, which does in the Southwest some of the shady deeds banned in California.

Ukiah. Saturday 16 January 2016.

Dennys's coffee shop "always open;" the Bargain Center, off the 101 expressway; and Friedman's, Rite Aid Pharmacy, Safeway—small town America. But the hills burst with the imminence of an early spring.

Everything speaks to me of a certain love I used to have, and the indescribable sweetness of this land.

The other day, we spoke about the idea of the soul. Mine, as I discover again, lingers in these forests of the Pacific range, sensing the end of a cycle. I needed to get away—alone. I picked up a suitcase in anger, throwing in a couple of shirts, and I headed north.

This is the time to think of death, not life. My mother died on January 17, sixteen years ago. Janine died in my arms on the same date, six years now. My father died on the 18th. I need to grieve, alone here, even in this standard American diner.

In the next booth, a middle-aged woman is helping her old mother to get up, her hair all frizzy and nice. I suppose her daughter took her to the salon this morning. She holds up her coat.

A sign says *Sizzling Fajitas – Breakfast served all day.*

Fort Bragg. Harbor Lite Lodge. Later the same day.

I'd planned to stop in Willits, but even under the Sun it looked sad, seedy, and dangerous, so I drove over the ridge to the coast and rediscovered this inn on the edge of a precipice, overlooking the fishing harbor. The Sun has just set in the ocean fog that drifted over the beach and the road. The sky offered a miraculously clear setting for a perfect half-moon. The hotel is built of massive redwood beams, black with wetness and sturdy with the experience of tides and storms. If I'm lucky, there'll be another tempest tomorrow, lifting my despair.

Flamine called in tears, catching me at a local shop. We spoke of our emotions, my anger, all the confusion. I told her I would return home tomorrow. Neither one of us understands.

Hummingbird. Sunday 17 January 2016.

John Schuessler sends along a note about an article describing "How to turn the brain off." He writes: "I found it interesting because it shows there is a mechanism, at least in mice, which can mimic what has been reported in many UFO cases: They found that making thalamus cells fire with a frequency of 10 hertz caused the mice to lose consciousness. If they fired the cells at between 450 and 100 hertz, the mice woke up again."

My notes reflect that it's the same with humans. It was first discovered by Andrija Puharich and José Delgado in secret CIA collaboration, along with the MK-Ultra Program, revealed a few years ago when some smart staffer discovered it, among the 100,000 pages declassified in 1986. Those two guys had Netter diagrams from an Atlas of both human and cat brains, where the Thalamus was clearly delineated. This was discussed before the Academy in 2020.

Now, I read that "early discoveries also suggested the cochlear microphonic as one mechanism... There are more places that could be tapped to insert words inside a person's head."

One CIA project called AccoustiKitty did the reverse, because they didn't know how to train a cat to understand English. Instead, they tapped the words the cat heard. Finally, at Los Alamos in 1983, a way was developed to send clear voice into someone's brain (posterior acoustic cortex directly, bypassing the 7th and 8th cranial nerves). It's all unclassified now. The first two projects were with small sentient smart mammals. The last one was with "large semi-comatose, somewhat dull humans..." All the same. All these and 45 additional neural projects were declassified... most in one fell swoop."

There may be many other clever projects, searchable using Isaac Koi's websites. I'm told it is tedious: 100,000 is a lot of pages. A similar pile deals with psychic research from CIA.

Hummingbird. Monday 18 January 2016.

A summary of current medical speculation around the group. My friends are impressed with one of the comments in the recently-released UK files about Rendlesham Forest, stating: "It may be postulated that several observers were probably exposed to UAP radiation for longer than normal UAP sighting periods. There may be other cases which remain unreported. It is clear that the recipients of these effects are not aware that their behavior/perception of what they are observing is being modified."

This is nothing new. You would find the same speculation, in almost identical words, in my correspondence with Aimé Michel and Pierre Guérin in the mid-1970s. In fact, that's exactly what was so discouraging to Aimé following the Doctor X observations, and his own, deep experience in Sisteron. But the UK report goes on:

"The reported effect of (presumed) UAP radiation on humans is that it is quick-acting and remembered–although, curiously, following the event there is little or no recall of events as a continuum. In short, the witness often reports apparent gaps, often for up to several hours. It is described as if the exposure causes a temporary memory erasure."

Again, these comments match things we've known from many cases and books about close encounters. But the anonymous author of the UK report provides more specific observations:

"Experiments (on animals) have shown that low (3mW per cm at 450 Mhz) exposures affect brain calcium ions. These are known to play an important role in the transmission of nerve impulses. Various modulations (e.g.5Hz, or 16Hz) were imposed on the radiation."

The analysis goes deeper: "Earlier experiments went on for tens of minutes—or even for hours. They were not representative of the very short UAP exposure times of a few seconds. In 1981 it was discovered that E fields at a modulation of 5Hz for only 5 to 10 seconds could increase the excitability of nerves for hours."

This again matches what Dr. Niemtzow found, and old experiments by Poher. The smart author of the report goes further: "It would seem that this effect can occur when close to the UAP, *either outdoors or indoors*. This causes the brain to interact in an unusual way with the imagination library, causing reports of visual activity *which are not in fact a true representation* of reality."

BINGO! Is that what happened to Aimé Michel, when he became distracted in the home of "Dr. X," unable to meet the Others?

Hummingbird. Tuesday 19 January 2016.

Pursuing the details of close encounters, I note an exchange about the death of UFO witness Arcesio Bermudez in Columbia, where a medical doctor from Bogota named Dr. Cesar Emerald attended the patient along with Dr. Luis Borda, at the Social Security hospital.

Emerald is quoted as saying that "NASA investigators came from the US and were all over the case," and "the patient was injured by gamma rays." In South America, the reference to NASA must be taken with a sense of theater and, preferably, comedy, the name "NASA" being a convenient placeholder for more discreet structures.

John Schuessler writes about his work on the Cash-Landrum case: "The basic gamma ray model doesn't cover these cases; there is a combined effect, like gamma+pulsed microwave." Claude Poher, in the 1980s, was working on similar models with Richard Niemtzow.

Hummingbird. Saturday 23 January 2016.

On my trip to the north coast, I understood why I won't mind dying: I don't need to cling to every last breath. I hope Death will be one more research expedition or, as Allen Hynek said, "my next assignment." If I died among the redwoods with the sound of the Pacific around me, everything would be nice and normal.

There is no greater beauty.

So, I simplify what I can. My friend Douglas Crosby had the same experience. "You have a light that shines," he writes. "I feel sorrow she's moving away from the warmth it provides, but that light is not dimmed."

14

Hummingbird. Sunday 7 February 2016.

The silence of the streets, under the quiet skies, catches the mind with its oddity. The Superbowl is playing in Santa Clara this afternoon, leaving San Francisco empty of cars and people, the avenues stretching in sunshine and the hills, for once, indifferent to the passing of the occasional car. I walked over to Polk Street for a veggie lunch, then coffee at Quetzal to work on the last pages of *Wonders*. The restaurant was empty. The waiters drifted indifferently.

On Thursday, I drove over to Federico Faggin's beautiful house in the luscious hills of Los Altos. I found him alone in the kitchen, enjoying a dish of spaghetti and a glass of red wine. We took his car for a visit to the gravity team in their new, unmarked lab, in the back of a non-descript, one-story converted warehouse building. Federico is familiar with inventors: He urged them to continue in their stealthy ways—and avoid government like the plague.

Hummingbird. Saturday 13 February 2016.

The divorce, polite as it was, has left me with mental distress and physical pain, but I was able to drive to the Valley yesterday to meet the founders of IonPath and to get an update on SpaceKnow, two of the startups I follow.

I should try modern dating sites, but they've turned the web into an infinite loop to share loneliness. So, I restructure the library, firming up plans to turn over my archives to Rice University in Houston. Hours in pain, flat on your back, do wonders to remind a man of mortality.

Among the books I'm reviewing is Arthur Versluis' treatise on *The Secret History of Western Sexual Mysticism* (10): a lucid overview of "sacred practices and spiritual marriage" that mentions many movements including those of Randolph, Thomas Lake Harris, Joscelyn Godwin, and even early contactee Jane Lead, one of the 'ancient witnesses' we quote in *Wonders*.

There's no great mystery any more behind these esoteric movements, although their secret is protected by the renewed pruderies of our high-tech age. The practices that couldn't be revealed in their time and forced Harris to leave Fountain Grove and California when rumors leaked out, had to do with *Karezza*. A woman correspondent of Alice Stockham wrote of it as "indescribable physical rapture, delightful interblending of spirit, soul and body."

Hummingbird. Sunday 14 February 2016.

In the ever-renewed turmoil of world economies, the relevant fact is that markets have fallen in value—by about 15% since the beginning of the year. Another sign of nervousness: gold and silver are up sharply (at $1237 and $16 an ounce, respectively) but still far from their peaks, indicating that some of the current alarm may be justified on economic parameters alone. The real problem is that trust in the key actors, such as central banks and major markets, has largely evaporated.

Hummingbird. Thursday 18 February 2016.

A major scientific conference has begun in San Francisco, dealing with photonic communications, so some friends from points east are in town for a few days. "A new technology is coming fast," they told me today as we sat down for lunch at Les Joulins. "Quantum entanglement is being applied to a new generation of networks. It will mean rebuilding the Internet from scratch."

While photonic devices today need to use several photons to send a bit, quantum technology can superimpose multiple pieces of data using different parameters (spin, frequency, phase, etc.) in a single photon. The work is still hosted in universities, particularly in Germany and Austria, but business applications are not far behind, with claims for attending improvements in cryptography.

We ended the conversation comparing notes about the MEDEA project and the 'Tic Tac' submerged in Virginia Beach. Is there a database about super-energetic devices that escape detection and pursuit, maneuvering across the atmosphere with impunity? The rumors are all over the web. But that file, too, is out of reach.

We-Ko-Pa Resort. Friday 19 February 2016.

The Open Minds organization, under Alejandro Rojas, holds its annual meeting on this grand Arizona site run by an Indian tribe near Fort McDowell, outside Scottsdale. As soon as I arrived, I saw old friends, including Ron Westrum and Lee Spiegel. Walking around the tables in the open-air café was Douglas Trumbull (with Marc d'Antonio, photo-analyst for MUFON), and David Jacobs, formal in a suit and red tie. I was introduced to Betty Hill's grand-daughter, now a paranormal investigator herself. If the mood is good-natured, there isn't much new. Conversation ranges from 'the messages inside crop circles' to abductions, the 'imminent' prospects for Disclosure, and the presence of hybrid Aliens among us, old hat, false promises with undefined meaning.

We-Ko-Pa Resort. Saturday 20 February 2016.

The day began with an interesting lecture on torsion fields by MIT-trained physicist Dr. Claude Swanson. Discovered in the USSR in the 1950s, torsion fields are presented as 'eddies in zero-point energy' caused by the acceleration of spinning particles. The names associated with the theory are those of Kozyrev and Laurentiev, also Lunev at Tomsk Polytechnik, Chernobrov…

Lee Spiegel and I held a debate after a discussion on Disclosure (that word again!) by Merrill Cook, an ex-Congressman from Utah. Our session broke attendance records as the organizers kept bringing in chairs into the large auditorium, an audience of about 1,200 people.

We-Ko-Pa Resort. Sunday 21 February 2016.

Everywhere, people want to say hello, have pictures taken, and talk about the experiences that brought them. I spoke to a microbiologist from the University of Michigan, Dr. Don Clewell, who has studied the skin marks and remains as puzzled as the rest of us. Chris Rutkowski, one of the Canadian scholars of the phenomenon, a tall fellow with a sense of humor and a frank, open smile, became an instant friend. People are eager to tell their own stories with passion.

Douglas Trumbull wants me involved in his radical effort to redesign movie cameras for the ultimate viewing. I tell him the plan will raise many questions for investors. Then at a break I meet Greg Bishop, one of the new freelance investigators in the field. A few years ago, he wrote *Project Beta* about the ordeal of Paul Bennewitz and his mental disintegration under the attacks of evil spooks (guided along by several cleared scientists from the shadows). When he decides to sort out the government lies, Bishop describes a dangerous swamp of false documents, fake projects, disinformation images, and ego wars that distract us from the real problem…or are they the real problem?

The lesson I draw from this is to bypass the tainted ufology channels and go straight to the witnesses. It's a sharp warning because, as Bishop writes, *"we have all been Bennewitzed."*

Hummingbird. Monday 22 February 2016.

Shocking news: Our close friend John Forge died of a heart attack over the weekend. He fell from his motorcycle near Santa Cruz when his heart failed. He broke eight ribs and one of them perforated the spleen. The medical team was unable to revive him. John (born "Jean-Eric") was a long-time companion from my early years in venture capital. He was fun, young in spirit, and made friends easily. His wife is a warm woman, a hard-working, courageous cancer survivor.

Hummingbird. Tuesday 23 February 2016.

This morning, I savored the privilege of driving down through the twisted hills of Los Altos; a fawn stared at me when I passed him on my way to Federico Faggin's house.

I wanted to give Federico my thick file about gravity research. He told me about the progress of his mathematical formulation of physical reality as created dynamically by consciousness. He made a cup of Italian coffee for me and we sat down in the kitchen with his wife, Elvia, who told us that Umberto Eco was being buried in Italy today. This launched us into a discussion of *Foucault's Pendulum* and *The Name of the Rose*, which I read as a deceptive book. Eco ridicules the antics of occultists without taking the trouble to understand the hidden vision. Federico, too, had put down *Foucault's Pendulum* before the end because it seemed a hodge-podge of superficial images gleaned from all the esoteric-sounding garbage on television.

Hummingbird. Thursday 25 February 2016.

Hard times: Peter Beren sinking into depression, Jack Katz on the edge (I encourage him to sell some paintings, one of which I've bought), Flamine struggling in Paris. For me, the next step is a visit to Rice University, whose motto is 'Unconventional Wisdom,' perhaps the future home of my archives. Then in September, a trip to Argentina to help a local movie director, Alan Stivelman, document the life of Juan-Oscar Perez, the boy who had an extraordinary experience at Venado Tuerto (11) in 1978 when he rode his horse into the morning fog and was confronted by two "robots" and a landed saucer.

Like other witnesses, Juan has become a recluse, ignored by his family. To see him again, and try to help him, would close an important spiritual circle for me. Thanks to Fabio Zerpa, we'd met the whole family. Now Alan has found out that Juan's mother, of Guarani heritage, had suffered her own painful experiences, never revealed of course, given the Tradition. I need those six months to relearn Spanish.

Hummingbird. Friday 26 February 2016.

Timothy Taylor, a remarkable entrepreneur and researcher who is in contact with Dr. Nolan, had agreed to meet with me this evening, at Bistro Vida. One of his patents has to do with metal-on-ceramics materials, which he plans to use on medical implants, his main field of investment. I'd backed Regeneration Technologies, now a public company making such implants in Florida, so we were in familiar grounds.

The ceramic-metal combination apparently comes from experiments he knew about at Cape Canaveral. I mentioned my interactions with Marcel Vogel of IBM research labs, who'd studied similar materials. Vogel did tell me his sample came from a recovered vehicle, and they had a psychic function. The inventor of thin-film digital storage, IBM owes a fortune in disk drives to his genius.

Hummingbird. Sunday 28 February 2016.

Eric notes the claim that "metamaterials came from Roswell," but scientists don't believe the allegedly unknown technology is esoteric.

"The fact that it took 30 years to find the analytics to identify the composites seems more related to technical abstruseness than ET cleverness, so I choose Black Aerospace," one expert writes, adding that today, Black Google and Black Apple also experiment with this in foreign labs, yet nothing gets published in academia. There is a sense that important discoveries are being made away from the government. No wonder: There's a wide impression that the Intelligence community is corrupt, using secrecy as a shield for evil within a structure that has become ineffective. The current legal fight between Apple and the FBI about encryption and privacy is an example of the deep breakdown in trust.

I drove up to Garry's modern home and we looked closely at my brain MRI. He taught me how to locate the structures of the striatum, namely the putamen and the caudate. In my case, they are heavily connected. That's also true for several remote viewers and some close encounter witnesses we know, but the research is only the beginning.

The putamen and the caudate are level with the top of the eyes. They belong to the basal ganglia, with the putamen linked to the rest of the brain. It is thought to serve as a collector of brain states: perceptions, emotions, executive functions. It mainly serves to plan and set up goals. It is also involved in the processing of fear.

Finally, we reviewed Tim Taylor's data about puzzling samples he'd analyzed. They came from two crashes: one in Mexico and one on the plains of Saint Augustine.

Hummingbird. Wednesday 9 March 2016.

I've begun sorting out the research archives into three sections: A for Analysis, partial Blue Files up to 2014, already organized; B for Background, secondary information culled from diverse sources, illustrating research; and C for Correspondence, also well-organized papers. In the future I intend to concentrate on the 'distillate' from this operation: the most exceptional events that deserve my continued study, beyond what was done at BAASS.

A professional writer named Annie Jacobsen is the author of a new book about Area 51. She knows about Project Monarch, an MK-Ultra mind control follow-on, a monstrosity that exploited children undergoing mental tests in Chicago and Canada. She was told by government experts that previously classified technology, and not magic, explained "most, if not all alleged real ufological data."

Of course, I've heard colleagues claim this since the 1970s. My best data would never be considered in such a study: Juan-Oscar Perez' experience (the 'Heavy Glove' case) or the cases Hynek, Fred Beckman and I studied in the Midwest would be disqualified even before witnesses could testify. The best way to get rid of a pesky problem in science is to disregard the data.

More to the point is the realization that medical ethics is broken in human research of the phenomenon. There's far more disinformation than we suspected, and the classified Establishment is deep into political damage control.

I was interviewed on the phone by Annie Jacobsen, interested in SRI's parapsychology work and in *pro bono* medical studies of UFO interference victims. She gave me an opportunity to expand on my own assessment and the role of information science in shaping the field. Annie leaves for Israel tomorrow. I asked her to give my regards to Uri. I had dinner tonight with Ed May, at his request (at Crouching Tiger).

Hummingbird. Saturday 12 March 2016.

When you start moving books around, you're bound to rediscover forgotten treasures. Among the latest jewels that have fallen on my lap from the shelves: *Mémoires d'un Ange Bâtard* by Harold Norse and *The Uses of the Past* by Herbert J. Muller. He writes, about medieval men, "We may be proud of them. We owe our being to their restless striving, their eagerness to experiment and adventure, in particular the ardor for learning, beauty and fullness of life that made the 12th century a profounder, more wonderful period than the official Renaissance."

Houston. Sunday 13 March 2016. Hotel Zaza.

Rice University must have a special rate here because I've been upgraded to Villa Champagne, all dark wood, chocolate wallpaper, and lurid velvet curtains. The artwork, splendid portraits of entertainers, is tasteful and fun in keeping with the modern spirit of Houston, liberated from its old Texas awkwardness.

The cab driver who brought me to the Zaza belonged to a company called Halleluiah. He shook my hand warmly and said, "God bless you." Tomorrow, Jeff Kripal's colleagues will show me the archives where the research will rest. I still see myself in the character of Pimen in Boris Godunov, the old monk who has kept the careful records of the follies of the age and the phantasies of destiny. Also, knowledge of a few forgotten crimes, their solution left to God.

Houston. Monday 14 March 2016.

A welcome break in the schedule: Jeff's magnificent office lined with books, guarded by a massive Iron Man, and scifi posters. I'll give my lecture before his graduate students and local members of Noetics.

The archives are a clean, airy suite of dignified rooms with every convenience for visiting researchers. What attracts me here is Jeff's commitment to active research. His magnum opus, over the next 15 years, will be a trilogy on the paranormal, starting with evolutionary biology (the role of mysticism in evolution, epigenetics, Wallace vs. Darwin, the Aliens as future humans...). Quite a program, in which he plans to enlist his graduate students in comparative religion, in the spirit of three great mentors: Scholem, who revealed to Academics the Kabala (considered as an obsolete Jewish oddity before him); Eliade with the revival of shamanism; and Pagels who showed that Gnosticism wasn't just a pseudo-Christian heresy. Jeff wants to expose how people use science today to express their paranormal and anomalous experiences.

Hummingbird. Saturday 19 March 2016.

In honor of J.S. Bach's birthday, a San Francisco group gave an organ concert at St. Mark's yesterday. I felt more at peace than I had in a long time. Lunch with Tania at the Fairmont. She has contacts for an early-stage medical fund and requests my help, but the plan is very preliminary.

"The Keep" near Santa Cruz. Sunday 20 March 2016.

James Lloyd is a sophisticated architect and builder, close friend of Oberon. He and his family live in a beautiful castle-like structure he built in the forested hills above Santa Cruz. They support the Academy of Arcana, with its storefront on a busy street of this University town. I gave an informal talk there this afternoon in a richly decorated room lined with bookcases and emotional memories of Morning Glory, with her Goddess collections of all countries and cultures. Vivian, one of their friends, joined us for a quiet dinner.

Oberon was in black velvet, striking with his flowering white hair and heard. We reflected on our long friendship and on the ongoing mourning of our companions. The group has moved away from the redwood forests of Mendocino, as I have, for similar reasons.

Hummingbird. Wednesday 23 March 2016.

Now a terrorist attack in Brussels morphs into the commonplace, pitiful images of the aftermath: people assembled to cry, bringing candles to a central square; listening to their king as he underlines the sense of dread, the loss of dignity at the hands of false worshippers of Islam. Where is the condemnation from leaders of their church?

Nobody bothers to ask why it happened, and what all those bodies mean, torn apart by sharp nails and the evil chemistry of hate. When the frustrations of humanity find their exclusive translation in this folly, how can we protect the Earth? Would a revelation of extraterrestrial life introduce a new order—or make things worse?

Hummingbird. Friday 25 March 2016.

Last night, a pleasant and erudite lecture on behalf of SETI by Andrew Krasnoi at Flint Center. I was greeted as a colleague, a welcome contrast to some of the vitriolic attacks against *Wonders* from the 'rationalists' in the group.

About Annie's stories about DARPA: she's been approached by a man who told her about seeing "a captured saucer and bodies, and the physician who did surgeries… One CIA guy said Roswell had to do with Operation Paperclip; the US continued the ugly Nazi work of Mengele." Where is the truth?

Hummingbird. Saturday 26 March 2016.

Annie Jacobsen flew up from L.A. to interview me over lunch at the Fairmont. We ordered their miserable Salade Niçoise and spent four hours together. We covered a lot of ground, in private. She was just back from Israel, where she'd seen Uri.

"He's put his UFO claims behind him," she said. "He no longer mentions the subject." Annie had Volume 3 of *Forbidden Science*, which she heavily annotated, filling four pages with questions.

Hummingbird. Monday 28 March 2016.

I picked up Annie Jacobsen's book about DARPA (*The Pentagon's Brain*) to read her account of the Internet, but the real Arpanet inventor, Paul Baran, is not listed in the index. Knowing Paul, he would shrug and move on, as he taught me to do. But Larry Roberts, the man who built the network with Paul's guidance, isn't mentioned anywhere either. In a world where technology threatens to overwhelm everything else, it is staggering to find that the most important events, still within human memory, are not only misunderstood but encrusted in misleading layers.

We don't even need the KGB 'active measures' to mess up our history. *We are misinforming ourselves about our greatest cultural achievements*, even in areas of technology we use all day long, where very little was ever secret. It didn't need to be because so few people understood the long-term potential. Paul Baran did, but he hated publicity hacks, and he's now beyond the historian's hungry scrutiny.

Brandan Fugal invites me to the Utah Ranch he's bought from Bigelow, but several of us have said we wouldn't set foot on the property.

Hummingbird. Tuesday 29 March 2016.

Over the long and lonely Easter weekend, I sorted the sections of my archives that will go to Houston, and I planned my second trip to Argentina. Now I'm reading again the French translation of Harold Norse's *Memoirs of a Bastard Angel* with as much pleasure as I did on first contact with his fresh, colorful narrative, evidently based on a well-observed, carefully documented journal.

Hummingbird. Thursday 7 April 2016.

My rare book expert friend, Todd Pratum, came over this afternoon to see the ufology collections. He examined the first part of the archives, the 18 boxes of the B (Background) section I'll soon transfer to Rice.

As he moved from bookcase to bookcase, he made colorful comments about some unique volumes or common tomes, relating them to little secrets of his trade. He even recognized the seller of a used book simply by the way the price had been penned on the overleaf!

After Todd left, I drove over to Point Richmond to see Jack Katz and take him to dinner. He'd had a fall, walked with a cane, and was in pain with a broken finger, a head wound, and twisted back, yet he managed to dazzle me with his vast vision of adventures beyond the dreaded borders of the cosmos.

Mabillon. Monday 11 April 2016.

The new chief of the GEIPAN group, Jean-Paul Aguttes, had requested to see me ahead of the general meeting tomorrow, so we had a long private lunch at Les Halles, close to CNES headquarters.

Aguttes is a balding man with glasses that give him the look of a solid French technical manager. Trained in physics and an expert on radar, he is at the end of his career and therefore immune to attacks from skeptical rationalists, but he's new to this subject. He has studied a few dossiers already, like Hessdalen, Haravilliers, and Lakenheath. He's also aware of our Pocantico meeting, through Louange.

Paris (Montgallet). Tuesday 12 April 2016.

The official French ufology group meets today at the massive headquarters of the ESA-CNES 'Launchers' group where the next version of Ariane is being designed. As large as two city blocks, the building is not identified on the maps of Paris.

The transition between Xavier Passot and Jean-Paul Aguttes takes place today. It represents a gain in visibility for the project since Aguttes is a deputy director of the agency, but he'll have much to learn. Pierre Bescond joined us. He chairs the steering committee.

Mabillon. Tuesday 19 April 2016.

Flamine proposed to see a play, so we took the Métro to Montmartre where a comedian had produced his one-man interpretation of the fantastic story *Le Horla*. I was struck at how Maupassant, in 1887, had anticipated our current puzzlements, including the possible role of cosmic beings intruding into our lives, beyond our ordinary senses. He even mirrored the modern anguish at the thought of galactic interlopers disturbing our bedrooms, as in the latest report from one of Garry's 'haunted' subjects that mentions videos of various lights (or objects reflecting the infrared sources attached to the cameras) dashing across a room.

The play brought back memories of late 19th century speculation about cosmic intelligence by Camille Flammarion, and the later work on hauntings by Commandant Tizané. Our experiments and hypotheses are nothing but echoes of research our grandfathers had covered. The only difference is our modern array of shiny gadgets that bring no new answer.

Mabillon. Thursday 21 April 2016.

The air of Paris, as soon as I came out of the train from Rochecour, assaulted me with polluted, gagging junk. All day, my lungs have struggled for oxygen and my brain for clarity.

The network brings news from John Schuessler about 'shadow people' in his basement library and storage area, again a scene worthy of *Le Horla*. As for Garry, he's on his way to New Mexico, hoping to dig up Alien artifacts in the Plains of Saint Augustine, his sense of humor intact, intellectual antennas fully deployed.

Hummingbird. Sunday 24 April 2016.

An interlude: I left behind the unanswered mail and the dreaded web to drive away to the Santa Cruz Mountains and visit Vivian and discuss psychic research. She's a wise, gray-haired wispy woman with enquiring eyes in a home full of books, in a redwood house up the hill, through the forest and across a little bridge, as in fairy tales.

Vivian told me she'd had two striking psychic episodes, visions of her estranged father who'd abandoned the family. Once, he came to apologize, an apparition at the foot of her bed; and a short time later he was seen again, begging her to care for her half-sister. He died a few days later in a car crash.

"What did you do?" I asked her.

"I moved to California," she replied quietly. "I met the girl. I raised her as my own daughter."

Hummingbird. Thursday 28 April 2016.

At Insight Editions this morning they handed me a heavy box, the first full proof of the Limited Edition of *Wonders*. It's a wonderful example of the printing art, and a great tribute to the skill of Chinese craftsmen. It is worthy of standing on a shelf next to Boaistuau, Lycosthenes, or the anonymous author of the Augsburg manuscript. Chris and I have set a new standard and given these old sightings their letters of nobility.

Hummingbird. Tuesday 10 May 2016.

The media, from the *Huffington Post* to *The New York Times*, vibrate with excitement at the news that Hillary Clinton will "reveal what the government knows about UFOs." I did call George Knapp today, and he put me in touch with Tom Delonge, whom he was helping with his own contacts. "The people in the group are high-tech execs," George said. "John Podesta, Hillary's adviser, is the only one who's been identified. The message I've crafted was, 'Let's assume this UFO stuff is real. Is there a way Tom can help you get out of the corner in which you've painted yourselves?' So far, they've responded positively." But will George's message reach its target?

Hummingbird. Friday 13 May 2016.

Today the Institute for the Future had invited about a hundred *Technorati* of Silicon Valley to meet the chief of an organization that uses powerful Internet search techniques to uncover the financial tricks of dictators, eastern potentates, sneaky bankers, and corrupt governments. A 'white hat hacker' and an idealist, he described the efforts of his group to link together databases across the globe and the resulting exposures, such as the now-famous Panama Papers.

His description of the criminal economy and underground financial circuits is chilling, from the structure of human trafficking in Eastern Europe to the sins of the oligarchs. But do the major governments, who use the same channels, really want to clean them up?

Las Vegas. Friday 20 May 2016.

My friends tell me the main event, every Friday afternoon in Vegas, is the arrival of the 'cougars in heat,' who've left their husbands to their male pursuits, like hunting and golf, to converge on America's capital of sin. I had other interests as George Knapp picked me up at the airport in a big black limo for a meeting with Delonge. Tom may be a punk rock star, but he's also well-organized and smart. A strong-built, likeable man in his early forties, he's the first to admit that he's getting way over his head amidst military intelligence folks. We all are.

In the process of promoting an idea for a UFO documentary, he approached the head of the Skunk Works with a pitch provided by George: Why not bring out the Pentagon's side of the story? Why not a series of books to go with the film, fiction and non-fiction? The brass fell in love with his idea and encouraged him with their own.

"They think the real story has to do with entities," he told me as we sat down in the back of a hamburger joint at the Mandalay Bay casino. "They call them 'the Others' rather than 'the Aliens.' They think these beings have been fueling religious strife on earth to force us to fight each other, and eventually to destroy mankind. So, I went to a general who works for NASA, and is interested in SETI…"

"Was that Pete Worden?" I asked.

"Yeah, you know him?"

"Our NASA venture fund my team created, Red Planet Capital, was under him at Ames. Pete is great."

"Good, so you see where this goes. After telling me three times that he was a skeptic, as they all do, he put me in touch with the general I really wanted to reach. Top-level guy. Can't tell you his name."

"Had you written something at that stage?"

"Yes, I'd been working with Peter Levenda, who looks at the cultist angle, and we told our new military friends that we wanted to help break the stranglehold of manipulation."

"Where do you think the real project is located?"

Tom grabbed his cap from his head, twisted it for a while, set it back, and said, "We got bumped around four facilities, trying to find it. Guards with machine guns. All decoys. Part of the project is corporate but it's run from inside the government."

"Isn't that a violation of some law?" I said. "Christopher Mellon just told Leslie Kean he'd looked at every secret project and couldn't find it."

"Well, whether it's legal or not rests on the interpretation given by a couple of guys who run it. I can't tell you who they are. One of my contacts is at Wright-Patterson, another is a high-level woman manager at NASA. They all want me to tell the story, but when I push them they only answer, 'We found a lifeform.' So..."

"Just like that?" I asked.

"Just like that," Tom said.

It was strange to have that conversation amidst the bustle of the restaurant, the skinny waitress in black with her tattooed skin, coming back periodically with the coffee pot. "And what about the role of the Nazi in the Roswell crash?" asks Tom, rhetorically.

Hummingbird. Saturday 21 May 2016.

I'm sending flowers for my daughter's birthday, "an old baby," Janine would have said with her wise, patient smile.

The discussions between Tom and his powerful advisors seem to have risen to the theological level. "Who are the gods?" he asked. His mentors pointed him to the Greek pantheon.

Other ufologists have come to him, brandishing legal papers and contracts to sign, asking about names of contacts. Tom didn't reveal anything: "My guys said I shouldn't have anything to do with them."

Later, as we rushed to the exit to catch my cab, Tom showed me a picture on his phone: "You should know about this. The general in question that Pete Worden put me in touch with is William N. McCasland, commander, Air Force Research Lab at Wright-Pat."

McCasland, Wow: MIT doctorate in astronautical engineering, responsible for a $2.2 billion budget in science and technology, with an equivalent budget for customer R&D. He's responsible for 10,800 employees, including the staff of AFOSR. During his career he served at all the right places, working at NRO before he commanded the Phillips site at Kirtland and oversaw special programs for the Acquisition, Technology & Logistics Agency. He retired in October 2013. Are all of Tom's advisors of that caliber, and retired? Another prominent Intelligence man, Christopher Mellon, was a contact. Mellon, who's just given a long interview to Leslie Kean in which he said he'd never found a serious UFO research project in secret government files, in contact with some space scientists in Europe, including one friend who recently interviewed me about monitoring techniques as he completed a technical report for the CNES. More unexpected connections, yet logical, and meaningful.

In that five-hour meeting with George and Tom at Mandalay Bay, I was struck by the fact that the names of the people mentioned had never appeared before, not even Pete Worden's. These men (and one woman) are unrelated to the Collins' Elite and seem immune to arguments about fundamentalism that have so worried the LoneStars.

Hummingbird. Friday 27 May 2016.

Here is a strange historical fact: As San Francisco plans to repurpose the prime real estate of Treasure Island, once a major Navy base, stories about the Cold War keep coming to light. There used to be a unique training ship on that island, a vessel that was never designed to go beyond the Golden Gate. It was called the USS *Pandemonium*.

Built out of scrap metal, it floated in a corner of the base and was purposely contaminated with cesium and radioactive dust to simulate what would happen to a ship in a nuclear war. Sailors in special suits used to be sent to practice cleaning techniques. Now my friends in City Hall need to rediscover the various locations where that 'hot' ship was used, in order to remove the lingering pollution.

Hummingbird. Saturday 28 May 2016.

Todd Pratum and Jeff Kripal will be here in an hour. They've never met, although Jeff once bought an esoteric library from Todd. This meeting is important. More than ever, I want to find a secure home for the archives, one where future scholars can study in peace while I focus on a few nuggets from the past, and put my energy into new cases.

I've read the manuscript of *Sekret Machines*, the first volume Tom Delonge plans to publish, and I liked it, so I just sent him the Foreword he asked me to write, curious to see where this new road may lead.

Hummingbird. Monday 30 May 2016.

Clas Svahn, the international director for a group of volunteer researchers in Stockholm (UFO-Sverige), was in San Francisco today with his family, so we went for a long lunch at Les Joulins. He's a tall, well-organized, fiftyish man who has managed to convince many groups and individual researchers around Europe to entrust their collections to his organization. In Stockholm, he's assembled a dedicated team of volunteers, including half a dozen full-time employees, to screen the documents, index them, catalogue them, and scan them to preserve everything on disks. The team includes a couple of retired librarians, and some unemployed young people supported by the State. The archival work is extraordinary.

Toronto Airport Sheraton. Wednesday 1 June 2016.

Eric is here but Hal has delayed his arrival by one day in order to get an update from Dave Jacobs (now retired from Temple University) about his abduction research. Garry is coming tomorrow evening, but John Schuessler can't make it for health reasons. Colm will be here with a free weekend now that Bigelow's ambitious BEAM module is firmly attached to the ISS. In two weeks, most of the group will meet with Brandon Fugal in Salt Lake City to discuss new research at the Utah ranch. I have politely declined, because of potential conflicts of interest with Silicon Valley startups, although my own information about the region may be useful to him later, if he wants it.

Toronto. Thursday 2 June 2016.

Since we had an unexpected free day, George Hathaway gave Eric and me a leisurely tour of his lab. I was much impressed by the seriousness of the research, speculative as it may be. He recounted how contact with his sponsor had begun after he performed a survey of the field of advanced energy physics and psychokinesis in 1988.

At 5 pm the rest of the team arrived and general discussion began, with queries about David Jacobs. In his latest book, *Walking Among Us*, he claims Aliens are indistinguishable from humans.

Toronto. Friday 3 June 2016.

George began the proceedings this morning with a recap of the eight prior meetings of the LoneStars. The topic to which we kept coming back is the Interference Syndrome, which we think is human made.

I thought we should focus on the nature of the channel, not simply on the content of what it carried. We're beyond technological facts now, but my friends don't agree, naturally; they are scientists. But even among them, there's agreement that the field is changing, as I keep telling them, becoming scary, influenced by invisible forces.

Hummingbird. Tuesday 14 June 2016.

Now, a text message out of the blue from Tom Delonge: "I'm having a face-to-face with Admiral (—) in Austin, so Hal Puthoff, Undersecretary Mellon, and I are going to meet afterwards. Would you like to come to Austin and have another meeting on July 19th?" The answer was easy.

This evening, Garry Nolan is having dinner with Richard Niemtzow in Washington, at the Army-Navy Club, at my instigation. He called me for more details. I told him Richard was a true pioneer; he had already researched what we now call the Interference Syndrome in the 1980s (including how UFO signals affected the permeability of cell membranes) well before others became seriously interested.

Hummingbird. Thursday 16 June 2016.

Reviewing my notes from the Toronto meeting, and going through archival material, I am seized once again by questions about the path we're following. Although I am fully onboard with medical studies of what we call the Interference Syndrome, the premise of the study raises deeper questions, as does the fact that so little information is available about the selection of subjects in our survey.

Take my own case. I score very high on the striatum connectivity scale, yet I am neither an experiencer in the sense of a close encounter victim, nor am I an exceptional psychic. The density of nervous fibers, in my case, could well come from intense experiences in childhood rather than anything related to paranormal phenomena. I recall Whitley Strieber quoting a study on rats subjected to extreme perils (random electric shocks) whose behavior was positively altered by the experience. Living my first five years in a high stress shooting/bombing war could be a much larger factor in the case of my brain than exposure to a UFO. Nor do I feel comfortable with the claim that UFOs are somehow screening witnesses based on their genome or the nerve bundles in their heads.

More to the point, I am concerned as I see us following the path of Budd Hopkins, becoming True Believers in our own hypotheses. A scathing article by Budd's wife Carol Rainey describes his gradual descent into credulity and obsession.

Hummingbird. Saturday 18 June 2016.

Five hundred copies of the Limited Edition of *Wonders in the Sky* have sailed under the Golden Gate at last and are now at the Port of Oakland, waiting for the trucker that will deliver them on Tuesday.

Internet blogs have already trashed the book, unsurprisingly, but I look forward to re-igniting the controversy about Vulcan and Neith, the two planetoids reported by some of the most trusted astronomers of the 18th and 19th century, summarily swept under the rug when theoreticians no longer needed them to complete the solar system.

Now some good news: Cogent, one of the surgical companies in my portfolio, is being acquired by Entellus Medical.

Hummingbird. Monday 20 June 2016.

Tom Delonge calls me about the latest 'Wow' revelations from his advisors. The new story line goes back to the Greek gods creating a hybrid race of 'Nordics' (hence the German esoteric traditions?) that led to German saucer prototypes in the 1930s. But the full power wasn't realized by the Nazi until they went to South America, gained the ancient Mayan knowledge…Those Nazi are in Antarctica…

None of this makes sense, but high-powered military leaders are openly involved, along with senior managers at Skunk Works. Tom's also been told that "the NRO has its own craft," something my friends suspected all along, but how does any supposed Nazi technology explain this body of data?

Hummingbird. Friday 24 June 2016.

Everybody woke up this morning to news of the UK's decision to leave the European Union. The markets are spinning out of control and pundits lecture about causes and effects, but for me the cause is primarily a *breach of trust*. People can go through tremendous hardships when they have a vision of the future or confidence in a leader. None of that exists in Europe now. Even those who, like me, have always hoped for the success of the European Union—intellectually, spiritually and economically—have to face that the fragile, fractal bureaucracy erected in Brussels is flawed.

Hummingbird. Saturday 25 June 2016.

Two meetings of interest in Menlo Park today. First, a long breakfast with Garry at the very busy Stacks coffee shop. I delivered the seven copies of *Wonders* he'd ordered, and we spoke of his meeting with Niemtzow; evidently, they got along well. They discussed the interesting position of Hillary Clinton on UFOs, and an hour later Garry practically bumped into Hillary herself in the hallway of the Hilton. She was looking pleased, emerging from a private meeting with Bernie Sanders, who went the other way, visibly upset.

Hummingbird. Wednesday 29 June 2016.

A producer for CNN called me this morning to explore the idea of a documentary, starting with "the mystery of the Black Knight." I told him I couldn't help much. The truth is that a weird media movement is rising up, looking for anything new to say, no matter how bizarre or distorted: no fact checking. As Tom's plans become clearer, I push back on his pronouncements about hypothetical Nazi geniuses in the wings. He replied by describing his interaction with one of his advisors: "We're talking about the way war seems to start so easily between 'tribes' these days, and then he mentions a classic science-fiction thread for my book..."

This made me even more skeptical, so I insisted: "Many witnesses are getting smart enough *not to report the cases* to people they don't trust, namely (1) the government and (2) the media, in that order. And they're not susceptible to common 'explanations' we see on the *X-Files* or in the movies. I'm left wondering where the advisors you have assembled are trying to lead us."

Tom wasn't disturbed by my rebellious spirit, writing again: "I understand your position. They're saying very clearly that the Nazis had contact with the phenomenon. Question is, given all the weird aspects of the phenomenon, are they all one lifeform...?"

I need to think about that, Tom has a serious point.

Hummingbird. Tuesday 5 July 2016.

Today, a call from one of the many podcasters around Tom Delonge: "There's a rumor that Lockheed Skunk Works had the SAP for the reverse-engineering of a craft that 'wasn't from this world.' Have you heard of it? It's all over Hollywood."

"News to me," I said, "Who's involved?"

"Supposedly, the man was in control was Glenn Gaffney, a deputy director of CIA and friend of Jim Semivan, but he won't talk to us. The project stopped in 2012. The material is classified."

"What was it?" I asked, a bit shaken.

"Well, among other things, a ceramics instrument panel. Could it be *esoteric, yet not ET*? Reportedly they also got tissue, slides, perhaps DNA. But that dated from the fifties and sixties, and now people don't give a damn, it just didn't get anywhere. They're not interested in Hal's physics, all that's old hat. They're more interested in you."

I had to let that last comment sink. I'm not doing physics.

"What about Delonge?"

"He's been told that the Roswell tale in Annie (Jacobsen's) book was true, about German technology dropped in the desert. Jim Semivan exploded when he heard that: 'Why in Hell are these four-star guys talking like this?'"

Obviously flustered and angry, the man went on: "It's also a violation of federal law to fool the American people by posting false data, classified or not. Targeting 12-year-old kids…"

"The believers are being used. We already knew that," I said.

Later the same day.

Over the Internet, Garry shows me his analysis of the object he picked up on the plains of Saint Augustin, with the strange knitting pattern of wires, one-fifth the thickness of a human hair. He tested it over the weekend with the prototype of the new MITI machine.

"The sample isn't radioactive; it seems to contain material with atomic mass around 210. We think it may contain polonium, not in natural form, and bismuth, astatine. We also suspect yttrium, zirconium, and strontium. There's a lot of aluminum and nickel, also silicium and calcium but those may be contaminants. The stuff was embedded in the dirt of the desert, for years."

I told him it was time to write this up, and to consult Federico again.

Hummingbird. Thursday 7 July 2016.

Breach of trust is in fashion, all over the world: Brexit, understandably rejecting the bureaucracy of Europe, and in the States the anti-Congress, anti-Establishment movement leading to Trump. Nobody went to jail for the subprime crisis that threw some eight million Americans out of their homes, and none of the white cops who blatantly executed innocent Black men in the last few months have been disciplined. Now, another bloodbath in Dallas, a discouraging scene.

Hummingbird. Friday 8 July 2016.

What's the meaning of the connectivity patterns in the brain? "Those with the connection form a distinct population from the norm," Garry said, reminding me that there was no bell-shaped curve in our findings. The effect is genetic, so it is pre-existing at birth, or at least pre-disposed, not what I'd thought.

He went home to pick up the honeycomb while I drove ahead to see Federico in Los Altos. This gave me time to report on the gravity experiments and to bring him up to date on our patent position.

Garry recalled what I'd told him about Federico's out-of-body experiences, so he probed the subject, eliciting a stunning description from Dr. Faggin:

"It was as if I'd been a fish at the bottom of the ocean all my life, unable to imagine anything else, then something picked me up and brought me above the surface. That little fish started seeing trees, mountains, strange animals, cities, and stars: he'd never be the same! That's what happened to me. It didn't last very long, only some 10 or 20 minutes, but that was enough. Since then, I haven't tried to leave my body, but it has happened spontaneously. I don't think people should train themselves to do it."

Hummingbird. Friday 15 July 2016.

Bloodshed again, in Nice this time. A mad Islamist driving a heavy truck has plowed through the Bastille Day celebration, killing 84 people, including ten children. As we try to process these images a military coup shakes Turkey. The machinery of human madness lurches forward. Now I see that Turkish soldiers have broken into the CNN studio in Istanbul. The last screen has gone dark.

Hummingbird. Sunday 17 July 2016.

The tragedy in Nice lingers horribly over the French summer. Jean-Claude Dufour tells me a family of his friends was on the path of the killer truck. They were thrown, dazed but unharmed, but the people with whom they were walking to the show were crushed and killed.

Flamine describes the secondary effects, the depression that has seized the French, the somber reality spreading over. Annick commented she was happy to have traveled around the world and seen the beauties of Turkey, Egypt, and other places thirty years ago, when those regions were at peace.

Today, we wouldn't think of wandering off among the bloodied marvels of the Middle East.

Austin. Tuesday 19 July 2016.

The only food United Airlines serves for breakfast in economy class is coffee with a *stroopwafel* in a plastic package. Opening the wrapper, I discovered that a *stroopwafel* was a round biscuit like a *snookerdoodle*, only flatter, with even more sugar, a diet disaster. In every other way the flight was unremarkable, and we landed on time.

The Omni Hotel, downtown. Tom arrived tall and relaxed, dressed all in black with his black FORD cap. We went out to dinner, and we spoke for two hours, impressed by his advisors.

The message interaction with these decorated officers, which Tom reviewed for me on his laptop, ranges over the territory we've covered before: Nazi saucers, Antarctica, claims of contact in South America, speculation about Roswell. Tom will publish non-fiction books, having hired Peter Levenda to help. I declined participation again.

Then I heard from Garry, who had just recovered some strange material during a New Mexico trip.*

Austin. Wednesday 20 July 2016.

Tom had a private meeting with an Admiral. It doesn't seem to have led to new contacts. His ventures have not done very well, although he invested in "over 30 startups" in 16 years. Venture Capital is harder than it looks.

Christopher Mellon, former Under Secretary of Defense for Intelligence, landed at lunchtime today, a tall Midwestern intellectual wearing a blue jacket, immediately comfortable among us. Peter Levenda arrived from Florida, a quiet man with a white goatee. He is the writer hired by Tom Delonge for the non-fiction trilogy.

We began discussing a 20-page statement Tom had assembled, but we learned little from the session, primarily devoted to finding a storyline for the fiction series. I gave up in mid-afternoon. The military officers want to get people to look at the world differently, starting with teenagers. To their credit, they're worried about the rise of fanaticism; they want to use a scifi framework to show how the future could be reshaped. The strategy is clear enough. The way it will influence the youth of America, less so. Some people even think it's against the law.

Hummingbird. Thursday 4 August 2016.

Now Tom Delonge calls to discuss progress with his advisors. His 'No.1' may have had direct contact with Aliens, he believes, because of a passage in a letter he sent: "Ask yourself if the group would be led to a contact through a formal protocol, or would it come outside any protocol?" But what about mind manipulation in the presence of Aliens? Could the contact happen at a national military command center?

---

* Garry showed Kit that material, with its white fibers, gray knots, and crosslatched chemical microscopy. "We found it in the dirt. Pristine, highly ordered. Artifact from the crash. Fiber-fabric?" he asked. It took Kit five hours at a science library terminal to find the solution among some 5,000 pictures. It came from two Copenhagen companies around 2010. Both failed to develop the material into radar reflecting fabric, and they went bankrupt. But how did it end up in New Mexico?

Hummingbird. Sunday 7 August 2016.

The Bach Soloists gave a concert last night at St. Mark's Lutheran, highlighting several 17th century composers I didn't know. It was a welcome diversion. My divorce with Flamine was effective on Tuesday, August 2nd. I am sad and tired with the end of that process.

I just mailed out some of the last *Wonders* books, and told my would-be partners that I wouldn't take part in any new biotech fund. I am bent on reclaiming my freedom, my time, and a little space to think. But the receding wave leaves a burden on my shoulders.

The first book issued by Tom Delonge's company (*Sekret Machines Book 1: Chasing Shadows*, written by a Shakespeare scholar named A. J. Hartley) follows Nazi scientists to Argentina, then to Antarctica where they are supposed to have a military base.

I told him that made no sense, and my Foreword is unrelated to that belief. Is that a *meme* that somebody is reviving, to be popularized by Delonge and—why not?—by Annie Jacobsen who suggested the object that crashed at Roswell was German, sent over courtesy of Stalin? Why is the same legend popping up in a novel inspired by American military brass?

In his introduction to the book, Tom writes truthfully that he was first contacted during "a summer morning in 2015 when I got the call from an old friend." The man was recently retired from the Skunk Works, which had an Open House where Tom was invited to introduce the top executive. He seized the opportunity to pitch an idea for "a project that could help the youth lose their cynical views of the Government and the Department of Defense."

Hummingbird. Monday 8 August 2016.

Ron Brinkley has dug up a fascinating old document about MK-Ultra and especially MKOFTEN, a covert project Sidney Gottfried initiated between the Pentagon and the CIA when he headed the Technical Services Branch. According to Gordon Thomas' book *Secrets & Lies* (12), the project was designed to "explore the world of black magic" and "harness the forces of darkness and challenge the concept that the inner reaches of the mind are beyond grasp."

As part of Project MKOFTEN, the CIA recruited fortune-tellers, palm-readers, clairvoyants, astrologers, mediums, psychics…and Satanists.

When I sent the link to the LoneStars, it was immediately noted that Project MKOFTEN focused on both humans and animals, overlapping with various biological warfare projects including the storage of prion-loaded human brains at Fort Detrick in the late 1950s-early 1960s.

One of our advisors confirmed: "Project Often was the host. The digraph MK means it was managed and funded at TSD, CIA. They were all 50s and 60s, and all ended *for cause* in March 1973."

He goes on: "There was one project that started in 1971 and was transferred to SAIC in 1973, called MK-Chameleo. The CIA had no scientists with qualifications to manage it. It was reported to have either died for lack of funds, or rumored to have been transferred."

Hal jumped in: "I recognize Sid Gottlieb, of course, he was the man to whom we reported on the Remote Viewing work."

Computer History Museum, Silicon Valley. Tuesday 9 August 2016.

I came here this afternoon with a heavy box of archives from InfoMedia and all my papers about social networking, which I'm donating to the Museum. For the occasion, Marc Weber had invited my friend Jake Feinler, the woman who developed the first Arpanet Network information center with me (the NIC) in the days of Doug Engelbart, so we happily reminisced about SRI days.

I keep reading about new warnings of possible Islamist strikes against France, in an atmosphere of continuing political decay and tourist unease. Yet Flamine talks about the glory of walking across the beautiful bridges of Seine as the setting sun throws its evening spray of gold onto the stones of the Louvre...

Hummingbird. Sunday 14 August 2016.

Garry Nolan, who attends a two-day symposium focused on epigenetic control, writes: "Epigenetics is the great unknown, basically a 3D encoding problem. DNA sequence coding, which we still don't fully appreciate, is trivial compared to epigenetics. It's a problem where the information content is dynamically transient, but 3D contextual at the atomic and molecular level. The 'states' are constantly (by the millisecond) transforming in a probability matrix that 'samples' opportunity and answers.

"*It makes me swoon to think about it. If there's anything that would make me believe in a higher power that 'designed' the universe to engender life, it would be the richness of our ignorance around how epigenetics operates.*

"Frankly, humans at our current state of evolution will never ever ever, **ever** be able to comprehend it. Superhuman AI will be needed to encompass an understanding. We'll understand pieces and rules, but there are many levels of unintended consequences...the sum is more than the parts...emergent behavior ... that becomes beautiful and scary in its own right."

Hummingbird. Wednesday 17 August 2016.

7:30 am. For a couple of weeks, in the morning, the hills of San Francisco have been wearing their white summer robes of pearly fog, fringed with the dust from the coastal fires near Big Sur. I am putting pressure on Todd Pratum to transfer my archives to Rice University, and on my Silicon Valley friends to enter the documents of our old InfoMedia Corporation (1976-1980) into the Computer History Museum in Mountain View. We anticipated full-scale social networking by 15 years; nobody remembers that.

Life moves on: I have my ticket for Buenos Aires to go and visit Juan-Oscar in September.

Hummingbird. Friday 19 August 2016.

In preparation for sending the "B" boxes to Houston, I've sorted out the index cards–thousands of them–left over from the statistical studies Janine and I ran in the sixties and seventies...and my eyes filled with tears when I saw them, covered by her intelligent, careful, fine writing... We worked so well, so closely! One more reason for shipping these records to Jeff Kripal, to keep them safe.

DelMar. Tuesday 23 August 2016.

Tom Delonge had decided we'd meet at L'Auberge, a fine hotel on the DelMar waterfront. I came over early, waiting for my friends in the sublime light of a Southern California morning, cup of coffee in hand, the Pacific visible beyond the pool and the hotel gardens, until Tom arrived and spoke about UFOs driven by Greek Gods and Atlanteans, reminding us that 'the Admiral' had mentioned their relevance.

"In the cockpit of the crashed disk at Roswell they found two bodies, an Atlantean and a German pilot," he said flatly.

I had to shake my head at this. Even the wildest rumors about Roswell (there are many, in conflict with one another) don't allude to anything of the sort.

"There's nothing especially 'ufological' and novel about the Greek Gods," I said. "They were inspired by Mesopotamian and Egyptian deities, themselves copied from earlier African gods. The Roman Mercury was the Greek Hermes who was the Egyptian Thoth who was Legba, son of goddess Oshun. Going the other way, Oshun became Isis who was Aphrodite who became Venus. If you track this, you'll end up studying the Yoruba pantheon."

DelMar. Wednesday 24 August 2016. L'Auberge.

Our second day of meeting, at a private table set away from the tourist crowd, with a plunging view on the beauties circling the pool.

For the aerospace industry to pick the rock music environment for its propaganda aimed at the Millennial generation makes perfect sense. The only universal language on Earth today is rock music, not English, or Hollywood scripts. Tom is a logical leader; a few years ago, he was camping in the China Lake area with three others when something flew over their tent. They "felt presences" and heard multiple voices drifting in the sky. Couldn't this have been a human device? They didn't get out to look at it.

This morning, Tom met with the Admiral, who brought up Ancient Greece, Athens and Sparta, the Atlanteans, and 'the Bugs.' I said Homer didn't write about any 'Bugs,' but Tom was unconcerned. He'd been told, "It's not what we know, it's how we came to know it."

Hummingbird. Thursday 25 August 2016.

Three meetings today, back in Silicon Valley, starting with an early-morning breakfast with John Hollar, chairman of the Computer History Museum. He wants me to come back and record a piece for the complete oral history of social networking and its business applications.

Next, another lunch with Garry Nolan. He had brought his laptop to show me the analysis of the material from the Plains of St. Augustine and especially the 'wires,' made up of aluminum and silicon, apparently doped with titanium and some organics. The resin substrate is of the same composition. He also said he'd experienced a terrifying episode a few nights ago, an attack on the subtle levels of consciousness.

Leaving Woodside, I drove over to Federico's house. He was clear-minded and happy. At Banff in Canada, he'd been able to speak with Arohonov and Paul Davies about the nature of time, retrocausality, and the dominant role of the mind, progress for his theory.

Hummingbird. Monday 29 August 2016.

The dust settles on the Del Mar meetings, with Jim Semivan taking a direct interest in Tom Delonge's projects. "We need at least $20 million," someone said, including funds for computer work.

Back in San Francisco, former NASA engineers Creon Levitt and Larry Lemke invited me for a tour of their impressive facilities at Planet Labs, in a funky converted warehouse South of Market. "More satellites have been built here than in the whole history of space technology," they told me proudly. We enjoyed a Japanese lunch and conversation with a friend of theirs interested in photographic analysis of 'unknown objects.'

Chris Mellon has responded warmly to my note. He must have heard about my skepticism about mixing up Greek Gods with Nazi saucers, so he pointed out that the group could be "providing a potentially useful inoculating function to *help prepare the public against what could someday be a shocking and disruptive revelation.*" Perhaps, but how does the Nazi connection help?

Noetics Institute. Earthrise. Friday 2 September 2016.

Flamine is back, so we took the truck and we drove up to Earthrise in Petaluma. We were first to arrive, finding the hills shiny with an early sheen of autumn in the setting sun. The staff of Noetics gave us directions to settle in the chalet where Jeff, Diana, Tim Taylor, and Whitley Strieber would also be staying. We met them warmly as they arrived. Bill Calvert was with them, as were Eric Ouellet from Quebec City and David Halperin from the east coast.

There was an orientation meeting after dinner as Garry came in, as well as Kary Mullis and his wife. Kary is the man who invented the polymerase chain reaction (PCR) in 1986 and was awarded the 1993 Nobel Prize in chemistry, shared with Michael Smith.

Earthrise, Petaluma. Saturday 3 September 2016.

Diana Pasulka gave the first presentation, centered on the origin of space technology in rituals and on the belief in extraterrestrials. When she finished *Heaven Can Wait,* her book on Purgatory, in 2012, she obtained a catalogue of observations of flying orbs or flying disks, and beams of light changing the material properties of objects, as preserved in the little Museum of the Holy Souls in Purgatory in Rome.

Diana somehow made a transition from these observations to relate the story of Jack Parsons and the hidden history of Space. In the discussion that followed, Whitley reminded the group that his uncle had been close to General Exon, who served as Air Force liaison with the science team that examined the Roswell material at Wright Field, well before 1963. Whitley thinks the secrecy can be traced to John von Neumann, who supposedly warned there could be a 'tripwire' in the human mind that would be triggered by contact with cosmic beings. Permanent denial must be maintained, he's supposed to have argued.

I spoke next, focusing on some lessons from *Wonders in the Sky,* especially two cases where the so-called evidence had been falsified. The MUFON claim of "the first flying disk in history!" is false; we know it was only an ordinary meteor seen in Kent in 1733, in the witnesses' own words, garbled by American editors. As for the amusing case of the attack of the French armies by flying beings at the siege of Angers in 842, it simply never happened. The latter case also mentioned the abduction of a French king, which never happened either!

My reason for extracting these two instances of negative conclusion was to cool the ardors of those who think they can simply dump the totality of UFO literature into a giant AI machine and extract the truth with clever software. It takes hard human research and hours of real study to clear the material.

Whitley, who sat facing the large windows opening towards the hillside, was fascinated by the wildlife roaming around close to the building. He interrupted me to point out strange beings were staring at my back through the glass, so I retorted a bit testily it wasn't my first experience lecturing before a bunch of turkeys. We all laughed.

Garry Nolan spoke after me, pointing out that a convenient way to study the data is to look at what happens to people: "Can we perturb the phenomenon to validate correlations?"

He described the instruments his lab was developing, notably the MIBI (Multiplexed Ion Beam Imaging) system producing near-atomic resolution of objects inside cells, including chromosomes, and leading to a new understanding of epigenetics. Then, he spoke of his own experiences as a child: faces in the window, and the human form that came into the bedroom in the mid-60s. In his early teens he saw a cluster of lights fly over. "There are non-human, sentient beings," he asserted, "and there's a brain structure for remote viewing. Is it special? Is it an antenna? How did it get there? From nature? From genetics?"

At the break, Bill Calvert spoke of the "Synaxis of the Primates" of the Orthodox churches held last January. Among the ten themes on the agenda of the Holy and Great Council (*which hadn't assembled for a thousand years*) two subjects didn't meet with consensus, namely "the question of autocephaly, and the question of the diptychs." The Synaxis decided they would be re-examined in another few centuries. No wonder such an organization can survive wars and revolutions. It even digested communism, now relegated under the gold of its altars and the sublime music of its five-hour long rituals.

Later the same day.

After our convivial lunch, I felt short of breath and extremely tired, so much so that I had to skip afternoon presentations and sleep for a couple of hours. Others were oddly brought down by strong allergies, notably Dr. Kary Mullis. In my case, the accumulated tension and hard work of the long summer came crashing around my shoulders.

Rejoining the group in the evening, I heard that I had missed a heated exchange between Garry and Dave Halperin when the latter began assigning witnesses' visions to psycho-sociological factors, as European skeptics are so fond of doing. Saucers with domes, for example, remind him of women's breasts with nipples!

I attended Whitley's after-dinner talk: "The phenomenon has to do with light," he asserted, citing Ezekiel, the flash seen by St. Paul, and the shining being that Mohammed met in a cave. But does it also have to do with the Dead? Whitley said he'd received new testimony from people who've had premonitions accompanied by visions of short beings. One child reported that his brother came into his room with little blue men and said he was OK; he had just died in a car accident.

Earthrise. Sunday 4 September 2016.

This morning, we heard from Eric Ouellet, who works with the Canadian military in Toronto. A young man with an open, congenial style, he gave an overview of the Belgian wave of flying triangular devices. I found it incomplete, but I have no special insight into the solution either. I was puzzled, however, when Eric blended together the UFO theories and ideas about poltergeists, ignoring the possibility that the flying triangles belonged to NATO. Jeff Kripal managed the debate masterfully, keeping all hypotheses in perspective.

Bill Calvert followed Patty Turrisi, a colleague of Diana who said the NSA had become interested in her work about Charles Sanders Peirce's theory of 'abductive reasoning.' Unfortunately, she didn't explain much about the theory, which I had researched on my own last year, finding it very relevant indeed, but I hadn't known of Peirce's occult connections.

There were continuous 'presences' in Peirce's home in Milford, Pennsylvania, and strange lights in his bedroom. His father was a math professor at Harvard, his brother was Secretary of State. His paramour, Julia, was a Gypsy priestess. Patty concluded: "UFOs pull the rug from under you... We don't know how to talk about things that are in the process of changing us."

Next, Tim Taylor spoke of his passion as an inventor of medical devices, especially for pediatric oncology, and his continuing work (a few days each month) as a controller of classified satellite launches.

As a break from the discussions, I'd suggested that we take a quick excursion to see the fabled Round Barn in Santa Rosa, so the group headed north in four cars and met in the well-preserved prairie where the huge redwood building stood, close to the luxury hotels and the hi-tech firms burgeoning in the area. Medtronic has already razed one older structure the Harris Community had built. I gave an overview of the Rosicrucian group's work on the physical and mystical planes.

We held a general discussion as we returned from the Round Barn, turning to Mullis. In the mid-1970s he had a secluded home on the banks of the Navarro River, not far from Spring Hill. One night in 1985, he met a glowing raccoon on the property who hailed him with "Good evening, Dr. Mullis." He has no recollection of what happened next, only regaining consciousness in the morning, still on the path.

The session ended with a passionate debate about strange events of the previous night. At about 4:10 am Whitley had felt a jolt, an explosion inside him. "The room became alive. I was on a college campus; I walked into a room where a strange instrument leaned against a door; there was a building with a cupola..." Eric Ouellet, too, had woken up with a start at 4 am, feeling an irrational fear. Tim Taylor had a terrible night, waking up at 4 am. He slept late after that, missed our first session.

Things became very interesting when Diana and Patti recognized the place in Whitley's vision: it matched a building called Schartz Hall on their East Coast campus, where a few students rebuilt ancient musical instruments. Dean Radin spoke up: "Think of this as normal," he said, "UFOs are not reducible to Psi, why shouldn't people have out-of-body experiences?"

Hummingbird. Thursday 8 September 2016.

Today's meeting of the Silicon Valley Aerospace 'angels' investment group took place at the SETI Institute. Thomas Dittler, the very dynamic German who chairs the group (I'm his co-chair), had arranged for a presentation by Nathalie Cabrol, director of the Carl Sagan Center and a principal in the Astrobiology Division at the NASA Ames Research Center.

Flamine and I had a good dinner with our friend Michèle, slowly emerging from the loss of John Forge, and last night we had a happy reunion with my daughter. Can we build a new life on all that?

"Of such threads do the Gods weave our destinies..." wrote A. Merritt in *The Ship of Ishtar*.

15

Buenos Aires. Thursday 15 September 2016. Hotel Esplendor.

My room in Palermo District is clean and classy without unnecessary fluff. It's also silent, which suits me. Alan Stivelman met me at Ezeiza with his friend Eduardo. During the long drive into the city, we spoke of Juan and his mother who still gets emotional when she recalls her own (unreported) experiences.

Alan tells me: "You'll find that Juan has grown into a simple but intelligent man with a single-minded interest in nature. He has a matter-of-fact approach to things, and no fantastic imagination."

That's very much the same psychological picture we saw when we met him in 1980. The new element is the Guarani lineage through his mother. He's had more recent observations of unknown objects, when he got lost in the forest of San Luis, and was guided to safety by a strong light. In that episode he had a very emotional vision, meeting his dead grandfather.

The Argentine Air Force has an official project, led by retired Commodore Ruben Lianza who gathers reports from the public through computer 'apps.' His first public lecture was this evening, so my friends asked me if I wanted to attend. I was tired from the long flight, but I didn't want to miss the occasion, so we walked over to the brand-new science and technology building where about twenty ufologists had convened. Ruben turned out to be dynamic and well-organized. Openly skeptical about UFOs, he showed slides from the French CNES to support his view that 98% of all cases were explained. He mentioned a bit of history, an early Intelligence study in October 1968 conducted by Liprandi, and a study at CNIE in 1979 by Graffigna. All led to the rejection of 95% of the reports, he said, adding that the CNES had similarly explained most of the cases in France.

People looked at me at that point, so I had to get up and stated diplomatically that more nuanced study was still needed…It would benefit from international cooperation.

I know for a fact that not all French cases are solved, since I attend the meetings when we study them, but I wasn't about to start a diplomatic incident just after landing in beautiful Argentina where I have many friends.

Ora Verde Observatory. Friday 16 September 2016.

We made the four-hour drive north to an observatory in the province of Entre Rios today. On the map of this enormous country, a four-hour drive is but a tiny jump. First surprise: several groups of ufologists who'd heard of our trip had driven over from other parts of the country. Greetings were exchanged and photographs taken as the shadows lengthened in the prairie, next to the observatory dome.

Discussions with them disclosed a long list of sightings the Argentine Air Force will simply never get. As in the US, people now understand it's useless to spend time with the bureaucracy.

There's a blackened metal disc on a strong support in front of the observatory: the heat shield of the soviet Salyut-7 station which crashed years ago in Entre Rios.

Las Margaritas, near Rosario; Saturday 17 September 2016.

Now everybody is waiting for Juan. Over breakfast, Alan and I reviewed the hypnosis sessions conducted by Dr. Berlanda, where Juan recalled the tall being pinching his arm in the 1978 incident. In tears, he also recalls his emotional reunion with his *abuelo*, his mother's full-blooded Guarani grandfather, in his later abduction or vision.

Juan smiled broadly as he stepped out of the car, and we greeted each other with open joy and laughter, as if we'd only been apart for a few weeks. It was easy to reconnect after these 36 years. 'Little Juan-Oscar' has grown into a full-bodied colossus, as tall as I am, his chest twice as large. I asked how Don Felipe and his mother were, the woman who probably holds the key to the enigma.

The rest of the day was spent filming under Alan's direction in the graceful setting of the estancia, loaded with history, its wide rooms with the high ceilings providing a congenial space for us and the team.

We had an intense dinner conversation long into the night, about the range of hypotheses and its implications for science and spirituality.

They asked about the Rosicrucians. For me, an identification with their tradition is not a creed or a religious statement, but an attitude towards the universe itself—a trust in science and the mind of man equipped with the tools of reason, principles that go back well before the Greeks. (They go back to the Three Kings, in my opinion.) They describe the place of man in space and time, his immersion in the universe, and the powers of his mind to comprehend nature and explore it; its limitations, too, and our many weaknesses.

Las Margaritas. Sunday 18 September 2016.

Slow morning, welcome rest in the beauty and silence of this hacienda.

My Spanish is coming back to me. I get better at understanding Juan and local parlance. This afternoon, as I was walking with him around the corral, we spoke of his peculiar dreams, of his concern because he doesn't speak Guarani well enough, of his estrangement from his family. In that respect, my coming from California to meet him will help. One of his sisters has just spoken to him of her own dreams, and her mother is now more willing to talk about her past.

Juan and I spoke alone, for a long time. (Fig.17) It was a special moment, the culmination of what I'd come to find, the closing of a circle. Juan needed to review the history, to express himself in full confidence. His family begins to understand what he's gone through.

Venado Tuerto. Monday 19 September 2016. Miro Park Hotel.

Dr. Berlanda, a psychiatrist, is driving Juan over to his parents' home, where we'll see him again tomorrow. The rest of us had a long drive along the Rio Paraná, which occasionally expands into a lake on both sides of the road. This is very flat, low country. It will be in major trouble if the level rises by just a few feet in coming decades. We arrived at dusk after fighting our way through long lines of 18-wheelers loaded with produce.

Alan had scheduled a side visit to the UFO museum in Victoria, a long, flat structure at a street corner, painted bright blue. Images of saucers in desert landscapes adorn the retaining wall. A businesswoman named Andrea Perez Simondini and her enthusiastic mother Victoria had assembled their group for our visit, so we were treated to a solemn *asado* of choice meats and salads.

Much of the material on the shelves was the standard fare of UFO books, magazines, and pictures, but two glass cages held special material, including sizeable chunks from the Ubatuba crash, so I spoke about our new isotope analysis at Stanford.

Fig.17: *Argentina. September 2016: Private conversation with Juan Oscar Perez. I gave him the photos I had taken in 1980.*

Fig.18: *The research group in Venado Tuerto, Argentina: Sept. 2016.*

Buenos Aires. Tuesday 20 September 2016. The Esplendor

We left Venado Tuerto this morning and headed for the countryside, turning into a dirt road that leads to Juan's family home. His father, Don Felipe, was there to greet us, looking very much the same as he did 36 years ago. I joked that he even wore the same Gaucho hat of brown felt with the wide brim I had on my photographs, that I showed to him. We laughed.

The wind from the fields sprinkles a super-fine dust over everything here. I can't blow it away, and my fingertips can feel its roughness on the pages on my notebook. Among the piles of accumulated wood and old pipes were a row of stalls for pigs and piglets, a box with three armadillos, and a number of dogs of all races, colors, styles, appetites, and vices. Also there, leader over all the activity, was Juan-Oscar with the blue and yellow knitted hat that never leaves his head.

No more ceremony, we embraced warmly, Latin style. Don Felipe, obviously moved to relive our encounter long ago, was very warm to me. As one of Juan's sisters joined us, smiling, I passed around the pictures from 1980, and I gave the negatives to Eduardo so they could make copies. Juan proudly gave me a private tour and again we spoke alone for a long time, going back over his old experiences and the new ones—he's seen new objects in the sky over the farm. I was happy to be able to have a long man-to-man conversation with him, my primary goal. I was happy, too, seeing how the family began to close around him.

As time came to leave, I presented Juan with a 17th-century token from Burgundy that shows a flying disk. He was so happy with the gift that he offered me his favorite hunting knife. He'd made it out of metal parts from a windmill, with the horn of a stag for a handle.

I told him I probably wouldn't be allowed to board a plane carrying a shiny 12-inch blade, no matter how unique and precious, so he gave me his water gourd instead, the *cantinplora* he carried the night he went hunting in the Selva de San Luis and was shown the way through the forest by a flying light. He'd made it out of the skin of a wild boar that he killed after it had slaughtered two of his dogs.

Buenos Aires. Wednesday 21 September 2016. Spring Equinox.

We're back at Hotel Esplendor again. Andrea Perez Simondini has driven down from Victoria to bring me the two Ubatuba samples in vials.

Commodore Ruben Lianza also came by for a longer, more relaxed talk about his role in the Air Force. Like most serious skeptics, he believes that, given improved technology like François Louange's latest photo-analysis, we'll be able to explain all the cases and even go back and 'clean up' the unidentified cases, once and for all. Then, instead of those irritating 2% unidentified, UFOs will go away completely! Unlike the US, we could talk as friends in spite of our differences, and we could see a constructive dialogue emerging.

The highlight of the day was a warm reunion with Fabio Zerpa, aged now and walking with a cane but accompanied by a much younger woman who introduced herself as a research assistant. His legendary memory, a historian's great gift, is intact. We reminisced about the case in Venado Tuerto, my lectures there, and his own research. He reminded me that he'd seen a UFO from an Air Force plane, back in 1959. That was the incident that convinced him that the sightings were real.

Later the same day.

As we drive along the streets of Buenos Aires, I marvel at the posters that acquire a different context in Spanish. The stupidity of the ads is on the same level as in France or the US, but it flares up brighter when you translate what they say, because it forces you to think about the message, rather than being subtly influenced by peripheral vision. One ad says: "Become part of a greater future!" You can do this by drinking more whisky with your friends. No wonder their economy is so shaky!

I did an online interview today, taking questions from the whole region. A Chilean researcher who follows my work, Patricio Abusleme Hoffman, from Santiago, quoted a man named Bosco Nedelcovic who claimed that as early as 1957 the CIA, together with Brazilian intelligence, simulated UFO encounters using experimental technology and psychoactive drugs in *Operation Mirage*. He strongly suspects that one of the incidents they fabricated was the Antonio Vilas Boas encounter with the "grunting" Alien female, the earliest report of sexually charged abduction. I have no way to verify that.

Asking questions about old Argentine history, it's obvious that its supposedly God-loving pioneers did terrible things to local Indians. In the 1870s, their government launched a campaign heralded as 'The Conquest of the Desert.' It boiled down to decimating the local population, particularly in Patagonia. The newly arrived immigrants were told by clerics that the native Guaranis didn't have a soul and therefore were not children of God and should be treated as vermin..

The invaders considered Patagonia a desert, never taking into consideration that it was carefully tended by a beautiful local culture. Discrimination only ended under President Alfonsin in 1988, if it ended at all, which I doubt after listening to Juan's experiences.

Knowing this dark history will be important in understanding his lineage and his tradition. He now embraces his Guarani heritage; next month, he'll go to Uruguay for a ceremony and a dance that will reveal the name he should have received when he was twelve years old.

In my last few hours in Argentina, Guarani psychologist Juan Acevedo made a special trip to say an emotional goodbye. He presented me with his ritual cup and a package of his personal *maté*.

Hummingbird. Saturday 24 September 2016.

Curse those long airplane flights, the dry air, the pollution onboard. Fortunately, it's warm in San Francisco, and peaceful, with a foretaste of Indian summer. At Dallas-Fort Worth the customs people were curious about the fur stitched around Juan-Oscar's canteen. I reassured them: "That particular pig has been dead a long time!" They laughed and let me pass. I also safely brought back the two vials with samples of the Ubatuba magnesium, from two different sources of which Dr. Sturrock and I had been unaware in his fundamental analysis. He'll be pleased with what I'm bringing back.

Hummingbird. Sunday 25 September 2016.

Jose Antonio Caravaca, a smart Spanish researcher I met in Madrid, sends me an expanded version of his Distortion Theory, proposing that many UFO incidents—including some of the most credible airship sightings of the 19th century—are the product of what Frank Salisbury would call "displays" rather than genuine encounters. (I think of them as exhibitions, but the concept is the same.) In summary, the sightings are designed to happen at a particular time and place, and details get built out of the witnesses' own mental images, like Betty Hill's famous star map. The problem, from my point of view, is that his model is incomplete because the UFO light or projection is *also* capable of precipitating as a physical machine that interacts with the environment, sometimes violently, and can outsmart any jet when it flies off into space.

I can do no more today, so I stayed home and read *The Ship of Ishtar*, a wonderfully prescient (1924) novel about good and evil written by Abraham Merritt. It was recommended by Jack Katz, who knows everything.

Hummingbird. Thursday 29 September 2016.

A positive step in print: the first paper by Kit and Garry about the unusual brain structures found by the LoneStars. The paper is titled "Incidental MRI and Genomic Findings in Human Striatum: Implications for Behavioral/Cognitive Research."

The abstract states: "Complex frontal cortical basal ganglia networks [with the dorsal striatum] are involved in behavioral and cognitive functions. These include executive, *rapid action/reaction decision-making, and information processing of massively parallel afferent sensorimotor data* [my emphasis-JV]. In this work, structural MRI studies of a unique clinical cohort showed ... enhanced bilateral white matter connectivity (visualized as fiber density) between the Caudate and Putamen. The findings were related to a retrospectively identified population subset sharing common cognitive functions."

The summary goes on: "The data also suggests these MRI findings are *non-pathological neural features* ... Given the location of these neural connections, and the attendant role of the caudate-putamen in deep processing of environmental context subject to executive function goals, it is worth speculating said connections might *afford affected individuals with novel cognitive or intellectual abilities.*"

Hummingbird. Saturday 1 October 2016.

This morning, Michael Angelo and Garry Nolan met me at the Stanford Blood Center in Palo Alto and showed me the prototype of the MITI cell scanning machine: One instrument at Stanford and the third one at Harvard, where a colleague of theirs is engaged in compiling a complete 'Cell Atlas' of the human body.

I turned over to Garry the two vials of Ubatuba magnesium. We discussed the follow-up to the striatum study.

Hummingbird. Sunday 2 October 2016.

Before leaving for Paris, I spoke to people who know Tom DeLonge's To The Stars Academy (TTSA), hoping to catch up. In a sobering phone call, one of them told me his skepticism had deepened: good intentions, but there's a shadow over the information. "Many of the memes are scary, not based on fact," my friend says, "or taken from misinterpreted classified writings." He now suspects there's an invisible hand, and it may be exposed in time.

Mabillon. Thursday 6 October 2016.

Paris was a delight of autumnal glory when I landed two days ago but the gray cold has reasserted itself to chill the boulevards. In Evry yesterday, I walked through quiet suburban mazes to reach the site of the Genopole. Everything there is designed around cars and drivers; the occasional visitor arriving on the train like me is left to find his way on foot along walkways bordered by ugly fences.

Once we assembled in the large square room, the expert committee session was a cheerful affair. Although I'd missed a couple of meetings, my colleagues greeted me warmly and we worked hard to review three cutting-edge medical startups with bright technology and interesting plans. We voted to finance them, so all three will be supported on campus with new facilities and real money.

Mabillon. Friday 7 October 2016.

Reality sets in with the creeping grayness, but life goes on and wine flows in smart brasseries (like the Roussillon, where I just had lunch with my son and his three-year-old son). The men still wear the obligatory dark suits and blue ties. They walk in and out of the portals of ministries, looking important. The women are slim, neatly coiffed, purposeful. The economy may be glum, but so far the French nomenklatura hardly feels the squeeze of unemployment and stagnation.

Mabillon. Monday 10 October 2016.

Turmoil in the high-stakes world of aerospace. François Hollande is crying foul in Warsaw because Poland just cancelled an order for 50 Airbus helicopters, shaking up diplomatic relations, while Qatar Airways just selected 19 billion dollars' worth of Boeing airliners instead of buying a fleet of A320neo, plagued with engine problems. Time will tell whether that was the best decision.

All this compounds the problems of Airbus with its military tankers and the huge A380. Meanwhile, SpaceX recovers from the loss of a fully loaded Falcon rocket that exploded on the pad during a cold static test. How could it blow up, when the kerosene was cold, and no engine was running? The inquiry is not expected to yield a clear answer.

Perhaps a single marksman with a good rifle and an exploding bullet hit the rocket from the top of the ULA building, 1.5 kilometer away?

A very good shot indeed, thank you very much. The site is a secure, guarded Air Force zone inside Cape Canaveral. The rumor says that one of the systems aboard was a non-US device that could have watched Iran to detect violations of the nuclear deal they just signed with Obama.

Mabillon. Tuesday 11 October 2016.

Suddenly, a storm is swirling around TTSA. Wikileaks has unveiled and published Delonge's email exchanges with John Podesta, Hillary Clinton's campaign director, even quoting Tom's offer of a meeting with "his Number One advisor, General McCasland," a name all of us had agreed to keep confidential.

Julien Assange, legendary hacker and major target for the Intelligence establishment, currently frozen in asylum at the Embassy of Ecuador in London, plans to use every sensational bit of the leak to insert a negative agenda into the American elections.

The information is splashed everywhere, from the *Wall Street Journal* to the web. Of even greater concern, mail directories have been pirated, exposing contacts and private interaction.

Hummingbird. Thursday 20 October 2016.

As soon as I got back, Garry and I met at Bucks and updated each other, discussing his frightening dream experience in London, the fallout from Wikileaks and my plan to secure my archives, and then move on. The analysis of my magnesium samples hasn't begun, but things will go faster once the lab gets its own copy of the MITI machine, which Michael Angelo is checking out in England.

Hummingbird. Wednesday 26 October 2016.

After the crisis with Delonge's top advisers was exposed on Wikileaks (AF Major Gen. William N. McCasland, and Rob Weiss), Tom remains active on media including *Rolling Stones*, and his project is alive again.

As I thought about future research on my metal samples, I came across a mention in *New Scientist* of a clean process to liberate hydrogen from methane. Invented in 1999 by Meyer Steinberg at Brookhaven, veteran of the Manhattan Project, it involves methane 'cracking' in a bath of molten metal that would presumably improve heat transfer and allow the soot which normally impedes the reaction to float to the surface. Could this explain the Council Bluffs precipitated metal Garry and I are investigating?

Hummingbird. Saturday 29 October 2016.

Tom Delonge organized a phone conference today with Hal, Chris, Jim, me, and Colonel Martin ('Marty') France of the Air Force Academy in Colorado Springs. We stayed on the line after the call and I told him how concerned I still was about the Nazi memes.

Hummingbird. Sunday 30 October 2016.

Well, it's Halloween in San Francisco. I'll meet my gifted artist friend, Karen, tonight at her funny house, and we'll roam the Castro district in costume like kids, at the risk of scaring her cats.

Hal and I compared notes this afternoon. I told him the TTSA plan looked good on paper, but my concerns remained. The idea of science serving a PR company worries me. Hal insists none of that will matter if the team keeps the process moving with the right funding. Like me, he doesn't believe that a Nazi super-science ever existed.

Rob Weiss, from the Skunk Works (and advisor to TTSA), has also publicly insisted that there was nothing real to the Nazi tales. As for the Horten brother who built a flying wing, he never achieved credible results. So why the lies? What purpose do they serve?

Hummingbird. Wednesday 9 November 2016.

California woke up this morning very groggy after the victory of Donald Trump as the next President of the United States. The entire West Coast, from the Canadian border to the Mexican frontier (soon to be equipped with an impassable wall?), voted heavily against him.

This is a quiet week for me, working on the material analysis of crash and implant samples. As for any hope that TTSA could force the release of special UFO data, that has been dashed for now.

Hummingbird. Sunday 13 November 2016.

On Friday evening, Tim Taylor, Diana Pasulka, Garry, and I had dinner at the Palo Alto Westin. They were on their way back from an Esalen meeting that included Whitley Strieber, Mike Murphy, and Jeff Kripal.

Arguments flared, unfortunately, between Tim and me about the Saint Augustine crash site, but Tim did agree to open some doors. The dinner ended on a friendly note, but I slept badly that night. In real terms, most of the information about the New Mexico cases has been known for a long time: Ron, who lives on-site, sends me data about what goes on there, including the coordinates of a specific ranch near Ground Zero.

Tim Taylor has written an interesting book about his early life. Entitled *Launch Fever*, it covers the Challenger and Columbia disasters in detail, and "the kind of struggles that lie ahead for every entrepreneur." He does know what he is talking about, and his insight is precious.

A new rumor says there used to be a study of certain medical files that referred to 'reverse transcriptase' applied to tissues from humanoids, labelled as non-human, originally located at Los Alamos around 1978... *before the words had ever been used in print*. They were related to autopsy tissue, but I have no way to verify any of it.

Tonight, Steven Greer gives a webinar in Vegas about false flag operations. This time he may touch a genuine topic. He claims that "Covert interests are laying the foundation for a false flag Alien threat, and a militaristic project: 1940s to the present."

Hummingbird. Wednesday 23 November 2016.

Manuel Terranova, the CEO of Peaxy (13) gave us hopeful news at an investors' briefing call this morning, reporting on progress. In the afternoon, Nathalie Cabrol, a French woman who now directs the Carl Sagan Center at SETI in Mountain View, invited me to come over. Once we sat in her office (I noticed with amusement the novel *Alien: Resurrection* by A.C. Crispin on her desk) I mentioned growing up in Pontoise. She's from Beaumont-sur-Oise, very close. When I explained I'd left because of the close-mindedness of the French science establishment, she said she'd suffered from the same lack of open enquiry.

Nathalie introduced me to her husband Bill Diamond, a seasoned fiberoptic entrepreneur who'd been president of Lucent. We had a relaxed conversation. To my surprise, it turned to UFO. "Many people here think we shouldn't get into that, but there must be a way to approach the subject scientifically," said Bill.

Hummingbird. Friday 2 December 2016.

When the LoneStars tried to contact the real reverse-engineering effort, which we know does go on, they failed. Now TTSA is trying to do the same thing, through a few generals putting their stars on the line, except for McCasland, who withdrew following the Wikileaks exposure. One trusted friend I consulted is skeptical. "I'm going dark," he said. "Washington is a zoo. I have no confidence in anything inside the Beltway. There's no integrity, no useful knowledge, no connectivity to the truth. It's downright dangerous."

He suspects the rumors about a 'genuine project' are just another layer of misdirection, built on some real data and actual teams.

Hummingbird. Monday 5 December 2016.

Projects are ending, projects are starting, and I am getting a bit more data about the San Augustin crash. Stanton Friedman, a good engineer who's been following the hardware for as long as I have followed the software, told me he thought the craft that crashed at Horse Springs was a companion (he called it a "wing man") to the one that crashed at Roswell. He was glad I was getting involved in the case, and kindly said he'd send me his data.

Next, I called Art Campbell, the retired teacher from Medford (Oregon) who's done early analyses. He, too, was ready to help and even to come down to San Francisco with a major artifact. It has been analyzed already through ICP (Inductively coupled mass spectrometry) and found to contain polyethylene. He told me the foil was aluminum on one side and a combination of other elements on the other. Art thinks the Roswell crash may have been a cover for the "real" crash at Horse Springs. No time to go there, however, as my next trip is to France, picking up the research in Europe.

Bellecour. Tuesday 20 December 2016.

"Every funeral is sad enough," says my son, "but it also brings to mind all the others..." The wife of Dr. Philippon was buried three days ago, and we learned of another death in a related family a day later, so we changed our plans for the holidays.

I did have a brief opportunity to visit the Institut Métapsychique, where I heard Walter von Lucadou recounting a series of psychic experiences. Méheust was there, and I saw Marie-Thérèse de Brosses; they said little is new in French paranormal research.

In Bellecour, the frost has come to this ancient town and its honored walls, carefully shored-up. The rich expanse of the fields greets us at the threshold of the last house of town. Olivier teaches me about the life of these medieval stones where artisans have sought refuge from the craziness of cities. They open dream shops, well-stocked in wooden toys, old books, strange objects. We walked back to the main road through alleys paved in mossy cobblestones.

The librarians at Rice University tell me that the last box of the "B" archives has arrived safely. The time has come to simplify, to focus on important subjects. We need a five-year horizon, not another phantasy team chasing ambulances and promising breakthroughs made-for-TV. Nor do we need another BAASS, paralyzed by obscure games around government temper tantrums.

Mabillon. Friday 23 December 2016.

Ever since we landed in Paris, we've felt the stinging fire of air pollution, needle pricks in the eyes and pains in the lungs fighting for oxygen. Also, the French continue to smoke heavily.

An ugly terrorist attack in Berlin, a truck charging a Christmas market, has further discouraged tourism in Europe once again. From Pigalle to Montparnasse, one can see empty bars and closed-up shops. I did have a chance to eat lunch with old friends at Siparex, to see Pascale Altier (at Les Editeurs, to talk about Bob Monroe's research), and a couple of days later I saw Dominique Weinstein, caught up in bureaucratic tensions.

The French don't seem to have heard about the turmoil around Podesta, whose ufological messages have been exposed by Julian Assange as part of the Russian software attack against Hillary. The whole thing wouldn't make too much sense in Europe anyway. Nor do they realize how much progress has been made in gathering hardware from crashed objects.

The CNES rejoices in self-congratulation: they have 'explained' all the cases submitted to them this year; but the files are curiously empty, simply because people are smart enough not to report anything to government agencies.

Hummingbird. Tuesday 2 January 2017.

The day after I got home, I had lunch with Federico Faggin to edit a white paper I'd drafted for SETI, which has been stuck in sterile research for years, finding no intelligent radio signal from the stars. That isn't surprising, Federico says: The most natural mechanism for contact would be out-of-body consciousness. There must be lifeforms in the cosmos that are not carbon-based and thrive in temperature ranges and atmospheres we haven't yet recognized.

Hummingbird. Sunday 8 January 2017.

A major storm sweeps the West Coast, torn trees, broken power lines. For California, this is only a return to a natural pattern, an interplay of the land and the elements marking the blessings of winter, the prelude to an early spring. A decade of drought had made us forget that we stand at the edge of a coastline exposed to a giant ocean with its own indomitable energies.

Hummingbird. Saturday 14 January 2017.

On his way back from a seminar in Napa, Chris Mellon came over for dinner with us tonight. We spoke of TTSA, which I'm still invited to rejoin in some capacity, but I can't align myself with it. Here lies the paradox: Tom's backers are among those who must know about the sequestered hardware, but what are they trying to do? Chris himself is leery of Roswell, preferring to believe the Air Force conclusions about balloons and classified detectors. I said I was comfortable continuing this research on my own. I mentioned the advice once given to me by Frank Pace: to only engage with organizations at the very highest level of integrity and reputation, *or not at all*. Indeed, there are urgent things to be done in the area of UFO information science, but they would be best housed within a Silicon Valley open model rather than yet another classified structure.

Chris recalled his days as a Senatorial staffer: a graduate of Colby College with a degree in economics, he'd gone on to study international relations at Yale, specializing in finance and management. He then served as a special assistant to the Secretary of Defense for Intelligence policy (Cohen) in 1998, then as minority Staff Director of the Senate Intelligence Committee for Senator John D. Rockefeller in 2002-2004, so his insight is precious.

Hummingbird. Sunday 15 January 2017.

My neighbor Dr. Kevin Starr died during the night: a heart attack, no warning and no pain. A big man with a booming voice and a formidable memory, he was a noted historian and served as the State Librarian for California. Sheila and Kevin, co-authors of a series of well-researched academic volumes about California, have been my wonderful, honored neighbors since 1992.

Hummingbird. Friday 20 January 2017.

Back at our primary Northern California site, the orbs are all over the backyard and even the home, combined with sightings of a 'vehicle' by three witnesses, and even a form of communication.

The Utah Ranch, too, is reportedly visited by 'vehicles,' even in daylight, with high terahertz readings on wide bandwidth RF devices used by

Erik Bard, still on the staff of Brandon Fugal. Personnel on the property have experienced dizziness, nausea, and worse. I'll try to understand this situation when I fly to Utah next month.

Hummingbird. Saturday 21 January 2017.

California woke up in shock today, faced with the reality of Donald Trump in the White House and the country in the grip of a reactionary Cabinet detested by half of the country and mistrusted by the very agencies they're supposed to start leading.

Enormous but pointless demonstrations erupted this afternoon. On the way to meet with Jack Katz tonight, I was caught in a massive traffic jam and had to turn around; people were packed along Market, blocking all the streets. America has never been so deeply, dangerously divided.

Hummingbird. Sunday 29 January 2017.

The Dow Jones index has climbed above 20,000, the result of a Republican alignment in Washington between Congress and the aggressively pro-big-business Trump White House. Then, the new government started implementing a populist, anti-immigrant, anti-Moslem policy, triggering new demonstrations against the US all over the world, as the balance of trade was shaken from Europe to Asia.

Personal privacy has been compromised beyond repair, a human disaster.

Hummingbird. Monday 30 January 2017.

Jim Semivan has taught us that 'Material Exploitation' from recovered hardware almost never starts with knowledge of sourcing. This is due to the National Technical Means they've applied to recover objects using Byeman-classified devices that maneuver in space or deep sea, he said.

Only one agency owns them, supposedly. In order to exploit, Federal Law demands that a DoD agency—not an Intelligence Agency—be willing to pay and to state they have jurisdiction for threat mitigation. Doesn't that almost ensure that no relevant correlation will be found?

Hummingbird. Thursday 2 February 2017.

On a visit to Planet Labs headquarters today, in a funky area South of Mission, next to a leather shop specialized in shockingly pink S&M outfits, and a block away from an X-rated bakery, I met again with Creon Levitt and Larry Lemke who, like me, had missed the last meeting with Dick Haines. They thought Dick was giving up the work because of health concerns (hopefully, temporary).

Creon grew up in Wilmette, Wisconsin, met Hynek as a kid, and later contributed to funding Dick Haines and NARCAP. As for Larry Lemke, he became interested through his father's experience working on esoteric materials likely to have derived from a craft. He's also met McCasland. Both expressed surprise that I didn't join TTSA.

Hummingbird. Wednesday 8 February 2017.

Today I picked up an article by Serge Kernbach (arXiv: 5 Dec.2013) entitled "Unconventional research in the USSR and Russia: short overview." The author, who works at an advanced robotics institute in Stuttgart, traces parapsychological research in Russia from 1921 to the present through cycles of abundance and famine. He explains the decline in 2003-2004 with the closing of many laboratories: "Unfortunately, most of the organizations that arose during the period of the early 90's—without proper funding and under ideological pressure from the Russian Academy of Sciences—no longer existed in the first decade of the 21st century."

Hummingbird. Tuesday 14 February 2017.

TTSA had hopes to raise $50 million, not just ten, for a combination of 'edutainment' with space research. A public offering would combine the company with Atlas, the new satellite firm backed by General France of Colorado Springs. (I know about Atlas, since they came to the Valley for funding.) Today, the deal is off. Christopher Mellon had an unexpected parting of ways with them yesterday.

There are other things on my plate, including real startups in technology and a three-day trip to Utah to meet Brandon Fugal and catch up with the work at Skinwalker Ranch.

Salt Lake City. Wednesday 15 February 2017.

"Waiting for the Panamera." That could be the title of a song about the busy roads of Silicon Slope, Utah. Brandon came to pick me up at the airport in that magnificent automobile, a black turbo Porsche. We stopped for lunch and spoke for an hour. Sadly, I learned of the death of Frank Salisbury, about a year ago. Then we shared stories about the Ranch: old ones for me, recent ones for him, including a couple of occurrences *at his own facility*. The office is a large, modern structure perched on the side of the mountains. Brandon introduced me to Erik Bard, a clever physicist with experience at IBM and deep knowledge of storage media. When they heard I'd brought a series of PowerPoint slides about similar sighting clusters I'd investigated, and occurrences of unpublished Skinwalker-like entities in Europe, we moved to the board room to go through the mutual presentations.

Salt Lake City. Thursday 16 February 2017.

A friend of Brandon's picked me up for the trip to the airport in Provo where the Citation was waiting, so we had more time to talk. She's a bright executive in a firm linked to Brandon's development model.

Brandon and Erik Bard met us at the hangar that used to serve as the lab for Motion Sciences, a company that sought to exploit the Puthoff theory about extracting energy from the zero-point of the vacuum. John Petersen was on the board, as well as George Hathaway and Kit.

The experiments lasted for years and consumed six million dollars, producing effects too elusive to reach the proof-of-concept stage. We landed in Vernal after a hop over the snow-covered peaks. The team drove to the Ute Crossing Grill run by the local tribe where I happily met Junior Hicks again. Smiling mischievously, his wife presented me with a scarf I'd forgotten at their home, six years earlier... It came in handy later that afternoon.

On the Ranch, we followed Brandon, in his white shirt, black coat, and black gloves, driving and sliding through the wet clay where Colm and Eric Davis had their sightings, and where local Indians have nailed coyote skins to wooden posts to ward off the Skinwalker.

We were followed everywhere by the two cameramen and by a friendly security man, also dressed in black garb and black gloves, carrying an AR-15 in case of any disturbance. Although interest in the Ranch has slowed down, scouts from the Militia are still known to sneak around, eager to catch UFO secrets while an astonished planet tries to grapple with Washington scandals. In Utah, many people regret, as I do, that Mitt Romney did not run again.

Salt Lake City. Friday 17 February 2017.

The Citation has flown back to its hangar in Provo and the time has come to reflect on what I've learned. At the Ranch yesterday, we revisited some old stories with Junior Hicks. He recalled wiring the place for electricity and the computer line that Colm and Eric used to report their observations to Bigelow. It went around the house to a connection where Junior recently found a splice. Someone had tapped into his bare wire and evidently captured the data, but where did it go? Deep mistrust by local people against 'the government' continues to this day.

We walked all over the damp property yesterday and even climbed the crumbling cliff to decipher some Masonic inscriptions from the late 19th century, carved into the rock face: A 33rd degree symbol and the initials "AW" left by Augustus Wally, a leader of the Buffalo Soldiers from the 1887 expedition garrisoned at Fort Duchesne, who had formed an alliance with the Ute tribe.

At sunset, the Ranch slides into the dying light of a dangerous orange sky. Then the land eagerly breathes its multiple mysteries where the Black troops of the western expeditions and their Masonic secrets meet the Navaho curses that unleashed the Skinwalkers— creatures like giant dogs that walk erect—that outrun patrol cars or sit pensively on piles of rocks, watching over cemeteries, like Anubis in High Egypt.

No wonder people get confused about the State of Utah.

Hummingbird. Monday 20 February 2017.

"Critical Mass," not a topic for the LoneStars this time, but the clever title of a concert by the Early Music Society playing Monteverdi. Alone, I walked down to St. Mark between two rain showers to listen to the madrigals of *Soave Libertate*, a poem by Gabriello Chiabrera (1595): *My soaring soul will return to its fair firmament, its sweetest dwelling.*

I have no such "fair firmament" but I do feel intrigued by some new data, so I fired up two of my computers and loaded up the memory sticks Brandon Fugal had given me. The PowerPoints I'd brought to Utah (slides about cryptids, information on clusters, and two presentations about the nature of the orbs) were still there, but he'd added a number of his own files: a photo survey of the Ranch taken by his drones, and a 218-page book, carefully edited, about the failure of Motion Sciences, Inc.

There was also a memorandum detailing his measurements with two new instruments he obtained from Infiltec, a company in Virginia. The first one is a microbarometer, measuring pressure changes due to infrasound frequencies. The second one measures seismic vibrations from distant quakes. Erik found a very sharp peak at 17.5 hertz, probably ground vibration related to oil drilling operations some 2 km away, as I had told them. Yet, researchers such as Persinger and John Derr have speculated that seismic stress could manifest as earthquake lights or deeply affect human perception (nausea, loss of balance, hallucinations) through temporal lobe effects.

Hummingbird. Friday 24 February 2017.

Five years ago, Flamine and I were married in San Francisco. We were happy; we looked at the future with confidence. Today we're apart, searching for new paths. Lauren Artress, the psychologist with passion for the labyrinth of Chartres, tried to help sort out my emotions but the result was only to send me into an even deeper funk. Crying Times?

As one beautiful song goes, "I can see that faraway look in your eyes…"

As the ideas of TTSA become public, the names of Tom's advisors are all over the Internet again, notably Rob Weiss at Lockheed Martin and General McCasland at Applied Technology Associates. Little is confidential anymore; people know that Industry has the hardware.

Hummingbird. Monday 27 February 2017.

At the Computer History Museum today, I turned over all my significant research records to archivist Mark Weber, with some feeling of accomplishment. I was happily surprised to meet my son's former boss, Charles House, who was there on a visit from San Diego.

Dinner with SETI executive Nathalie Cabrol at the very modern, geeky Scratch restaurant in Mountain View, the new center of gravity of Silicon Valley. I brought her together with Garry and Federico, co-authors in our White Paper about multi-disciplinary SETI research. The depth of experience around the table was quite obvious.

Hummingbird. Saturday 11 March 2017.

Jim Semivan has written a revealing, luminous foreword for Tom Delonge's paperback edition of *Sekret Machines*: "We may stumble across something that we should not have disturbed: then again, we may just fall headlong into a new realm of existence that has been hidden from us and has always been our birthright."

There was a grave incident on the Ranch. It seemed to be exactly what they'd feared: they interpret it as a direct beam attack. On March 5th, a staff member working on the security perimeter and the new gate was hit by what seems to be a narrow-band radiofrequency beam.

I offered the help of the physics team at Accuray, the directed radiation surgery company where I've served as a board member, but I was told confidentiality demanded otherwise. I discussed it over lunch with Garry in Woodside, going over my New Mexico crash data, plans about estimating the isotope ratios of ejecta, and the bizarre proportions of elements in the knitted wire of the suspicious material picked up at Horse Springs. Ron Brinkley, too, has made progress with Chuck Wade, getting up to speed on all the crash information in New Mexico.

Now to focus again on packing up my archives to be sent to Rice University before I fly off to Paris.

Mabillon. Sunday 19 March 2017. Brasserie Le Rostand.

Gray windy Paris has reset all the clocks by one hour, so many cafés were still shuttered when I came out, but I feel on vacation now, although America continues to intrude and new changes loom ahead. The Presidential elections pit smart young Emmanuel Macron, technocrat and banker (Rothschild) who's never run for office, against Marine Le Pen, who advocates leaving Europe and abandoning the Euro, an extreme idea that could only benefit Putin.

On my bedside table is the autobiography of Edith Cresson, France's first female prime minister. It's an honest book about her projects and the many mistakes in the declining years of the Mitterrand era. I remember my hopeful meetings with her in the days when technical innovation was a hot topic in Paris, but nothing came of it.

Mabillon. Wednesday 22 March 2017.

The Equinox came with a little rain, a little wind, and much anticipation of political changes. At CNES, Jean-Paul Aguttes has sent around five, well-analyzed cases for review and discussion next week. The documents are detailed, the interviews are very professional.
 Apart from the government project, private research goes on. Jean-François Boëdec brings me up to date about the principal case we've agreed to embargo: too much, and too many implications.

Lyon. Saturday 25 March 2017. Radisson-Blu Hotel.

I took an afternoon train to meet Maxim for dinner. For old time' sake I booked a room at this modern hotel that looks like a fat pencil, near the train station, its proud illuminated top cone dominating the great city. The atmosphere is tense after another terror attack in London. I found Max smart, lively, adapting well to this busy world in spite of the complexities of his first year at the university. The trip was good for me, too, a refreshing perspective.

Mabillon. Tuesday 28 March 2017.

Over lunch with Alain Dupas, we were joined by one of his friends from Astrium who was leaving his job after many years. After the loss of the military helicopter project in Poland and the accumulating problems over the refueling tanker, the employment climate is frigid. The new technology leader, US scientist Paul Eremenko (whom I met a few months ago in Mountain View), had to close down the Suresnes site, eliminate 308 jobs, and reshape the R&D. This is taking place in a generally poor environment for French hi-tech, contrary to all the hype about entrepreneurship and startups on TV and in Macron's speeches.

Mabillon. Wednesday 29 March 2017.

A full day of work at CNES HQ reviewing recent sightings, weighing hypotheses, and planning for increased use of the advanced database. The larger problem is that the most meaningful sightings don't get reported through the official channels.

There were thirteen of us in the room, including three full-time staff. My long-term colleagues Weinstein, Louange, and Lagrange were there. The workstyle and the intellectual discipline set a very different tone, more serious than what I've seen in the US.

The real pleasure is to walk home across Pont-Neuf in the glory of spring. The sunshine is glorious as it kisses the monuments. Even the twin towers of Saint-Sulpice look lovely, enlivened by the statues recently cleaned up, now re-attached to the superstructures.

Hummingbird. Tuesday 11 April 2017.

Jack Katz works on page 201 of *Beyond the Beyond,* his ultimate opus. He can no longer afford the rent for his one-bedroom apartment, but his admirer in Colorado is purchasing his unit, so he'll be able to stay.

Historians of Popular Culture at Ohio State have discovered that Jack did originate the *genre* of the Graphic Novel, four years before anyone else, so at age 89 he's being feted by academics, to his surprise, and he feels like a monument.

"They've found an ancient relic!" he snickered, bright as ever, as he exuded prolixity anew. I have remained close to Jack as he transcends and redefines what it means to be human. His paranormal experiences, long repressed, are coming out through his art, one by one.

Hummingbird. Friday 21 April 2017.

I had lunch at Hotel Mac with Peter Beren, six weeks in the hospital, intellectually sharp but walking slowly, with a cane. We spoke of Jack Katz. Peter recalled taking a call from him in his car once, as he was driving with a friend. Jack went into one of his rambling tirades, his voice booming over the car speaker.

"I couldn't interrupt him," said Peter, "so we listened for half an hour, after which I said, 'OK Jack, I'll see you later,' and I hung up. My friend jumped: 'That guy was real? I thought it was the radio!'"

Hummingbird. Thursday 27 April 2017.

The French Presidential elections have turned into a complex conflict, pitting two incompatible systems against one another. I went to the Consulate last Saturday, finding only three people ahead of me. The Consul was talking to employees in a corner and didn't say hello.

In that sedate atmosphere, oozing with the staleness of bureaucracy, I grabbed papers and voted for Macron with little enthusiasm. Democracy must go on, and new leaders are needed, but Mr. Macron is trained to manage finance, not to inspire a country to renewed greatness. He speaks like the product of PR agencies, but his opponent is a direct danger to France.

In the controversy about Annie Jacobsen's book, I've learned from a high-level participant in the remote viewing work that "At the start of the [remote viewing] program at SRI, the direct CIA link was the DCI at Richard Helms direct request, and a month later, John McMahon, who later became Associate DCI in 1982. The program, from Day One, was managed deep black in the Operations Directorate, and what Russ and Ed May and Ed Dames and others have been saying for years is flat wrong. Annie's got it right, but sadly, it takes reading the whole book to figure it out." In 1986 John McMahon resigned from CIA and joined Lockheed, and in 1994 became the President of Lockheed Missiles and Space Company.

When I pointed out I was bothered by Annie's emphasis on Puharich, my friend responded: "The Puharich spin bothered me, too. But Annie's intention in bookending him turned my confusion into respect for her brilliance. The way you write about your consternation is valid. She was almost too subtle. He was the progenitor of it all ... without what he did, there would have been no DCI Helms in this story, no Ed Gregor, no Gottlieb, no Dave Boston, no Kress, no Peter Tompkins, man with the Bowie knife, D/DCI to later become Chair of Lockheed, no SRI program. Annie figured all that out."

Hummingbird. Sunday 7 May 2017.

Emmanuel Macron has won the Presidency over Marine LePen by a wide margin (66%) and he's already succeeded in changing the discourse. This evening, Karen came with me to the Unitarian Church to hear the Bach Choir in Vespers by Mozart and Beethoven's Mass in C. We spoke of her projects, which include marrying her friend Paul, and the plight of so many lost and homeless people on the streets that she tries to help.

Joshua Tree Retreat. Mojave Desert. Friday 19 May 2017.

Tired by hours of analysis and corporate issues. There are many recent changes in my small portfolio: Taulia and Materna, generally going in a positive direction, while others struggle.

Flying to Palm Springs. Cars everywhere, typical American camp, good humor. About 3,000 people have arrived; they're expecting more tomorrow. Linda Howe reminded me of her Tungsten-Magnesium-zinc layered object, studied by Puthoff and Hathaway. "It moved spontaneously," she claimed, "when zapped with a 7Hz frequency while held in a Van de Graaff accelerator," but nothing's been published. Conversations now revolve around the multi-dimension concept I detailed in the 1980s. Parallel universes and tunnel loops in time-space, once pooh-poohed by the cognoscenti, are the rage.

Well, I materialized the multi-universe theory in *Le Sub-Espace* in 1961, and many other places since then, but it's hard to change established thought patterns in science.

Joshua Tree Retreat Center. Sunday 21 May 2017.

Yesterday, one presenter spoke eloquently of disinformation and the control of news. He told the wonderful story of the 'Dog-walkers Rebellion' in the Soviet Union, an episode I hadn't heard about. Then my friend Janine Taicher showed up, so we spoke at length, of Hawaii and the Menehune.

One of the new people I've met here is Jeremy Corbell, a video artist who makes documentaries based on his interviews. He took me to his ranch, a fine modern structure built on a bluff overlooking the desert at Frontier Town. I also met his artist wife, fashionably tattooed but not inclined to discuss the phenomenon. Night fell over the desert, we had a quick sandwich at a nearby bar once built as part of a movie set, and Jeremy went off to record a conversation with George Knapp, leaving me with two local women friends of his who kindly drove me back to Joshua Tree. They run a good local restaurant, La Copine.

Hummingbird. Tuesday 23 May 2017.

There's something sweet and disarming in a grown woman who will chase a wounded bird around an airport with a broom and a plastic bag in hopes of healing (and freeing) the poor thing. That's the scene I found when I reached the gates at Palm Springs airport yesterday afternoon. Her name was Ann. She'd heard my lecture, so she gave up on the bird, put down the broom, and came over to discuss the conference.

While waiting for the plane home, I discovered that she used to work at HP and Amdahl, specialized in human development. She also, by the way, spent five years in Washington on the staff of a three-letter agency, got disgusted with the financial waste and shady deals, and retired. Her main interest is archaeology, so we ended up speaking about the unexplained giants.

New York City. Friday 26 May 2017. The Highline hotel.

Lieutenant Colonel Tom McNear has been invited, along with Hal and I, to speak about the life of Ingo Swann at the Philip Dick film festival here in Chelsea. Marianne Bilham Knight, with her photographer husband Robert, has produced a wonderful movie about Ingo, his achievements in psychical research, and his art. The hotel itself is something of a monument to Ingo, whose friends own the place. His old sofas are in the lobby, bringing sweet memories of intimate soirées. "Gloria Swanson sat in this sofa," Robert tells me.

New York City. Saturday 27 May 2017.

The White House continues to unravel. It's fascinating to watch, as Trump's son-in-law and confidant admits to begging the Russians for a secret phone line "to unnamed persons." In the meantime, Trump himself travels in Europe, insulting people at every step. American politics at its most grotesque.

I have several companies to visit before I fly off to Paris in less than two weeks, and we face a heavy experimental plan to run samples from New Mexico when Chuck Wade and Nancy come over next week with a collection of UFO fragments.

Hummingbird. Friday 2 June 2017.

Chris Mellon has called me with news of Tom Delonge, after spending a couple of days in San Diego in the company of Hal and Jim. They'd made progress in DC. "The military doesn't know much, but work continues to evaluate the performance characteristics (of crafts)," he said, which led to questions about the isotope ratios I'd mentioned at the conference. Then we spent some time talking about Trump's decision to withdraw from the Paris climate accord, seen as an abdication of responsibility on the part of the US.

Hal had heard my talk about the ejecta and wanted access to them so that Tom could show something exciting and raise serious money. Colm and Garry are shareholders, as are Hal, Chris, and Jim who sit on the board, along with a couple of DoD folks I didn't recognize.

Hummingbird. Saturday 3 June 2017.

This morning, I voted for Lefebvre, the moderate right candidate in the French parliamentary elections. An hour-long drive through the eastern suburbs took me to Livermore, where Ann and I finished the conversation we'd begun in Palm Springs. New information and stunning photographs about early civilizations of the Pacific, one of my private interests, keep coming to light, and more surprises! She gave me the transcript of a conversation where Dr. Jim Harder believed he had contact with an Alien when he hypnotized one of his witnesses. His estranged wife is said to have sent Jim off to a hospice, reportedly asserting that he had Alzheimer's.

Now I watch the aftermath of another series of terrorist attacks near the London Bridge: continuing horror, six dead, thirty wounded.

Hummingbird. Wednesday 7 June 2017.

Chuck Wade, 77, and Nancy, his wife of 52 years, drove into town yesterday, bringing two suitcases of precious samples from the Plains of Saint Augustine. Nancy is a smart, gray-haired woman, who once worked at White Sands as a mathematician. They later built a service company as contractors in New Mexico, in the Gallup-Corona area. I'll introduce them to Garry tomorrow.

Hummingbird. Thursday 8 June 2017.

Second day of work on the samples, while television networks are wholly concerned with the anti-Trump testimony of James Comey, the former director of the FBI, summarily fired for refusing to stop investigation into Putin's meddling into US elections.

We went to the lab of IonPath, in an incubator facility close to the mudflats of the Bay. First results from selected samples. Garry speaks of "the lust in his heart" for keeping some of them as treasures. 'Relics' would be more appropriate, but is that the way to study them?

Hummingbird. Friday 9 June 2017.

The results we're getting from the samples are so striking that I called Tim Jenks, the CEO of NeoPhotonics, so that he could see the honeycomb. In my brief introduction I told him we needed advice as we encountered "a technology we didn't understand." We're not finding what we thought we would find, perhaps indicating a capability to re-engineer matter at the atomic level.

There is drizzle in the morning, turning to a fine, refreshing rain. I only sleep 4 or 5 hours a night, and I pay the price in memory lapses and a general feeling of exhaustion.

Mabillon. Tuesday 13 June 2017.

The luxury of deep jet-lag sleep in the warm atmosphere of an early Parisian summer. Five busy weeks ahead of me, starting with a review of the science environment and our recent findings. On the plane I read most of Lou Marincovich's book *True North* and particularly his OBE in the midst of an Alaskan storm.

Several discoveries were made recently. Fossils found in Jebel Irhoud, Morocco, analyzed at the Max Planck Institute for evolutionary anthropology in Leipzig, have revealed that *Homo sapiens* did not emerge in East Africa 200,000 years ago, as previously thought, but *much earlier* in Saharan regions and perhaps as early as 500,000 years ago, when our lineage separated from the Neanderthals.

Another discovery is the third detection of gravitational waves from a pair of merging black holes at the LIGO instruments. According to the team at MIT, the second detection (in 2015) came from black holes that had been orbiting each other, but this new pair is thought to have formed independently and to be much farther away, at 3 billion light-years, confirming Einstein's contention that all gravitational waves must travel at the same speed. Then there is confirmation of the Higgs boson, discovered in 2013, thought to be the particle that gives mass or inertia to all other particles, and perhaps to the entire universe.

Mabillon. Thursday 15 June 2017.

Lou Marinovich's autobiography, *True North*, describes a dramatic out-of-body experience in Alaska with the clarity and precision only geologists can achieve. Lou is the man who discovered that the earliest opening of the Bering Strait was 5.5 million years ago, *some two million years earlier than previously thought*. I am stunned to learn that this breakthrough happened 'simply' when he noticed a small Astarte clam from Alaska lying in a museum collection at the California Academy in San Francisco. It had been there for two decades before anyone recognized it. Only one or two other paleontologists would have known that Astarte had been dwelling in the Arctic and Atlantic oceans for 100 million years.

"However," he writes (p.153), "I was the only paleontologist in the world who had ever been to Sandy Ridge, knew how old the rocks were, and that 'Astarte' was an unambiguous signal of an open Bering Strait. Experiences like this were a reassurance that life is not simply a random arrangement of molecules moving chaotically: *it has an underlying pattern.*"

Mabillon. Saturday 17 June 2017.

After a simple dinner last night near the Sorbonne, Flamine and I walked home across the Luxembourg Gardens in the sublime sunset. Now she works with a patient while I read (or rather, savor) Philip Dick's *VALIS*, one more time. We had the same insight, in the early 1970s, about a universe of information, except that he received it in a form he describes as an intense beam of pink light directed at his brain, while it only came to me in scattered experiences and coincidences.

Mabillon. Tuesday 20 June 2017.

Days begin in sunshine but quickly turn into the sticky oven the French call *canicule*. Sweat drips down your back and your mind turns to mush. In spite of this, I was able to attend a presentation at Ecole des Mines, and to renew an old friendship with a professor there, Thierry Weil. I came out in time to meet Olivier with 5-year-old Raphaël. Now I read *VALIS* for the third time, amazed again.

Mabillon. Friday 30 June 2017.

On Monday, I made an informal presentation about our isotope study before the Institut Métapsychique in Paris. Then yesterday, Flamine came with me to meet Philippe Favre, and I taught the masterclass he'd asked me to give to graduating MBA students at the HEC business school. The audience of some 60 graduates was a clever international mix of dynamic students, half of them women. It was an opportunity to renew an interesting link with the tony, high-tech Western suburbs around Saclay, Gif-sur-Yvette...

Le Mans. Sunday 2 July 2017.

This old city, on Sunday morning, lies idle and silent. There are construction sites at every corner, big and small, idle as well, contemplated from above by pensive cranes, pennants waving.
    Flamine and I came here yesterday for the wedding of her niece, a very fancy affair in the lovely medieval village of Sainte Suzanne. I felt uneasy with the ladies' incredible hats, the singing of hymns, the long lines of Mercedes and Maseratis looking for a place to park; it stunned the local folks staring from stone benches.
    The reception that followed the ritual, led with steely precision, took place in the perfectly groomed gardens of a nearby castle. The whinnying of racehorses could be heard beyond the tailored hedges.

Nice. Thursday 6 July 2017. Hotel de Verdun.

We took the train to Nice and went to consult Jean-Claude Dufour. He was waiting for us at the tram stop in the warm sunshine. Soon we were sitting in his living room. He told us of the recent death of Pierre Gueymard, alias 'Doctor X,' Aimé Michel's major test case.

Events in the area are as mysterious as those of Utah: men have disappeared, strange beasts continue to be seen... Dufour gave me a striking map where he'd marked those events.

Good reactions continue to pour in after my Paris presentation on the isotope study. I even heard that the department of astrophysics of materials at Saclay had a mass spectrometer (heavily used, however) and two people have offered new samples of material.

Following the IMI-Isotope conference, Flamine and I invited a small group to Mabillon: Méheust and Rospars, along with Laurent Chabin, Alexis Champion of IRIS Intuition, and Patrice Galacteros. The latter turned out to be well-informed (ex-IBM) and a source of good ideas.

Méheust was surprised to learn about the IRIS Intuition and its success applying remote viewing in industry and government, at a time when the academic community remained so closed-up. Our freemason friend was eloquent as usual, speaking of African traditions among hunters and 'forgerons' (blacksmiths). We discussed a Greek island, which he knows well. It was good preparation for the LoneStars meeting.

Vienna. Wednesday 12 July 2017. Hotel Steingenberger.

Our group assembled at 7 pm, without Colm retained by Bigelow and John Schuesler too ill to travel. Eric spoke about Boltzman Brains and the possibility that stars were conscious, especially those that can be tracked, drifting fast on irregular trajectories. He wrote the foreword for a book by Greg Matloff, *Starlight, Starbright*.

The meetings themselves were private, and so will my notes be. Vienna was splendid, the weather fair, sun and passing cloudbursts.

Vienna. Thursday 13 July 2017.

The tenth session of the LoneStars was called to order by George Hathaway this morning, in a modern meeting room of a local palace, adorned by three large photo-realist paintings of grand landscapes.

The first topic was a review of disclosure effort. Steven Greer now argues publicly that "Disclosure serves Secrecy," which reflects my own thinking. He points to false flag operations and the disinformation of the public. Someone strongly objected: "I'm not aware of any false flag operations," he said forcefully. It was pointed out that Rick Doty was involved, "stimulating new physics," and monitoring what everyone did, including Greer.

I won't report on the proceedings, covered under the secrecy rules, but 'interference' obviously came up in open conversation over dinner. The idea that groups of individuals (or entities?) prevent understanding of how the phenomenon interacts with humanity is a favorite topic for some of us, actively involved. I asked for clarification between Alien interference, vs. human (social) interference, or what I've called The Undercurrent. The group remains confused in terminology.

Vienna. Friday 14 July 2017.

The phenomena at the Utah Ranch follow people through their personal life, so the interference problem is much larger than we thought. This was discussed at length today.

Where does *To The Stars* (aka TTSA) fit in all this? Jim Semivan has joined their team, along with two consultants (Dr. Norm Kahn and Dr. Paul Rapp from Uniformed Services). Reportedly, there was a meeting with two Skunk Works members, and meetings with the Lockheed-Martin director of future technology programs in April and May.

Someone laughed: "All that's needed now is a chicken, and an egg..." Along with Drs. Kelleher and Nolan, Tom Delonge is forming a Public Benefit Corporation under the new, flexible SEC rules.

Others involved as distinguished Advisors are an impressive group:

- Rob Weiss, exec. VP, advanced development, Lockheed
- Steve Justice, director of advanced systems, Lockheed
- Eric Schrock, director of air vehicles, Lockheed
- Major General Mike McCarey, ex-Space Command
- Major General Neil McCasland, ex-commander of WPAFB.

For the moment, I'm waiting to read the fine print, as I was taught by my elders. John Schuessler and Eric Davis are not joining. In a rare display of emotional reaction, deeply troubled, George Hathaway, the advisor to our Sponsor, only said that something was *extremely* wrong. Real investors, so far, are staying away.

Mabillon. Saturday 15 July 2017.

As we assembled our suitcases for the return trip on Friday, we learned that our Sponsor had decided his funding would come to an end in one year, "as his age made it necessary to change his role." It's a clear signal, from someone who knows the cards.

Flamine and I flew back to Paris where the weather was delightful. I began answering emails, then a short walk through the city. To dispel the somber impressions of looming dangers left over from the Vienna meetings, I continue reading the life of Barbey d'Aurevilly, with much pleasure.

Four more days here, then I return to California.

Hummingbird. Sunday 23 July 2017.

7 am. Home again, and alone, I wake up to a desk piled high with letters to answer and bills to pay, but the room is flooded with sunshine and San Francisco spreads its generous landscape from the ocean to the hazy depth of the Bay, embracing it all, lifting my mood, still shaky from the bustle of airports, the trains, the crowded flights, the bulging suitcases.

The field of ufology has entered a new phase in the United States, to the tune of screaming commercials and unfocused media frenzy. Various UFO groups claim hundreds of cases each month, but most melt like summer snow as soon as someone investigates the details.

In Northern Utah the talk is still of mysterious injuries, orbs, and monsters, not flying saucers. Brandon Fugal plans a big review. I declined to attend; I said it would have no impact.

Another important initiative is that of Chris Mellon, in close touch with two subordinates to General Mattis. We spoke over the web: "There's so much going on!" he said... "There's a new team in the Defense Department declassifying some gun camera footage, and actual materials. My friends are skeptical of the government's ability to treat the issue properly, until it's public."

Clearly realistic about Congress, he adds: "Nobody's in charge!" and he suggests briefing the Academy of Sciences. Good idea, but I recall the failure of the elite science group of the JASONS. His other idea, more likely to have an impact, was to push for new Hearings.

Hummingbird. Thursday 27 July 2017.

After three days of paper-shuffling, I've answered all the letters and paid all the bills, but I feel lonelier than ever, unsure of directions.

Chris Mellon is coming over for dinner on Tuesday, Linda Howe is sending me the bizarre bismuth-magnesium sample that the Puthoff-Hathaway team couldn't fully validate, and time will be set aside on the mass spectrometer next weekend.

Hummingbird. Friday 28 July 2017.

It's taking me a very long time to recover strength and sleep after this long stay in Europe. I did start on the indexing of the research correspondence (11 folders for Allen Hynek alone), putting that mass of data behind me, preserved in good order. My correspondence with Aimé Michel, too, will always evoke cherished and occasionally amusing memories. But my mind wanders, unsettled after what happened in Vienna.

Hummingbird. Wednesday 9 August 2017.

I went and bought three pots of zinnias today, then planted them on the deck where I've cleaned up the two small raised gardens. There were zinnias in Pontoise when I was a child, simple flowers my mother liked for their ruggedness and color.

I also cultivate my friends, now that I've been reconnected with Vint Cerf, Houria Iderkou, and Karen Breschi (we watched *The Circle* together at her house last week), and Jack Katz of course, with his friends from Colorado who keep getting very boring psychic messages from Heaven, which they relate with enthusiasm.

Hummingbird. Sunday 13 August 2017.

I visited Jack Katz on Thursday. He's mellowed somewhat, his home situation now secure. Our friendship deepens in surprising ways, given his propensity to interminable soliloquies, but he stops talking long enough to point out sublime music (*Transfigured Night* by Schoenberg, and Moeran's *Lonely Waters*, which I'd heard on the radio on the way over, by another, annoying or wondrous coincidence). He also told me about three remarkable old movies: *Enchantment, World Without End* and *Somewhere in Time*, all having to do with the time dimension and profound disorientations of the human condition.

As for his drawing ability, it remains unequalled, but he never brags, speaking warmly of other artists who were his friends, comparing their talents. He points out he was the only one who could draw human anatomy in motion accurately without a model, or a photograph.

Hummingbird. Tuesday 15 August 2017.

I've received the box from Linda Howe with the bismuth-magnesium sample, plus two letters: one from an exquisitely polite Arab gentleman in Paris, with a dissertation about his research on the Jinn (he's an expert on pre-10th century semitic languages), and another from a doctor in New York, a Columbia geneticist who invites me to Krasnoyarsk on behalf of his father, a parapsychologist from southern Siberia. (I can't find the time, unfortunately.) Then two of my little startups decided to raise new rounds of financing with the usual complicated terms and grabbed my attention.

Hummingbird. Sunday 20 August 2017.

A surgical operation on my ears has kept me close to home, with the wonderful added benefit of Catherine's company. She took me to the hospital and back, cooked for me, cleaned the kitchen so thoroughly that I can't find the spoons, and generally was a delight.

There's intense activity, the kind I love, with innovative companies in the midst of new financing and the flow of research ideas. A plan for sample testing is taking shape, too.

Yesterday I had lunch with Federico. "I get up at 3 am to write," he said, "my book has transformed itself into a trilogy, I've never felt so alive in my life!"

He took a sip of Italian red wine and reflected on his career: "When Zilog was sold, I had enough money to stop working, but it took a while to realize that I could decide what I really wanted to do, and to choose the topic of my work—to make an impact on what's important…"

Hummingbird. Saturday 26 August 2017.

In the ever-changing landscape of California, the roads around Mount Tamalpais are a stimulating challenge. Winter rains have redesigned the cliffs and closed the highway, so I discovered a new way to reach Stinson Beach on Thursday, as Lee Spiegel and I were trying to join James Fox at his Bolinas home. James, who has produced two previous films on UFOs, wanted me to check his recordings of Air Force PR man Bill Coleman, and of Clinton associate John Podesta.

Bolinas, hidden away on its isolated peninsula beyond the lagoon facing the legendary surfing spots of Stinson Beach, remains fiercely estranged from the American mainstream, and aloof from the very California that makes its unique lifestyle possible.

I returned for lunch with Lee and James to discuss things we didn't want to put on the Internet, including our approaches to material sample data. James has a few of them yet to be analyzed, including intriguing material from Socorro. I accompanied him up the hill to the home of Marion Rockefeller, Laurance's daughter.

Hummingbird. Sunday 27 August 2017.

I just learned of the death, three months ago, of Simonne Servais, with great sadness. My last visit to her was with Flamine, at a nursing home. She is buried at Le Heaulme, an odd little village I know well, near Gisors, notable for the incident witnessed by Mr. Delangle and his friends at Haravilliers.

There was a 3-line, classy notice in *Le Figaro*. It cited her numerous decorations, except for the medal of the Résistance (the only one she wore), but it was silent about her prominent role in government. She evidently intended it that way. The Pompidou Foundation briefly expressed sympathy. General De Gaulle was never mentioned.

Hummingbird. Saturday 2 September 2017.

Thinking again of the episode in *The Odyssey* when a shipwrecked, naked Ulysses is forced to hide in the bushes, I discover some new lessons in Homer.

Recall the scene: Ulysses addresses Nausicaa on her way to the beach to wash clothes with her maids. Seeing him, the girls scatter in fear, but their mistress calls them back: "Haven't you ever seen a man? This person hasn't harmed us, he needs our help…"

The scene is eminently modern at a time when so many families flee from war and terror and get washed up on the shores of the Mediterranean, hoping to be picked up by a few generous souls.

Nausicaa, a practical girl, gives Ulysses a piece of cloth and some advice to seek help in the town. (I notice another modern touch in the story: the girls have been playing with a ball, the very first instance of a ball game in human literature.)

The interest of the scene, apart from its underlying sexual tension, comes from the predicament of Ulysses. He is a king, after all, forced to cover his nakedness with branches, hurting from the salt, and dirty from the sea and the sand. He has to borrow a piece of cloth from a mere girl. As always in Greek writing, there's a reason for all that, at a much higher level: Ulysses' ship didn't just sink; it was destroyed by a storm whipped up by Poseidon, mad at him for killing one of his sons in battle. He only escaped with his life because Athena admired him and extended her protection while Poseidon thought he was dead.

So, how did we go from these beautiful accounts of human fate, complex but exquisitely sculpted, to the dreary, sadistic stories told by modern religions? The bland remonstrations of what passes for spirituality today in the West, the contradictions of our warlike God, give a stunningly backdrop to our 'modern' times.

Our aspirations end up in contradictions: We preach against meanness and stupidity, often acting mean and stupid with the best intentions. We become furious when this is pointed out.

I think of it because as they age, I watch my friends turn into hardliners: They can't accept that the Three Kings came to adore Krishna, not Jesus. There's a tacit betrayal of the text, by a thousand years.

Even in Vienna last week, one member read to us an apology praising Catholic history for the concept of interplanetary life, yet the ecclesiastical judicial record long condemned that idea and forbade to research it under pain of death. Why sweep ten centuries of falsehood under the rug of a few well-chosen quotations, along with the blood of innocents? Has that ever helped the Church?

I keep closing circles, as does Jack Katz, whom I saw tonight. He recommended a couple of beautiful old movies: *The Four Feathers* (1939) and *World Without End* (1956). Now the Bay Area suffocates under heat and smoke: 106 F, a level never seen since the City began keeping records.

Hummingbird. Sunday 3 September 2017.

A conference on political science has brought to San Francisco the eldest son of my dear cousin Francis, a tall strong fellow. He and his partner teach at a college in Geneva, New York. She has been interested in UFOs during her research and has read my books, so we had lunch and spoke for four hours. When we came out, a breeze was finally cooling the air a bit, dispersing the evil smoke.

Hummingbird. Thursday 7 September 2017.

Alan Stivelman has just sent me a link to *Witness of Another World*. Juan Oscar's personality and the mystery attached to him drive the action to its spiritual conclusion. In Paraguay where the Guaranis hold their initiations, the elders state the problem simply: The white man keeps getting distracted (by money, silly secrets, technology) and *doesn't know how to look for what's important, or secret.*

As if to underline this observation, a monster storm is bearing down on the wealthy resorts of Florida, only a couple of weeks after the hurricane that devastated Houston. This time, the Earth is angry and shows it with disasters, one on top of another, on a planetary scale.

Hummingbird. Sunday 10 September 2017.

Hurricane Irma devours Florida; all one can do is watch in amazement. It runs up the spine of the state, much of it evacuated. But Irma is a portent of future storms: more violent, more energetic, and more unpredictable. The new administration, unconcerned with danger, now promises to allow uncontrolled building along the coast. Yet this isn't just another flood: the swamps are being replaced with an ocean. In 50 years, half of the state could vanish.

Chris Mellon has called me on the web. He's obtained three official videos from friends in DoD showing UFOs, including recordings from the *Nimitz* encounter. But none of that will constitute definitive proof for stubborn scientists, I told him. Where's the deeper research?

Hummingbird. Tuesday 12 September 2017.

Handling bureaucracy all morning, I feel beaten up by irrelevant paperwork. I drove to Menlo Park for lunch with Jessica Fullmer, the CEO of a network-edge video distribution company. She commented about seeing me with Flamine: "You looked so radiant when you were with her." But Flamine is in Paris now, in the bustle of the streets, the Métro, her patients' problems, the coziness of coffeeshops when she meets her sister's kids—all the things of which I was depriving her.

I learned more today about phenomena inside the homes of friends. People are getting overwhelmed with a sense of urgency I'd never detected. Days follow one another. I lead the quiet life of the bachelor, but I've become reconciled to the pace of knowledge, endlessly destroyed and re-ignited at unexpected moments by the wisdom of hermits or the cleverness of infants.

Hummingbird. Sunday 24 September 2017.

A whole day to myself. No appointments, only a walk to the bank to send money to my grandson, then a ham and Swiss sandwich on Polk Street where one of the waitresses, Elisha, has a wonderful smile. Most of my time is spent scrubbing the archives, mailing out a fourth box of carefully sorted acid-free folders to Jeff Kripal. There are gems in every box I've sent out, but I don't need to own them anymore.

Now this magnificent City kneels gracefully before the splendor of the setting sun. The newer skyscrapers of the waterfront reflect the glow against the eastern sky, a pale blue that will soon turn to gray.

I've got a lot done. Reorganizing the documents for long-term preservation and research is active work, not at all an end of the road; rather, it's a new beginning, staging future projects, freeing my mind. Those are the cases the BAASS project never saw, details the Pentagon ignores. This will have long-term effects, a long fuse. There is time.

Hummingbird. Sunday 1 October 2017.

There was a 90th birthday celebration for Jack Katz on Thursday. Karen came with me to Point Richmond, guiding me around a huge traffic jam that blocked the Bay Bridge and the east shore freeway.

We had a warm reunion with our unstoppable prophet, increasingly convinced he has access to transcendent truths transmitted through the "impervious cowling" of the giant Black Hole where our universe has unfortunately sunk. The language has become so abstruse as to be incomprehensible, but the graphics, patiently inscribed (Jack works ten hours a day, and often at night) remain the product of a master.

Douglas Trumbull, in touch with Alan Stivelman, has now seen *Witness of Another World* and loves it. He writes: "Thank you for this unique and memorable film. I think it is the closest depiction I have seen of your philosophy about the phenomenon. We share a deep passion to continue to understand what is happening."

Hummingbird. Friday 6 October 2017.

Garry and I finally found the time to get together at IonPath, planning to run some tests on the Ubatuba material, but the machine was inoperable (something to do with leaks under pressure), so we were only able to pre-load the material. The 'Sierra' metals will have to be tested at Stanford on the French machine.*

Hummingbird. Monday 9 October 2017.

Winds have been blowing all night, swirling through my window, lifting the curtains. They have triggered a catastrophe in Santa Rosa where the grass was dry. Trees fell on power lines, starting fires that still burn uncontrolled. All the systems we take for granted in a modern city—sprinklers and fire alarms, quick communications, access to help—were overwhelmed in a few hours. Phone calls no longer go through, and flames are raging. When I got through to Oberon, he told me the Round Barn where I had taken Strieber, Kripal, and Diana last year was gone. The fires engulfed Potter Valley, Redwood Valley, and the Black Bart Canyon up to Willits. I am afraid for Spring Hill Ranch.

---

* Seven years later, they still haven't been fully exploited; science takes a long time, under the best conditions. Yet these samples could unlock a mystery.

Hummingbird. Tuesday 10 October 2017.

The formal announcement by TTSA went out today, promoted by Leslie Kean in the *Huffington Post*. The team is impressive; around Tom Delonge are Mellon and Semivan, and Hal describing physics breakthroughs to come. Garry Nolan is there with Colm Kelleher and two space technology heavyweights: Steve Justice, a VP for advanced development at Lockheed, and Luis Elizondo, head of the program to track unidentified global threats in the Pentagon for the last 10 years, a media exaggeration.

The company plans to go public immediately, with no operational experience, under a special high-risk procedure using the glitter of science as a hook. The scheme can work if the 'science team' holds together. They do represent a pool of talent.

Santa Rosa keeps burning, 100,000 people are in shelters, a dozen dead, communications impossible in some areas, a modern city in shambles, brought to its knees in just four hours.

Hummingbird. Wednesday 11 October 2017.

Images of the Round Barn in Fountain Grove burning helplessly still keep me awake and sad. Evidently, it wasn't a priority to the firemen, "just a barn." A time of fire and doom has arrived, and I don't know what it means. Only that we don't care for the wisdom of the past.
Yet a civilization that doesn't care for the past has no future, does it?

Over the Internet, boosted by a flurry of articles by Leslie, the Delonge initiative feeds a new torrent of news that only enhances the failure to speak clearly to scientists and the public. There are stories I won't repeat, of a secret project that failed, and those who knew about it said nothing.*

Hummingbird. Friday 13 October 2017.

After days of mismatched agendas, I found a window to catch up on several topics with friends who remain hopeful to restart research. Yet another worker had been harmed at the Utah Ranch. Cattle may be getting hurt as well, and reportedly there are daylight sightings now.

One correspondent found the book by David Booher (*No Return*)

---

* The project in question was KONA BLUE.

very troubling "because it connects all this to the history of MK-Ultra, but there's another level, and it's thematic…it may not even be related to UFOs; it's very big, it involves the sightings of 'Tic Tacs.' It's hardware-related and it's very big. I shudder when I think kids will be exposed." (14)

He went on: "We're at the intersection of several technologies with unintended consequences, managed by people who don't care."

Now *The Sun* in the UK reveals that a large National Intelligence program has been tracking USOs all over the world; that the physics was beyond our understanding: 150 to 200 knots travel with no cavitation…in the ocean… One of their reporters was even allowed to board a sub. Courageously, Lue Elizondo says the article is 100% true: he was in charge of that SAP as the OCA (Original Classified Authority), he claims.

Interference cases continue, in one husband-and-wife case, two kids, it happened inside the home. "Five days ago, there was a creature in the kitchen, a Grey. It could be touched, maintaining eye contact." Catching up with my notes: The wife, separately, saw a figure that vanished, dog-like creatures walking on their hind legs in the yard.

The kids don't go out to play anymore. The man himself saw the creature near the bed at 2 am. In a scene reminiscent of Whitley Strieber's experiences, the entity pivoted, went into the bathroom, lifting its knees in a stilted way. The dog went nuts and jumped on the bed. The man has a wound since that night (as in the cases I saw in Brazil) but no implant. The injury looks like an infinity sign.

On the positive side, the caudate-putamen study is moving ahead slowly. The MRIs are in Detroit, Garry has the genomics, and Collin is securing the double-blind storage. The DNA is analyzed separately, under a study funded in Europe, a multi-year project.

Chris Mellon, who's joined TTSA, verifies that the new company expects two stock offerings, at $10 million each: "They've taken over the idea others were going to pursue, about research on the brain and telepathy, the antenna in the brain…Someone passed on the idea to them. A few people are upset about that."

Hummingbird. Sunday 15 October 2017.

The Redwood Empire keeps burning. Over forty dead, thousands of homes reduced to ashes, large portions of Santa Rosa and Sonoma destroyed, more threatened. Some horrible videos of Tomki Road are posted on the web, showing burned trees, devastated ranches, flames still flickering in many places where gas is escaping. I can't tell if the Spring Hill Ranch was touched, but I recognize the road on TV: all black, dust, ashes, and soot, with dead branches scattered on the pavement.

Exchanges among the LoneStars have turned sour. Even within our collegial communications, things are tense. Even Hal chides me for calling TTSA an "entertainment company," while it designs spacecraft to cross space-time in seconds!

"Let's go back to fundamentals," I told him. "I am prone to mistakes and errors of judgment like anyone else, but the Public Offering Memorandum for TTSA, which I've read, calls for money from the public, and very clearly states *it will be making movies and videotapes*, and perhaps research later. Where's the due diligence for a technology base? And, by the way, what's wrong with making movies?"

Hummingbird. Saturday 21 October 2017.

Last night I had a chance to brush away negative thoughts for a concert at St. Mark: *Missa Brevis* and *Cantata 21*, enchanting and powerful. I was amused to read that in composing his *Kyrie* for the *Missa Brevis* Bach had borrowed some inspiration from classic works but decided to change the form to a three-themed fugue to communicate that this was a *Lutheran* Mass, certainly not Roman Catholic!

My Passavant ancestors, chased away from France because they'd worshipped the Lord under the wrong music, must have approved.

Hummingbird. Sunday 22 October 2017.

Over a long lunch with Garry at Café Flore, I clarified my position on TTSA, and we went over the results from the latest run on Sturrock's Ubatuba samples, making lists of everything yet to be done.

We also agreed that we wouldn't attend the SETI conference, where we were invited based on the paper we submitted with Federico Faggin. "SETI is trying to build a science on a single equation that's sixty years old," I said, "and they haven't found anything to support their hypotheses. They only define themselves by rejecting what Aliens could be, instead of opening up their lexicon."

Quite sad, because they have fine scientists, trying to change things.

Hummingbird. Thursday 26 October 2017.

Tomorrow, I fly to Albuquerque, where Ron Brinkley will meet my plane. The change of scenery will be good for me because my email was compromised last night, leaving me with a sickening feeling. A correspondent from CNES had sent links to recent podcasts, notably an interview by Grant Cameron, skeptical of TTSA. One link (using "RoguePlanet.tv" as a channel) seems to have planted a bug and lifted the contents of two folders. One contained messages about *Sekret Machines*, my own interactions about Delonge; the other was a file of Fugal's remarks. Was this targeted by Counterintelligence, eager to know if TTSA was leaking classified tidbits?

There was nothing objectionable on my system, but business records I'm bound to keep in trust were there, so whoever did this could be brought to Court. I'm determined to renew my life away from this junk. Ufology disgusts me more today than it ever did because some of the attacks are likely to come from 'agents' befriended by the LoneStars themselves.

Datil, New Mexico. Friday 27 October 2017.

Ron picked me up at Albuquerque airport and we drove through Socorro to San Antonio near the Trinity Site (there was a reported crash there in 1945, to study separately), and past the impressive alignment of dishes of the Very Large Array radioastronomy facility to our own base of operations in Datil, where Chuck and Nancy Wade met us. Tomorrow we drive to the crash site at Old Horse Springs, but before we do, a word about the incident is in order: The crashed object was first seen by a soil conservation engineer named Barney Barnett, early one morning in the first days of July 1947, possibly coincident with the more popularized Corona (and Roswell) crash.

This tiny motel also serves as grocery store, restaurant, and social club, so we were greeted by a jolly group of elders singing old Spanish ballads, led by a violin and an accordion. Some of the informal dinner talk touched on politics (Chuck detests Obama) and the Martian conspiracy theories inspired by Doty, which he buys enthusiastically.

The Plains are a beautiful old lakebed or seabed, framed by dark mountains where elk and deer hide. Chuck thinks the object was shot down (by what means, he doesn't know, suspecting the radars). Loose pieces were left behind, so they came back with garbage trucks and spread metal pieces, cans, and old car parts over the landscape. Tomorrow, under Chuck's capable direction, we start digging.

Datil, New Mexico. Saturday 28 October 2017.

We spent the day on site, Ron and I, with Chuck and Nancy. The crash was discovered by a soil conservation engineer named Barney Barnett who witnessed the shattered ovoid craft on a warm July morning early in 1947; he kept the secret until February 1950 when he confided to Vern Maltais. There are other witnesses: archeologists who were studying Bat Cave, a few miles to the south, and the Anderson family who came over with their five-year old son Gerald. It's Gerald who befriended Chuck Wade and told him where the site was.

I approach the place with the respect due to an accident scene (whoever the victims were), grateful for Chuck and Nancy who've kept the secret. Ron has become a trusted friend to them because he comes from a New Mexico family, knows the state so well.

The land impressed me with its vastness and beauty, its gentle wilderness too. The ranch lies between two ridges with tall grass, manzanitas, and madrones. Higher mountains, well above 7000 feet, look down from a few miles away. The craft came from the northeast, hit the first ridge, dug a trench across a mile-wide pasture, and came to rest against a row of large rocks, throwing off fragments of metal.

We had an effective metal detector. We must have dug up 30 rusty beer cans, part of the junk the military had spread, but the deception encouraged us. We found good samples of the foil, including one complex piece. We're coming back tomorrow.

Albuquerque. Sunday 29 October 2017.

We spent another five hours on-site before I had to drive back to Albuquerque. Chuck wanted to concentrate on an area where he thinks the military buried the craft to hide it while they summoned the right equipment to carry it away. Overall, we must have dug a hundred holes, and four of them yielded metallic parts, pieces of foil.

Tomorrow I'll ship my correspondence of the last half-century to Rice University: some good data there. I'm happy to know this will be safe with

Jeff Kripal's team. The archives of Hewlett and Packard kept by a high-tech company at Fountain Grove burned down two weeks ago in the Santa Rosa fire, along with the Round Barn.

Hummingbird. Thursday 2 November 2017.

On a recent two-hour Joe Rogan interview of Tom Delonge, we heard renewed claims about the Nazi. Garry is supposed to have discovered that the human genome was tampered with by Aliens, the Roswell object was manufactured by the Nazi in Argentina, and 'TTSA scientists' will soon reveal a spacecraft based on reverse-engineering. Hal himself supports the idea as "more that media hype."

Robert Bigelow called me today, denying he'd joined Delonge's project. He felt the group underestimated the difficulty of achieving its objectives, that they were unrealistic in their approach to 'disclosure' (a slippery concept), and there are the ITAR (International Traffic in Arms Regulations) implications…Without new support the project can't get traction, unless a rich uncle is hidden somewhere, but the Defense community isn't amused. Some believe there is indeed a rich uncle, but he's been silent since Wikileaks issued its exposé.

Hummingbird. Thursday 9 November 2017.

Mill Valley after the autumn rain is a quiet town encased in hills and multicolored trees, with the peaceful feeling of a provincial town time has forgotten and modernity has bypassed. It's all a front, of course.

I had a colorful and amusing lunch there with Marion (Rockefeller) Weber and James Fox, discussing his project for a follow-on to *Out of the Blue*. It started raining again in the evening. My hope for new research stayed with me like an echo of some long-forgotten truth.

Hummingbird. Friday 10 November 2017.

Karen Breschi came with me tonight to hear violinist Elizabeth Blumenstock in a Vivaldi concert that included a wonderful piece by Tartini: a long, hopeful expectation of higher truth, a patient search.

When we walked out to the street, we could see crowds streaming out of the Symphony, the Opera, and Herbst Theater, all splendidly illuminated in the autumn evening. I felt honored and privileged to be part of a community that cares so much about excellence, a city capable of filling three stately buildings dedicated to art and civilization, night after night, and many more around the whole Bay Area, from Berkeley to San Jose.

Hummingbird. Monday 13 November 2017.

Jack Katz has been ill, and I've been a bit sick myself, so I haven't seen his latest work, but I've had reports about its scope and beauty. Jack's rambles about the need to "break out of the impervious cowling of our Black Hole." Fortunately for us, he has been "endowed to reveal" all this, which he does by drawing his marvelous adventures warning us about the Great Calamity. His pencil is precise and his characters are as sharply drawn as ever, transcending our puny universe.

Now the rain has come. There's a veil between Pacific Heights and the vagueness of Telegraph Hill, and Berkeley beyond.

Hummingbird. Tuesday 14 November 2017.

I've seen Garry, just back from Tokyo, ready to fly to Zurich and Berlin for Stanford. I brought him up to date about my work in New Mexico and we discussed the advisability of launching a small research company dedicated to materials research.

Now, one researcher is said to have compiled an unpublished list with some 250 new cases of medical interferences with UFOs. I have my own lists, still unprocessed.

Hummingbird. Thursday 16 November 2017.

This afternoon I had a long conference call about an automobile networking company, after which I thought to call a knowledgeable expert back east for advice. I found him at home. We spoke about cars, and then I mentioned the pursuit of UFO hardware retrieval.

"What hardware?" he asked. "At TTSA, Steve has made it clear he was never connected to that project." As for Lou, when he hand-carried a letter of resignation to aides of General Mattis, I'm told he was debriefed and escorted out of the building. People are upset about it, but no hardware is mentioned.

"He does have photos," my friend said, "and the 38 papers produced by Hal and Eric, but only one was classified, and it may be obsolete by now. It won't take you to Alpha Centauri."

"So, what did Lou have in mind when he joined TTSA?" I asked. "Directed energy is interesting but many people are working on it, and it takes tons of money. I've spoken to startups in Silicon Valley who study new applications, they're real but in very early stages."

"I'm as puzzled as you are, the people who've taken over Lou's project in DC have little knowledge of the phenomenon."

Hummingbird. Sunday 19 November 2017.

Passing another milestone, I finished packing four more boxes of documents to Rice today, and completed an index of the research correspondence. It has been a long, careful project for most of the year: over 700 correspondents in 11 boxes of documents, a fairly complete picture of the field over 50 years.

Hummingbird. Black Friday 24 November 2017.

Last month's firestorm keeps burning in my mind, so I drove up to Ukiah and on to Redwood Valley, eager and anxious about what I might find at Spring Hill. The tragedy was obvious when I reached Calpella, where multiple red signs warned about hazards, falling debris, and heavy equipment ahead.

Disaster: the entire eastern side of Tomki road had burned, leaving only a few chimneys and twisted metal next to half-melted pickup trucks. Ruins in piles of burned-up objects were covered in white ash, so the remains of my former neighbors' homes were oddly shiny amidst black tortured trees.

Freys' vineyard was dark as well, their metal tanks oddly twisted, but Spring Hill was miraculously untouched. It had its usual look, only marred by all the junk piles and old cars that remain a hallmark of the whole stoned-out, doped-up region.

I drove up the hill on the Cleveland property where all the trees had burned and found their home in ruins. Next to the house was a big motorhome, a light showing through the windshield, so I called out and a friendly, blond fellow came out, barefoot. We shook hands: "How bad was it?" I asked. "How is Mrs. Cleveland?

He motioned to the rubble. "Nobody was inside. The old lady is away, she's got cancer."

"How long have you lived here?"

"We bought the place back in 2000. We had to move out quickly when the flames came over the ridge, then the firemen wouldn't let anyone come back. We sneaked in from the north."

From Willits, I thought. Fording all the creeks. I had done it a few times, slowly, in my four-wheel-drive truck. But in the middle of a firestorm, that took guts. He gestured towards the hills: "There was a wall of fire. We fought it for two days and one night, but we lost the vineyard."

"Who lives at Spring Hill now?" I asked, my attitude changing to admiration. His matter-of-fact account was heroic.

"My son lives in the castle." I had to chuckle; my simple one-bedroom tower had been ennobled. He was happy to meet me. He'd read my books. In Santa Rosa, Fountain Grove is a war zone. Of the Round Barn, only a thick mound of very black ash is left, to be dispersed by the wind and the rains. So much for the Rosicrucians, their secrets blown in a cosmic fire, as befits their silent wisdom.

Hummingbird. Saturday 25 November 2017.

Reconnecting with George Knapp today, I found him puzzled about TTSA's plans after a Vegas meeting that included Lou on one side, and Bigelow on the other. He was impressed by Lou, his letter to General Mattis and his plans for bringing out those videos.

George asked if I thought Steve Justice, under Tom, could build a hyperspace craft. I said he probably could obtain funds to extend the theoretical work, which would be an achievement, but claims of galactic travel would require quite a different approach to funding…

Hummingbird. Sunday 26 November 2017.

A contact who recently attended the debriefing of one of the *Nimitz* pilots with Lou and Brennan McKernan tells me that ufologists will publicize the incident as a 'definite proof.' This led to heated exchanges. My contact is "too upset to talk about it anymore."

Fortunately, we have those amazing sunrises in San Francisco. At 7 am the sky is a technicolor orgy of pinks and mauves on blue-gray volumes of changing vapors with great cavalcades of blackness riding on the top of it. My only sorrow is that I am alone to watch it.

Hummingbird. Monday 27 November 2017.

More news about the firestorms, from a local friend: "Yes, I am safe in Santa Rosa, just barely out of the evacuation zone. I am with my elderly mother and, of course, several members of other species. We are prepared to stuff ourselves into my car and hightail it out of here, if necessary, but it's looking hopeful that we will not. My son lives nearby within an evacuation zone. His house has not been touched but a few blocks away it looks like a post-apocalyptic survival scene: rubble, stumps of bushes, blasted trees, tall chimneys standing alone, and jets of gas ten feet high, punctuating every block. I think of you also, often recalling that time there were two comets with the full moon in the sky: so magical! What a lovely time that was!"

A note from my former neighbor: "There were nine fatalities in Redwood Valley. Considering the magnitude and speed of the firestorm, it's a wonder there weren't more. One of my brothers and three of our employees narrowly escaped. One employee drove with his ten-year old son through a tunnel of fire north up Tomki Road. Another with her family got to the top of Rattlesnake Hill when the super-heated air melted the air filter in their van, killing the engine. They coasted through flames on to West Road and were rescued by a truck. I heard they put a new air filter in it, and it started!

"One 87-year-old man ran out his back door, slid down the bank, crossed the Russian River, dug through hot embers with his hands, and buried himself in the cool soil beneath. The next day he crawled over to Tomki Road, and a Sheriff picked him up with minor burns.

"Most of the vineyards hit hardest only suffered fire damage along the edges. Further into the center, the vines were OK. The edge vines never burned, browning out the leaves. These vines will recover. They can be cut down low and still come back, sprouting a new shoot the next year as long as it's alive above the graft.

"We hadn't picked our Tomki vineyards before the fire. Any smoke-tainted wine will have to be sold to an alcohol distillery. That's where the heavy monetary loss comes."

Hummingbird. Tuesday 28 November 2017.

Jack Katz now puts it in clear terms: "We're born into an allegorical fantasy," he told me on the phone as I was walking back from a meeting downtown, fighting the morning crowds, my mind on other things. But he's right. What people call the real world is a quantum illusion teased out of an infinite soup of random vibrations.

Hummingbird. Tuesday 5 December 2017.

We're making some progress, thanks to the instruments at IonPath. Garry has summarized the findings: "My feeling, and direct experimentation, shows that what Jacques says is true. We paint with around 80 elements but we don't fully appreciate (yet) what the various flavors of elements (isotopes) do.

"We simply don't have the ability to put together what this material is made of. Even the measurements I'm doing are far too crude to understand the local structure of what I am looking at. What I'm doing now is like trying to discover where the family portrait was hung in the living room, using a ten-ton wrecking ball as my scalpel."

Yes, I said, but we don't even know if what we're looking at was done by a Surrealist, a madman, Picasso, or a 5-year-old toddler.

Yesterday, we met with a lawyer to incorporate a testing company.

Hummingbird. Saturday 9 December 2017.

We've reviewed the first isotopic results, those from the IonPath instrument. Garry had just returned from Washington with a cold, so we only spoke briefly, about his meeting with Jim and Lou.

Chris Mellon did not attend but Hal showed up, in town for a presentation. Garry was shown the second video of an object being tracked by the military. The group believes it will be dismissed by the science community for lack of proof. They had "IP worth 30 million dollars," but in the form of what? Garry asked if they had access to hardware and the answer was no. "There's a left foot and a right foot," Jim Semivan observed, obviously concerned. "The left foot doesn't know what the right foot is doing. There's a strong possibility we'll be shut down without knowing why, or by whom."

There was also discussion about research on data structures, and I was happy to hear that Hal and Jim had suggested that I run it, if conditions were ever right for me to join them.

Mabillon. Wednesday 13 December 2017.

In the evening, at home, I transferred small pieces of the 'Gateau' samples that Michaël Vaillant had saved, with Jean-Pierre Rospars, Flamine, and François Louange as witnesses.

I told Michaël we didn't need the big piece, given the fine capabilities of Stanford's machines.

Mabillon. Sunday 17 December 2017.

The New York Times article is out, under the signature of Leslie Kean and a couple of others (15). It reveals the existence of the $22 million secret UFO study at the Pentagon, which ran between 2007 and 2012 "by a military intelligence official, Luis Elizondo," the money went to an aerospace research company run by Robert Bigelow at BAASS at the initiative of Harry Reid. Associated projects were: Project Abel Gray, Project Sierra Doors-KIM, Project Orion, Project Flight Mock. "It's all public now," I'm told.

I'd never heard those details. Didn't Marshall McLuhan remark that "Only puny secrets need protection; big secrets are protected by public incredulity"?

Mabillon. Tuesday 19 December 2017.

The press now claims that the Pentagon has metal alloys scientists don't recognize. "They have material from these objects that is being studied, so that scientists can try to figure out what accounts for their amazing properties," Ralph Blumenthal told *MSNBC* (16). Mr. Blumenthal said the DoD is puzzled. "It's some sort of compound they do not recognize," he added. We've known this for twenty years.

Mabillon. Friday 22 December 2017.

TTSA's hype of unscientific themes is tied to a perception of the public's appetite. It may well be factual, if we believe a book called *Hitler's Monsters: A Supernatural History of the Third Reich.* (2017): "Captain America contains all the elements of Nazi supernaturalism in the popular mind: the connection to occult forces, mad scientists, fantastical weapons, a superhuman master race, a preoccupation with pagan religions, and magical relics supposed to grant the Nazis unlimited power.

"From comic books produced already during the Second World War era to 21st century video games like Castle Wolfenstein, from classic science fiction and adventure films such as *Raiders of the Lost Ark* and *The Boys from Brazil* to contemporary horror movies like *Dead Snow* or superhero franchises such as *Captain America*, popular culture is awash with images of the Nazi supernatural."

TTSA resonates with these themes: implants and crashed disks are the new 'magical relics,' zero-point energy produces the new 'fantastical weapons,' and David Jacobs' hybrids hint at master races. All the elements are there. Guess who the mad scientists are?

Mabillon. Wednesday 27 December 2017.

We spent a warm Christmas at Bellecour with Olivier, Claire, and the kids, and then a second Christmas in Paris, a quieter affair with Flamine, her sister, and the family. In the midst of celebrations, we continue to catch the waves of surprise raised by the phenomenon and its secrets. Whatever happened to us and our friends, swept away in the wake of the *Nimitz*?

Part Twenty-Four

SOLITUDES

16

Hummingbird. Monday 1 January 2018

The night is festive. Alone on the wooden deck, watching the fireworks over San Francisco, I feel no compulsion to go out and celebrate. Better to stay here and dwell on pleasant memories of Normandy and Paris.

Not everything speaks of sweet leisure, however. Garry and I have formed a small company to study our UFO materials, as we review the characterization of our sample set (1).

Other issues grab my attention. Peaxy, the industrial optimization company, deals as its prime client with an increasingly shaky General Electric, which has trouble staying in business, a sign of the times. As for my best physics friend, he's now in his seventies, and his remarkable mind vacillates just as the team's experiments begin to show tangible results. Yet his gravity theory appears far more solid than all the speculations I hear.

Hummingbird. Tuesday 9 January 2018.

An unexpected message from Gilda Moura in Rio: "I am a Brazilian psychologist active in abduction research, a friend of Dr. John Mack."

She has performed a study of alternate states of consciousness with brain mapping in Brazil, and another study of abductee groups. With Irene Granchi (2), she has participated in her study groups and published three books on the subject. She proposes to meet and discuss the phenomena in Brazil and the abductees for whom she's performed hypnotic regression.

Another hopeful startup company, dear to my heart: I just attended their board today, by teleconference from Prague (3). Their work is the culmination of remarkable efforts in planetary image analysis. They have signed up a few significant users for their images, but I need to know how strong their claims of AI results are, so I plan to visit them in two months.

Hummingbird. Monday 15 January 2018.

Physicist Jean-Pierre Petit, *enfant terrible* of French ufology, long perceived as a painful pebble in Claude Poher's shoe, writes to me in kinder tones: "We've known each other for a long time. At the beginning I didn't understand how there could be a link between the UFO phenomenon and the world of the paranormal. But after everything I've gone through, it's obvious there's a link. The phenomenon isn't either nuts-and-bolts, or the paranormal, but both.

"A lot of things have happened since our first meeting in 1976. I've been given everything: physical meeting, phone calls, non-stop scientific information..."

Hummingbird. Saturday 27 January 2018.

Two events in the high-tech neighborhood: a board meeting of AlterG, then what I project as my last venture, a small check for a company that proposes to modernize the management of fleets. I visited them yesterday in Marin, at the Toyota dealership that is testing their software, deployed on top of a blockchain, the latest breakthrough in information technology. (4)

Echoing Jean-Pierre Petit, Claude Poher has answered my Christmas greetings. "No encouragement from French investors for the moment," he reports, "I'm starting to wonder if they even exist! Either they don't know what I do, or they don't understand, and I'm no way going on a sales campaign."

He tells me I shouldn't describe his work as 'antigravity propulsion' because "that doesn't mean anything. One should say *propulsion without ejection of matter*. In reality, what I do is *pro-gravitational,* the only way to move to the stars. The point is to give an enormous amount of motion that is the product of its mass by its velocity. For the trip to be compatible with the lifetime of the crew one must reach a relativistic speed at least equal to 80% of the speed of light to benefit from time compression."

Observing that one can never use more matter than the initial mass of the vehicle, he says one has to move without ejecting any matter, using the 'emitters' he's invented.

Their motion doesn't come from mass emission but from quanta from the general field. His experiments have been replicated by NASA-MSFC in 2013, followed by a nice letter of thanks, but there was no further progress: "They've sent me their video recording at 250,000 images per second, where one observes a 50-gram emitter move 3 mm in 300 microseconds under a 1,500-volt (900 Joules) discharge. The total displaced mass is about 400 grams, immersed in liquid nitrogen. This gives a speed of 10 m/second, the top possible hydrodynamic velocity in liquid nitrogen, with motion of 4 kg. meter/second for 900 Joules, and acceleration of 3400 g. They've understood nothing! Yet that's in line with what I've published."

Dr. Poher now works on direct electrical production from the gravity field, a prerequisite to such propulsion. The physics is hard, whether one follows relativity (as Poher and Puthoff are doing), or going beyond it, as the Palo Alto team does.

Hummingbird. Monday 29 January 2018.

Dinner with Garry and Kit at the airport Marriott. We spent most of the time discussing the failure of the LoneStars to work as a real team (my complaint), and the biased approaches of TTSA with its military and intelligence entrenchment. When I restated my legal concerns about the structure, I learned that my confidential memo to the team had been shown to some journalists at *The New York Times*, which upset me.

Hummingbird. Tuesday 30 January 2018.

I spent the day on research while Garry had to stay at Stanford, worried about yet another drifter who threatens the labs at the Medical School. The field is increasingly dangerous, bordering on the criminal. Chris Mellon has wisely sequestered the tapes from AATIP. Only three of them will be broadcast: the *Nimitz* tape and two others.

My daughter's home. Sunday 4 February 2018.

This is Superbowl Sunday, so I'm relaxing here, munching on cookies and watching the game among a formidable array of TV commercials. Yesterday morning was spent at Garry's home, again discussing isotopes and the many complex ways to identify them with machines based on different principles, including his team's latest inventions. Things are not going as well with the study of propulsion. Over a recent lunch in Palo Alto with the current CEO of our gravity company, we concluded that all we could do now was to fold the company and try to preserve the intellectual property.

Hummingbird. Tuesday 6 February 2018.

Gilda Moura and her family invited me for lunch at the Fairmont, and we spoke about Brazil. She is a psychologist from Rio who urged me to make plans to travel with her group to the new sites she identified beyond Manaus, which will be difficult for me given our work here.

Much of the discussion centered on a certain ranch where a strange craft was seen every night during her trip. She later remarked that she was "very happy with your calm in handling such questions." She must be used to rabid arguments with American ufologists...

Although its research team is mocked by skeptics as "a rock star with a guitar," TTSA has turned an important page in this field. Who knows? The National Academy of Sciences may well analyze the *Nimitz* FLIR data, congratulate the pilots for their gallantry, and conclude there is nothing worth a study, but this time, research will go on. In the process, a number of new people will rise up and enter the field for reasons ranging from wisdom to folly, egotism to greed. They will find backers, if only because the publicity is wonderful and can be fed into eyeballs for the cable channels.

Some effect has already been achieved: the intensity has been cranked up. So have the expectations. Yet nothing else has changed: The physics narrative is as lame as it was before, the important facts are still swept under the rug, and the public has zero perspective on the history of the underlying phenomenon.

Later the same day.

James Fox had asked me to call Alan Sandler, the producer for a series of old documentaries featuring Allen Hynek and me, for help with his own documentary. I reached him at home in Oregon. He remembered me and the conversation was cordial, but his style was as abrupt as ever. He won't release anything unless he sees our script to make sure "we're nice to the government."

At my urging, he spoke about the days when the Defense Department was trying to get the Blue Book story out. He'd worked with Secretary Weinbrenner and a former Under Secretary of State who opened doors for the team. "We did try to get it out, and nothing happened. The Air Force was attacked with all kinds of conspiracy stuff. It was all negative, so they got upset and withdrew."

"I don't remember it that way at all," I said. "As you know, the stuff they'd promised was withdrawn before we could see it. As for the people we'd spoken to—Chartle, Miller, and Scott—they were all quickly reassigned. Our Holloman movie was a simulation of a sort of war game: I'm glad we didn't go off on a limb, telling the public it was real. To me, that whole thing was a hall of mirrors." (5)

He disagreed, continuing to speak of "his friends in the Pentagon." He discounted the Tom Delonge effort, saying "there's nothing new, those pictures (from *Nimitz*) don't show any convincing details." He's right, in this case at least, but the pilots had visual contact.

He did tell me he had kept a film of a UFO circling a rocket, and a photograph of an object taken from Skylab, that would be of interest for James.

Hummingbird. Sunday 11 February 2018.

On Friday I had dinner with Thomas and Andrea Dittler and their son at Slanted Door. We spoke of Europe and America, of art and technology. Thomas is an aviation expert based in Munich and an investor with our Band of Angels. We're trying to restart early-stage venture capital for aerospace and for aviation.

Then today, for old time's sake, I drove over to Ukiah and visited Spring Hill with Diane. We both wanted to see the damage from the recent fires, which turned out to be horrible, even though Spring Hill itself was saved, thanks to the pond we had dug up. Ukiah itself has grown: it's cleaner and friendlier.

Hummingbird. Monday 12 February 2018.

Writing to Jeff Kripal, I pointed out the 'threat' implicit in the *Nimitz* reports. Anything able to fly with impunity so close to a nuclear carrier, and to play games with a squadron of F-18s, is a game-changer, strategically and economically. Also, scientifically. Also, mentally: Where did it go? Where is it now? What about the other object that remained underwater? Is it a reconnaissance device from the Navy?

Some of the mystery has to do, very simply, with the fact that the TV presentation to the world was badly bungled.

Jeff responds: "I have not said anything, partly because I have not been asked, but also because I remain deeply skeptical of the context and the 'dripper.' Who is releasing these videos? Why the Pentagon 'threat' story? Why the cold war narrative? I don't see anything threatening in the videos, other than the fact that we are allegedly looking out from a fighter jet—the threat lies entirely with us, so far. I also remain deeply troubled by how it is all controlled and hidden from real researchers. That is not what you do when you want to understand something. That is what you do when you want to control what we all understand." Jeff is right. Shame on us for not seeing it.

Hummingbird. Wednesday 14 February 2018.

My friends at the Band of Angels organized a fascinating private conversation today with mathematician Whitfield Diffie, Stanford professor and legendary co-inventor of public-key cryptography. A typical academic with longish white hair and shaggy dress, he spoke of the nature and care of such esoteric technologies, explained why the old 'socket layer' of the Internet was upgraded with the new notion of 'transport layer security,' and he described the breakthroughs he'd initiated, where even knowing the codes gives your enemy no advantage in deciphering messages.

All that had happened to the dismay of the Intelligence boys who woke up too late to catch him, although they would have loved to throw him in jail for a very long time—without his computer!

Highly respected around campus, Diffie was clearly distraught at the recent loss of his wife. He spoke of his computational exploits as one might recall a hard jungle trek far away, long ago…

Hummingbird. Sunday 18 February 2018.

The spring sun gives an illusion of warmth, soon contradicted by a blast of cold wind that I fought last night to have dinner at the Big4 with Jeff Kripal and grad students Timothy Grieve-Carlson, Simon Cox, and Learned Foote for an evening of fine food and a wide review of everything ufological.

It was time for dessert when a waitress who'd recognized me came over with a note. She wanted to know why releases of reliable UFO data were always mysteriously held back by the government, an easy answer when you think of what 'the government' consists of.

Hummingbird. Tuesday 27 February 2018.

The gold gun at IonPath isn't working, so Garry's team has returned to an oxygen gun, less precise and susceptible to recombination with the elements to be tested. This will require another phase of analysis, and further delays in testing my samples for unusual isotope ratios. Our discussion goes on about AI neural networks.

Hummingbird. Saturday 3 March 2018.

The Computer Museum maintains its archives in Fremont, where I went yesterday with my friend Houria. My donated documents now have an official Lot number: X8145.2017.
 Psychologically and pragmatically, I must prepare myself for a critical cancer test tomorrow morning, and for a trip to Europe in one week. My financial affairs are in order, but I have fallen behind in capturing events in this Journal. What events? The same arguments continue to fly among my friends about threats (overt and covert) from a retinue of scoundrels, and about exposés by the brilliant but enigmatic Isaac Koi on European websites.
 What we'd feared is happening: The popular revival of interest in UFOs has spurred Hollywood craziness, and the field becomes increasingly weird. Garry, too, is assembling a series of projects and companies, even as he fights the Stanford bureaucracy, as difficult to deal with as it was when I worked at the Computation Center, decades ago. Our focus is simply to improve material analysis, with renewed intensity in the resolution of the samples' enigma. We hope the oxygen gun will be working this week, so we can soon put the first phase of the work behind us.

Mabillon. Thursday 15 March 2018.

The city is pleasant, spring sunshine alternating with periods of rain that keep the air fresh, and my daily lunchtime walks to Saint Michel an easy, entertaining little trip. Flamine has left California behind. She has rebuilt an active, productive life here, with her colleagues and patients. Our contradictions are intact.

Mabillon. Friday 16 March 2018.

Lunch with Dominique Weinstein today, at 'the usual place,' as we enjoy writing in our faux-conspiracy tone. More seriously, he told me that the CNES was leery of what they've heard from the US. The videos transmitted to *The New York Times* by Lue Elizondo remain controversial, given the huge amount of missing data. According to Nate Jones, director of the Freedom of Information Act Project at the National Security Archive, the videos seem to have been altered and haven't gone through formal declassification. Also, French researchers already know the *Nimitz* data, although presented as new by the NYT; it circulates on the web (and the dark web) since 2007.

Wired magazine finds nothing to indicate the videos actually come from a UFO program, and imaging experts speculate that the objects seen on the infrared video could be advanced drones. That wouldn't account for what the pilots saw with their own eyes, of course. It's just the old negative bias, but the story is indeed incomplete.

For the financing of its projects, TTSA has succeeded in 'circling' some $2.5 million but it may get less than half when the checks clear.

Mabillon. Saturday 17 March 2018.

A colleague who follows the field has drilled into the activities of Intelligence super-skeptics and their tattered band of 'sock puppets.' He's not impressed: "When the Cold War ended, ufology stopped being a dedicated tool of the IC. What you see now isn't a coordinated program within US intelligence to manipulate opinion. It's just a show to get even with a few people." The real opposition to scientific study is elsewhere. It consists of scattered bureaucrats and political appointees driven by fear, careerism, extreme fundamentalism, and incompetence.

Mabillon. Monday 19 March 2018.

Bruno Mancusi, a Swiss chemist interested in private, scientific UFO investigations, was kind enough to come all the way to Paris by train from Lausanne today. He brought us three sealed bags of samples from a case of 'home invasion' by small luminous balls.

The witness, a 40-ish hospital nurse, has assembled evidence left by the exploded orbs on her window and on a wooden beam. The collection includes spectacular videos of a large, bright object in the sky filmed at dawn, bouncing around luminous beams.

I am reading *The Beautiful Cure* by Daniel M. Davis (about progress in cancer research at the Parker Institute), and I bought Camus' *La Chute* to read on the way to Czechoslovakia on a company visit.

Prague. Penta Hotel. Friday 23 March 2018.

Our hotel, with its cavern-like, warm modern lobby and great comfort, is just around the corner from the offices of the company, so Flamine and I simply walked over to meet the chief scientist. He brought me up to date on new software that tracks industrial and economic developments for clients around the world.

We went to lunch with part of the team and heard about their skills and dedication, thrilled as ever among young mathematicians and software designers with this level of talent. The restaurant looked like a factory. We ate in the massive kitchen and brewery, filled with busy young people, vibrant and smart.

It rained a bit as we came out, the weather was as cold as Paris. So, we took the subway to the Old Town and walked around the astronomical *horloge* tower with the tourists. We went all over, covering a dozen kilometers on foot, happy to be together, discovering this land.

The Internet brings echoes of sharp words exchanged among my LoneStars friends. I only catch echoes of it from Prague, on the tiny screen of my phone. One group complains about the medical data accumulated under BAASS but withheld; it was critical in any assessment of the situation in Brazil. I find this ironic since I tried for many years to interest colleagues in those same Brazilian injuries, which they kept explaining as insect bites and nonsensical anthropological factors as far back as thirty years ago.

The team has now briefed some Senate staffers in a four-hour classified setting, expanding on my Brazilian data, but I'm not invited, although I still regularly get first-hand reports of the matter. As for Bigelow's own data on South America, it remains wisely locked up as well.

Everybody has started to speak excitedly about "thousands of pages of revelations," but either those didn't survive, or Brazil is not willing to release them. More likely, it's the old morass I knew, of reports in official

Portuguese, incomprehensible unless you go there and talk to the people in charge, but of course most have died by now. I turned off the Internet connection for the rest of the day.

Prague. Penta Hotel. Saturday 24 March 2018.

On the way to the castle today, we admired one impressive church after another, but the places we loved were the paths of the Old Town under the ramparts and the intensely vibrant Golden Lane with its row of small houses, more like cabins built by the guardsmen, now turned into precious small museums. One of them in particular held a meticulously reconstructed alchemy laboratory: three rooms filled with retorts, globes, glassware, candles, and athanors.

I wondered whether I might sit at the desk and retrace the experiments I had started there in the sixteenth century.

We walked all over in the cold air, had a light lunch in a vaulted medieval cellar, and crossed the river on Charles Bridge, back to the areas we'd already enjoyed yesterday. I was tired, so Flamine wandered on her own throughout the city, returning with a gift of a remarkable book on Esoteric Prague, chock-full of historical details about Alchemy, Kepler, Tycho Brahé, and the history of the Rosicrucians that I had not seen anywhere else.

Prague. Penta Hotel. Sunday 25 March 2018.

Prague is beautiful and discovering it with Flamine is a lover's delight. We've located the house of Johannes Kepler, so we went to pay our respects in the sadly deserted courtyard with its simple armillary. The place is neglected amidst the tourist shops selling trinkets and the glamorous museums displaying royal treasures. Afterwards, amusingly, we passed by a bright shop called the Hynek Cibulka.

Mabillon. Monday 26 March 2018.

As soon as we got back, Flamine rushed to assemble an informal dinner party for our Parisian friends. Michaël Vaillant and Morvan Salez arrived first, and we were soon joined by Marie-Thérèse de Brosses, to whom Michaël returned the Gateau material sample. She's had her own troubles, a series of break-ins at her apartment and even physical attacks that have left one of her arms nearly paralyzed. (She'd found some papers at a restaurant, left behind by stock market crooks. They were not amused and tracked her down.) She seemed to have recovered her strong mind, fortunately. A trusted, long-term friend.

Mabillon. Tuesday 27 March 2018.

The all-day meeting of the expert committee of GEIPAN was held today at CNES Headquarters in Paris. There were 16 of us, led by Jean-Paul Aguttes with his assistant Brigitte Vergé. Bertrand Méheust attended but Rospars was away. I met or rediscovered some faces: Luc Dini of the Sigma2 commission of the International Astronautics Federation, Jacques Zlotnicki of the Earth Physics Observatory in Clermont-Ferrand, cognitive psychologist Jacques Py, and investigator Gilles Munsch. Our committee went through a detailed series of updates on cases reported in France and elsewhere.

After the obligatory lunch at the colorful Pied de Cochon (Pig's Foot) across the street, we devoted the afternoon to the review of five unsolved French cases to be re-evaluated in detail. I did learn about their circumstances and about aviation issues I hadn't thought of. Next, a presentation by Zlotnicki about the Hessdalen phenomenon left me puzzled: Is there anything paranormal in the Norwegian observations?

It was late when I left the building in Les Halles. Paris was dark and moody under the rain. There are street works and building sites everywhere, open trenches filled with grayish water, and the square in front of the building still misses its trees and bushes. More ominously, somber thoughts linger about recent terrorist episodes and a looming three-month (!) train strike.

Mabillon. Wednesday 28 March 2018.

I woke up to a pleasant childhood memory, the song of a dove perched on a nearby roof. Back in Pontoise, our house was surrounded with large gardens, home to birds of all types and their pitches, tones, and colors.

John Schuessler has just sent me a recent article about mind-reading technologies flowing from breakthroughs in brain science. If this now appears in print, it means the covert studies must be far beyond what the public sees. (*New Scientist,* 10 March, p.3)

Now we're finally getting to the point where UFO data can be studied in the lab, but the main incentive among US researchers is to understand how people get paralyzed or killed when they come close to these objects, so they can build new weapons. Does that really make sense, in the full picture? The same people don't ask how the knowledge can help improve or heal mankind. Or what the real, long-term social consequences of their secret ways will be.

Hummingbird. Saturday 31 March 2018.

The first item I found in my mailbox as I came home was the catalog of my correspondence for the archives at Rice. My most useful contribution now is to continue this archival work.

While I was away, some improvements were made here, so I found the apartment bathed in an inspiring high-tech glow. The place feels like a space station hovering over this magnificent city, and the harbor beyond it, an orbiting capsule I inhabit alone, unfortunately, lost in space and in my emotions.

Silly skirmishes in research: With the support of a Russian YouTube channel, Steven Greer has launched an Internet video attacking Garry Nolan and Stanford University because 'his' extraterrestrial specimen from Chile has now been explained as a very unfortunate stillborn female child suffering from various genetic diseases (6).

Greer complains the CIA must have influenced Stanford, and he uses the case as a steppingstone to launch verbal missiles at TTSA, calling DeLonge an "aging rock star," and building on supposed revelations from his newfound friend Rick Doty who confirms the government once sent him "carrying bags of cash to the media" to disinform the public about UFOs (7).

Greer adds random historical references to Battelle and claims that TTSA, Hal Puthoff, Senator Warner and Bigelow ("the loony billionaire rat pack") have "set up a scam to convince America of an imminent Alien threat." How does he know? Where's the research?

Hummingbird. Friday 6 April 2018.

There's a French channel on my TV, so I watch the crowds in Paris, the latest strikes, and the ever-present spectacle of social upset that never changed. What scares me, beyond the appearance of control Macron tries to display, isn't so much the nastiness of their daily turmoil but *their inability to name it.*

When I wake up in the night and wonder if I should be making plans to spend more time in Europe, I recoil at what I saw in France, an existence drained of any willingness to transcend it. The TV shows crowds crushed together on the quays of railroad stations from Lyons to Normandy, waiting for trains that won't come, in part because nobody knows where the locomotives are. Once the strike is settled, bringing all the machines back to the barn will take three days.

Everything has turned green here. Polk Street is silly as ever, small town America with the delicacies of its Thai restaurants, oddly spiced up by the naughtiness of Good Vibrations next door and the rumbling cable cars from another age. Deplorably, I don't have anyone with whom I can share its oddly poetic life.

Flamine has rushed back to the safety of Parisian confinement and another train strike. Solitude or not, my optimism stays high. I sleep without the anguish that plagued me years ago. I look forward to a healthy California breakfast of cinnamon-raisin bread and those marvelous pears in syrup grown in the coastal orchards.

Hummingbird. Saturday 7 April 2018.

Dr. Green called me today with a frank endorsement of *Forbidden Science 4*, in spite of the record of occasional friction it contains. "It's the first book in decades that I couldn't put down," he said warmly. "I made myself a cup of coffee, turned off the news, and I didn't stop reading for seven hours except for a quick lunch. You mention several important conversations I just didn't remember. I think I may have experienced some fugue states, but everything else is fascinating."

We spoke of the 'soap operas' that took place in the nineties, the absurd extrapolations from abduction stories: "I'm glad our thinking has now turned to more solid data, although I'm afraid of what we're going to find as our hypotheses become crisper. My own thinking became clearer after I left the Agency, in 1985."

We reviewed our recent research, and Kit did correct me about the facility Pat Price once remote-viewed on Mount Perdido in the Pyrénées. The CIA has its own 'Imagery Analysis Service' attached to the National Photographic Interpretation Center (NPIC), the organization set up by Art Lundahl. Kit went there to check the graphic slides and saw the image of a facility, not previously reported, that straddled the Franco-Spanish border, but the coordinates were a bit different. It's a long building with a dome, not military, with housing for about 50 people. Price had thought it had to do with UFOs when he remote viewed the facility. Did he just pick up the dome?

Hummingbird. Sunday 8 April 2018.

Garry has returned from another whirlwind lecture trip, so we were able to review our latest samples to prepare them for isotope analysis. I also told him about the videos from Lausanne, and we watched the bizarre phenomena on the big screen. I gave him the 'shard' that Morvan Salez had brought to me in Paris, from the collections of the late Dr. Roger Leir. The instrument at IonPath (our main machine, from now on) has been repaired; the tests are looking good, so our first samples should be able to run later this month.

An amusing piece of news: There is a new rock music group in the Midwest calling itself Major Murphy, based on the character I created in *Messengers of Deception*.

Hummingbird. Wednesday 11 April 2018.

Long-time friend and renowned executive recruiter Paul Gomory invited me to lunch today in the magnificent setting of the St. Francis Yacht Club. Like me, Paul contemplates spending more time in Paris.

He's a clear thinker and a knowledgeable colleague (What stories couldn't we tell about Silicon Valley lifestyles if we wrote a book together!) so the conversation boosted my resolution.

Paul had brought the old index card under my name from his former Rolodex, a prehistoric device, with my address from our first meeting in 1986. Ironically, all his earlier computer records have become unreadable on today's equipment, so that piece of cardboard is the only key to past information. In these heady days of AI hype and transhuman memory, the obsolescence of common objects keeps calling itself to our attention like the siren of an approaching fire truck, but we choose to ignore the signal.

This is a decisive week. The team at UCSF tells me it is time for the prostate operation. I've been followed in 'active surveillance' since the cancer was first detected in 2015. I do trust the Accuray robot that I helped finance many years ago.

Hummingbird. Monday 16 April 2018.

Lunch in Menlo Park today with Tim Jenks and one of his top optical engineers at NeoPhotonics. They returned to me Linda Howe's Tungsten-Magnesium-Zinc layered sample, with an apology for being unable to run the specific tests in the way we'd anticipated.

All is not lost, however, because they have a design for an original laser-based device that would generate terahertz modulation in a fixed plane—enabling very precise measurements of any gravitational alteration. Is it worth it?

Hummingbird. Saturday 21 April 2018.

A fine weekend is ahead of me, but I'm wasting it, sleeping deeply again in the afternoon, the undisturbed sleep of an animal in search of new bearings, sweeping the emotional horizon. A call from Garry woke me up. He'd just arrived in Hawaii and was sorry not to join me for a dinner with Chris Mellon, but the instruments at IonPath are running well again. Later it was Jack Katz, proud of his latest graphic page, inspired by the music of Romeo and Juliet he plays in the background. He spoke of the *State before Existence*, and how it got shattered. He promised to reveal events from his childhood that he's beginning to insert into the graphics, as they finally emerge from his unconscious like long-awaited creatures from meta-reality.

My daughter's home. Monday 30 April 2018.

I'm her guest for a brief interlude, a quiet night away from the City in this peaceful retreat. I've enjoyed dinner with her and her companion, as we took the opportunity to discuss the next senatorial elections, and to joke about local politics.

My dinner with Chris Mellon on Tuesday was most cordial. He offered support for our projects and said he might finance TTSA. I note that Lou Elizondo has just been invited to lecture by Italian and Brazilian specialists, a nice form of official recognition.

Hummingbird. Saturday 5 May 2018.

Cinco de Mayo…San Francisco is dancing the whole weekend away.

I've never completely known how the Bigelow research efforts on UFOs terminated. In response to my inquiry, I've only heard that General Burgess may have blocked third-year money, already appropriated.

The simple fact is that the senior leadership, including Robert Cardillo, were concerned that BAASS or any extension of it would end up on the front pages of *The New York Times*. Along with General Clapper, they took the risk-averse path.* I'd naively thought that Harry Reid, the most powerful member of the Senate, would ensure continuity of the program. The answer I hear is that he was bluntly told by General Burgess to go away; he was yelled at by Jane Holl Lute and rebuffed by Ashton Carter. Then he was told, "Sorry, my hands are tied" by Leon Panetta.

There must be more, however. For the senior Senator to be treated this way, there must be something at an even higher level that maintains an incredibly tight curtain over the entire phenomenon. As the program was ending, Senator Reid forwarded new efforts to 'SAP' the program, elevating it *because of the progress.*

Who were the risk-averse bureaucrats, or the Deep State actors concerned about exposure? What are we missing? The answer, I hear, is typical Washington: Ruth Davis helped kill the Army Stargate program; John Gannon re-established it in another Agency; Dr. Davis then confirmed it. Some in the leadership still supported BAASS, but "undisclosed events" blocked it.

---

* The proposed KONA BLUE classified program, proposed by our group in 2011, had been denied by Homeland Security the following year. It would have been built upon the data from BAASS.

Fig.19: *Visiting a satellite imagery company in Prague. April 2018.*

Fig.20: *AI case retrieval demonstration, with Paul Hynek, 2019.*

17

Toronto. Monday 7 May 2018. Four Points Sheraton Hotel.

The hotel is in Mississauga. The group assembled in the afternoon, and I joined them for a late dinner (minus Garry who has a big milestone at Roche, Colm who's at work, John Schuessler who is ill, and Bob Bigelow who declined). Jim Segala (8) joined us, as well as Collin Puthoff. The conversation began informally. We discussed Jim Semivan's dramatic home encounter; he'd told me that he'd consulted Dr. Ron Pandolfi following an episode when he developed bleeding in the back of his neck. Pandolfi told him that, if he wanted to know what was going on, he should get cleared for some additional SAPs. But Jim and his wife decided to take a leave of absence instead, judging the situation too dangerous.

We've also heard that TTSA would soon be "restructured into new legal entities." Dr. Segala now works at George's lab on radiation models, and Dr. Green has investigated events at the Ranch. As for EarthTech, it has been funded since 1989 but the support may terminate as their European supporter retires.

So far, TTSA has drawn down one million dollars from the public, as Chris Mellon once told me. Their most visible member now is Lou Elizondo, who advocates the Delonge-Puthoff technical theories. Elizondo was a GS15 at the Pentagon, then elevated to an SIS-1 with one star when he was sent to Gitmo (Guantanamo). After that assignment, there was apparently no job for him in DC, so he reportedly went back to his former rank. He now speaks openly about UFOs, as he recently did on Russian TV, and will carry his message to Brazil in a few weeks.

I still find much uncertainty in the current data, even in the *Nimitz* images. Raytheon has firmly reminded us of the optical artefacts in their equipment, and of the failure of radar correlation. There are uncertainties, too, in the scientific argument. The objects make sudden moves, but those could be an effect of cloaking and uncloaking. When the pilots flew under the objects, they saw the underside as fuzzy, not solid. We've heard all that before.

Toronto. Tuesday 8 May 2018.

The main presentations began with the recent sea change in credibility because of the revelation of the Government's own work on the subject. Harry Reid had stated on 16 December 2017: "We don't know the answers, but *we have plenty of evidence* to support the questions. This is about science and national security. If we don't take the lead, others will."

Yet I don't hear the obvious questions: What 'others'? Russians? Chinese? What's the nature of the threat? Are we facing the drones of a foreign power, possibly off-world? The object tracked by the *Nimitz* measured 46 feet in length and was stealthy: radar couldn't attain a lock. But this was no isolated incident: Over the last five years there have been dozens of such sightings, always with 'Tic Tacs,' hence the Pentagon's decision to only partially release the data. Scientific response, unfortunately, has been nil. The subject scares everybody.

Hummingbird. Wednesday 9 May 2018.

In my discussions with Dr. Kit Green (after the formal meetings) about Brazilian phenomena, his hypothetical pronouncements surprise me, so I try hard to see the situation through his eyes. He'd begun his presentation with an appropriate citation from Nabokov: "Queer, how I misinterpreted the designations of doom…"

Said Kit: "In Colares there were 37 victims in the period 1976 to 1978, and they fall into three subsets. There were also 73 non-Colares medical cases, from 30 years before to 10 years later. We thought that the same interference cases we knew applied there, with acute and subacute injuries, leading to the notion that we dealt with the same phenomenon. That hypothesis was false."

So, what's his conclusion? "The cases are homogenous," he said, "but they aren't related to demographics or cultural anthropology, and they don't match the interference syndrome. They result from a false flag operation, along the lines of the Iron Mountain deception." (9) He laughed: "Somebody is after us!"

This left me confused. How can the hypothesis that "we deal with the same phenomenon" be false if "the cases are homogenous"? He never separated the non-Colares cases from the primary ones. He may well be right, but he provided no access to that data.

I continue to believe the cases are very much related to culture and demographics. Perhaps what we've called the Interference Syndrome is the problem. I've never heard a clear operational definition of it, but I did review the injuries from Colares, and I've spent time with the officers who had to deal with them.*

Later the same day.

I can't comment on much of the proceedings, under the strict Chatham House rules we follow. One exception today was George's presentation about Alien beings in antiquity, quoting the 'Watchers' of the *Book of Enoch* and Michael Heiser's *Reversing Hermon*. Were the Sons of Heaven in Enoch the same entities as the Sons of God in the King James Version of the Bible? There were supposedly 20 leaders of groups of ten beings, and the Nephilim were the children of these 'giants.' The discussion was fascinating.

Hummingbird. Wednesday 16 May 2018.

Reflecting on the vagaries of journalistic reports about the Pentagon's programs on UFOs, Kit recalls that "BAASS was pretty much dead by 2011; and my view is that the continuation of AATIP since then up to now is a fiction, wrapped in a partially true story. The terrible exaggeration with the spin toward ET is a side-effect."

I think it's a brilliant observation. Who should own the medical information in the Capella files I designed? I wasn't aware of the full operation while I was away from Las Vegas, so another two hundred medical cases of mine never made it into Bigelow's files.

Lou Elizondo had to cancel a foreign trip after someone asked, "What do you think you're doing, getting mixed up with UFO promoters in Brazil, and those people in Italy, the fake cases…" But when he goes back, he will take a film crew with him.†

---

* At this date of writing, in 2024, I have found the answer to this: Kit and I are both right, but about different places and times.

† The Italian trip did take place in 2019, resulting in the deeply flawed Vetruccio story in June 2019.

Hummingbird. Thursday 17 May 2018.

A happy evening with much nostalgia as we celebrated my daughter's 50th birthday at Alioto's famous restaurant by the Bay, two blocks away from our old Hyde Street home. Just Becky, Cathy, and me, and the setting sun over the Bay. An icy fist held my heart when I thought of those who were gone, and those who were away.

Hummingbird. Sunday 20 May 2018.

Vociferous arguments, frayed nerves, pointed rejoinders among my friends, following the release of a report about the *Nimitz* by George Knapp. Is it really "the true report extracted from the Pentagon files"? Or simply a fraction of it, fascinating but partial?

The argument is important because the scientific community has yet to react to the major revelations, and the current hype on TV isn't having any impact among scientists. The pilots have described some amazing facts in exact terms, with precise data. Were they ever made aware of the full range of phenomena being tracked from the surface vessels, at the same time?

Hummingbird. Monday 21 May 2018.

A powerful dream woke me up at dawn. That hasn't happened in some time. I was driving away from a meeting and a woman said: *There's a new wind within the wind, there are new worlds between the worlds, there's an unseen universe crawling in.* She told me this so sternly, and it was so clear that I woke up on the spot, the memory crystal sharp.

Hummingbird. Sunday 27 May 2018.

The Sun has come out at last, draped in a double veil of high clouds and ground haze. This is the Memorial Day weekend, so computer networks have calmed down and quarrels have been set aside.

There is an unusual asteroid in the solar system, captured by Jupiter. Discovered three years ago, it moves in the opposite way from the planets, so *New Scientist* jumps in with a cheap headline: "Alien among us!" Skeptical academics are both fascinated and repelled by the idea of space visitors, and they cannot resist the temptation to make awkward jokes, like frustrated adolescents.

Hummingbird. Friday 1 June 2018.

The preliminaries for my cancer operation are scheduled, first with the implantation of gold markers for the Cyberknife, then in early July for simulation planning and imaging. The radiation treatment may extend to August, when Flamine will be here.

Robert Bigelow called me a few days ago to discuss *Forbidden Science 4*. I had already heard from Kit and Hal, who requested no significant change, so I was eager to hear from Bob.

He was pleased with the book. "It's given me better insight into the territory we covered in our research," he said. "Some parts even made me laugh, remembering what went on and seeing how you described it. I've been away from all that; I've had no contact with the folks from the old project, except for Jim Lacatski, who seems to be OK. I'm going to the SSE convention in a few weeks, so I'll probably see John Alexander there."

"Did you follow the discussions around TTSA, the *Nimitz* video and Delonge's latest announcements?" I asked.

His response surprised me: "No, I haven't had time to pay attention. The public is saturated with competing topics, so I don't expect we'll see much significance. Even the *Nimitz* case is already old, I'm not sure it matters. The public sees this as just another series of stories, like the Phoenix Lights observed by thousands of people including the Governor of Arizona: it's had no impact… It will take a lot more to get people's attention."

I can only agree with his assessment.

Hummingbird. Sunday 3 June 2018.

This is the first weekend of warmth and sunshine this year, so I went over to the lake and rowed alone. Despite my concerns, I feel alive and strong, walking every day. Last night I went over to Point Richmond for dinner with Jack Katz, a bit more bent and walking with a cane but always kind, intellectually aware, and often brilliant. He even wore his *bérêt* in my honor. (Americans think that the French wear bérêts. The last time I did, I must have been 12 years old, but movie stereotypes always win.)

We sat outside at a small Vietnamese restaurant by the Bay. We launched into a complicated mythological story, a legal trial of Destiny arguing against Fate. Jack's productions involve dark forebodings, the knowledge that several civilizations arose and died on the earth before we came along, and a colorful theory that our DNA originated from the Crab Nebula (astrophysics be damned). He writes a lot, courageously fighting dyslexia and getting lost in philosophical minutiae that delight him.

Back at his apartment he played an old movie for me, *Tales of Manhattan*, with a stunning cast: Charles Boyer, Rita Hayworth, Ginger Rogers, and other stars of that era! "It's a serious movie, made in 1942 at a time when we thought we were losing the war," he said, "You can feel the drama and the intensity in every scene."

Later the same day.

The mass spectrometer is still broken at Stanford, so Garry and I are making no progress in the analysis of our 20-odd UFO samples loaded on the plate. We spoke about his childhood. In a thinkpiece he wrote some months ago, he notes: "It started when I was five or six. Little people in my bedroom, next to me. Faces looking in the window. I know I was awake each time. I couldn't move. I complained to my parents, and I was told I was just having bad dreams.

"I forgot about it until my early teens and was a paperboy...had to get Mr. Pritchard's paper to him by 5:30 am or he would complain to the paper distribution office, and I might lose my route! Lights in the sky behind Mr. Pritchard's house. I had to pass behind to take a shortcut to the next street. It was through a copse of woods away from other houses.

"The lights passed overhead—no more than 20 feet above me. Soundless. Probably 30 feet across. No shape I could discern—just the lights... I can see them in my mind's eye to this day. I hated having to go through that area of the route every morning since that time.

"Nothing happens again until I'm 33. I wake up and there's something tall at the bottom of the bed. Paralyzed again. I can see it—I'm not afraid. It's just tall and thin, kind of smoky and see-through. 'Go to sleep' I hear clearly in my head. And I do."

Hummingbird. Friday 8 June 2018.

I work hard on many fronts, the main one an update of all venture projects, a secondary effort reviewing the four extant volumes of *Forbidden Science*, an eye to the long-term. At day's end, before a glass of wine and some microwaved food, listening to classical music or to a debate on radio about some new farce in Trump's Washingtonian wilderness, I find some happiness in my solitude, and I don't understand why. I am free from the artificial constraints most people put on themselves. Staying away from the public hoopla hyped up in the name of 'research' means antagonizing some of my friends but it may turn out to be one of my best decisions.

Fred Adler taught me never to compromise deeper standards. I watch people throwing away their compass to chase after pride, ego, or plain greed, and that's their choice.

A harder decision is the disposition of my main research files (Section A, the analysis 'Blue Files' with the field investigations). I have assembled the data in the first box of such records destined to go to Rice, when they would join section B for 'Background' and C for 'Correspondence.' But these Blue Files represent something unique, something many pundits of ufology haven't done: *actual travel to the sites, long hours with witnesses*, careful reconstruction of events, and in some cases, professionally designed series of measurements and follow-ups. Only the MUFON group has done something similar in the U.S., and the Bigelow team for the cases in Utah.

I had plans to develop all that information, homogenous in methodology with a single focus, which isn't the case elsewhere. But that opportunity never presented itself realistically under Bigelow, and the new visible groups on the scene, including Delonge's, are unlikely to do it, consumed as they must remain with politics and entertainment. Thus, the work of 50 years will quietly pass into the future, under new angles. Just as well. I have learned what I needed.

Hummingbird. Monday 18 June 2018.

Garry's clever students have fixed the device, so we were able to run the analysis a week ago.

Dinner with James Fox and Lee Spiegel at the Big4 last night. Most of the discussion was about the *Nimitz* (Kevin Day's revelations), and James' trip to China where only one UFO research group has the indispensable official blessing. James also said he'd found a letter from Dr. Hynek at the National Archives, dated Sept. 1964, with the authentic Socorro symbol.

Hummingbird. Thursday 21 June 2018.

George Hathaway, in two impeccable reports on the Bismuth samples, has negated earlier claims made before the Society for Scientific Exploration, about an Alien origin for the material. Others still go on with the claim in public and before Congressional budget committees. Could they be right?

Now, an exchange with an East Coast expert, who laughs: "The days of the CIA's Weird Desk have long been over!"

"That's what they said 20 years ago," I replied. "There's no way they can be out of the business, for at least two good reasons, and the USAF for a third." I had to add, realistically: "Come on, the real projects can't be little $22 million explorations like BAASS. They've got to be ongoing, full-scale, with the higher clearance you need to touch the hardware."

Now serious, he concurred: "That's correct. But I was addressing only the Weird Desk, which was only one office within the Intelligence Directorate as compared to *the actual UFO program*."

Indeed, I recalled that in May 2015, DCI John Brennan initiated widespread reform, so the program could have gotten disbanded and its long-time members reassigned. Or a new one created.

Hummingbird. Tuesday 3 July 2018.

All our initial samples have now been run, including both my pieces from Europe and Brazil, and the materials from New Mexico, but the analytical program is so abstruse that Garry himself has trouble running it, so our progress has been slow. He did reorganize the output, however, and agreed we should publish initial results soon.

Kit didn't attend the recent SASC (Senate Armed Services Committee) with Hal, but he went there later with Jim Segala, Brandon Fugal, and Erik Bard. "Hal gave us his slides from the first briefing," he told me. "They were about EarthTech, including some medical stuff. His second briefing had no slides. Kirk [the staffer] and the SASC committee on Emerging Threats were told explicitly that I had no participation with TTSA."

Hummingbird. Wednesday 4 July 2018. Independence Day.

Yesterday was spent at two medical centers for calibration of my radiation treatment in two weeks. I walked all the way between the two sites, enjoying Haight-Ashbury and the greenery of the park.

Now, I have confirmation about what we'd heard from radar officer Kevin Day, namely that some appropriation has been obtained for further research on the *Nimitz* case. Our team, however, was kept out of it, further evidence that the LoneStars were an old page in a big, growing book.

My friends think there's more, but they don't know how much money is involved. Even the undersecretary for intelligence gave out inconsistent stories. "Something's going on, Jacques," I was told, "something very unfair." I do agree, for my own reasons.

I made some inquiries on my own through a financial expert who follows space-related budgets: "Actually, there's not one dime for BAASS or EarthTech in that proposed appropriation," he said on the secure line.

"It's all going to a massive, well-connected company, the one that had the last SAP for UFOs. There's 40 years' worth of data, somewhere... What did you expect? It's mostly politics, money directed to just a few people. There were two SAPs, not one, but both are moribund. People have looked at camouflage, stealth, materials, RF-modulated lasers, and waves going through walls, but not UFOs themselves, and no 'Tic Tacs.' They can project a platform in a CCD environment, so some systems will make images appear as solid objects. Remember, there are many SAPs in activity, just from NSA, the USAF, and Navy. Read the novels by Robert Gaffey."

"What did the pilots see, then, those guys from the *Nimitz*?"

"They really don't know. Possibly, drones from a fleet flying around for six days, taking evasive action, from Catalina to the island south of Mazatlán, where they landed. Yes, there was data on the disks..."

"The pilots saw something visually, and consistently," I objected. "That wasn't a bunch of drones churning at 100 knots."

"True again. The super-Hornets had visual, plus TV and Raytheon contact; and the *Princeton* saw something, and captured radar, but they couldn't lock. Visually, the pilots did see something at the surface of the water. Electronically, the objects moved at high speed, up to 20,000 feet in 0.065 seconds. But that time interval *was the same for every change of altitude.*"

"That could simply be the refresh rate of the screens, then," I said, "rather than real object data."

"You got it! And Lou was in Naval Intelligence, he knows that."

He paused, then he added, "Keep in mind, too, that the same Commanding Officer was in charge in the *Nimitz* case in the Pacific and in the 'Gimbal' case in Norfolk. Support your DERFUM theory." (10)

"Are you saying it's all gone dark?" I asked.

"No, but somebody may want to make it look 'Alien,' artificial, an unintended result from *something else*; then peddle the Tic Tac as non-human technology: just another hypothesis."

Very interesting, I thought, but I can't jump to that conclusion yet. I've seen very real cases, like the Tic Tac, from decades before.

Hummingbird. Thursday 5 July 2018.

The initial plan to build up a research company with Diana and Tyler is falling apart. We hadn't planned it very well, so Tyler went out and Diana declined as well. I will sleep better knowing that I don't have to help manage one more structure with such a challenging product environment. Any intellectual property we might derive from isotope analysis of UFO fragments remains attached to a mystery, even at the basic molecular level. Everybody is going too fast, talking about things they don't really understand yet.

Hummingbird. Sunday 8 July 2018.

The day was spent at Garry's house, deciphering the first results from the analysis of UFO samples. So far, all the isotope frequencies appear very close to nominal values. I told Garry that some of the brochures circulating from researchers who claimed extraordinary figures and very long lists of components should be regarded with skepticism: poor sample preparation, or problems in calibration of the lab instruments? I just can't believe their results without replication.

Hummingbird. Saturday 14 July 2018. Bastille Day.

At a French Consulate reception this morning I had a chance to speak with Stanford Professor Elisabeth Paté-Cornell and her new husband, Retired

Admiral James O. Ellis, who has an office at the Hoover Institution. In a very distinguished career, Admiral Ellis, now 71, four stars, served as head of the Strategic Command at Offutt AFB in Nebraska and as head of INPO until May 2012. He's on the Board of Lockheed-Martin, where he took his wife's seat. They were both open to further discussion. So was the new Consul, Emmanuel Lebrun-Damiens, who suggested a private meeting to catch up with my investment outlook.

I'd be happy to reconnect with them and the French community when the time is right. Previous consuls only seemed to focus on general representation as hosts for distinguished French visitors and didn't seem to realize their assigned territory included all of Silicon Valley, so I've had no regular contact for ten years, except for routine administrative papers.

I left the consul's residence shortly after "La Marseillaise," and I happily joined Catherine and Rebecca at the sun-splashed Embarcadero Plaza, where a crowd was gathering around food stands and an improvised orchestra.

Hummingbird. Thursday 19 July 2018.

Whitley Strieber was in town today. He suggested dinner together (at the Big4 again, where Julie was delighted to meet him and served us with her usual good cheer). He said he was on his way to meet with Garry and Hal to catch up with 'things.' First, he wanted to know "What I thought 'IT' was."

I took a deep breath and told him I'd have to ramble a bit. I mentioned the things that had changed, TTSA serving unknown designs. We agreed there was manipulation on a big scale, and the media were compromised. Then it was my turn to ask about his recent experiences with entities. "They're still around," he said. "But I rarely see them, yet other people do. The main thing they say is that there's some secret I must absolutely keep."

Coming back to the government he said, "I worked for them when I wrote *Black Magic*. That book was inspired by the Intel guys, to confuse the Russians; it mentioned the remote viewing of a site in the USSR, with many details," he added, leaving me puzzled.

"When my son was two years old, I decided to get out of that relationship because others were killed, you see, and I didn't want that to happen to my family."

"How did the connection take place?" I asked.

"My uncle Mickey and General Exon used to work together at Wright-Pat," he replied. "They handled the Roswell material and the bodies. They thought the beings were artificial."

That would match my own tentative conclusions from the Trinity case.

He went on to tell me about something in France: "Two men from the French Services came to my hotel room. They told me the US was completely controlled; if I wanted to be free, I should move to France. They put me in a trance to recall more of my experience, which they recorded, but all I know is that I was elated after that session." He added: "I'm convinced the French know a lot."

"It couldn't have been the regular police, or French Intelligence," I countered (I am skeptical that "the French knows a lot," although they have much data). "The bureaucracy would not work that way. Perhaps it was a parallel organization, very hush-hush, outside normal channels? As in the US?"

"Maybe, but Mitterrand's family did have a close encounter, did you know it? That was the reason for his interest. Mrs. Reagan, too."

That remark was meaningful. I recalled the hush-hush story I'd investigated, the episode of the manuscript notes inspected in Nice. I had not solved the case, really.

Yet Whitley's book had no initial success in France; everyone thought it was science fiction because of his previous works. His Paris publisher had placed *Communion* in a fantasy collection.

I had to ask: "Where do you think the hardware would be?" He didn't give me an answer, but he obviously knew about 'Tyler' because he mentioned who'd collected samples. He said he'd been approached by Intelligence people playing the usual games: "What we're about to tell you is highly classified, but we think you should know…" He concluded: "It's their trick to get you hooked, to make you follow what they feed you. Then, you become their little toy."

Hyatt Regency La Jolla at Aventine. Saturday 21 July 2018.

As I continue to read Peter Levenda's *The Tantric Alchemist*, an excellent book, I understand better why I don't resonate with the occult groups in the Crowley tradition and the alchemists' interpretation of sensual metaphors. The symbolism seems stale to me, even when tied to colorful references to

Mercury Influences, Sulphuric Emanations and the 'Chemykhal' panoply of inscrutable Archetypes. Love is alive in its urgency, its flooding of our senses and our imagination. Even the basic, physical experience of sex is never twice the same, nor should it be. To ritualize it, as Crowley and the OTO did, is to reduce its power. Codified into incantations and obligatory movements, its virtue is weakened. Celebration of life can be ritualized, but it should be for change and discovery, not liturgy. Levenda recognizes this because he writes that "Tantric texts are indeed fortified with warnings to students who would wish to practice these extreme forms of meditation without a reliable guide."

Hummingbird. Sunday 29 July 2018.

Jim Semivan congratulates me for a recent presentation on UFO materials back in Paris. The *Nimitz* affair remains at the center of other enigmas. In one case, about a hundred discrete vehicles were tracked over six days, in clusters of eight, transiting at unsustainable speeds of 100 km/h. A group of coordinated drones could do that, I thought, given what I'd seen on the drawing board years ago in Berkeley labs. Also, those strange triangles over Belgium, decades ago, never caught, never revealed.

Noetics Institute, near Petaluma. Saturday 4 August 2018.

Dean Radin, the science director of IONS, had invited me to give the J.B. Rhine lecture at the annual conference of the American Parapsychological Association at the banquet tonight. It was a fun occasion, among researchers from the whole country, and a few from Europe (11).

Hummingbird. Thursday 9 August 2018.

I am resting after my first encounter with the Accuray robot that will destroy the growing tumor. The modern Cyberknife, based upon the AI-driven model I financed with the Euro-America funds in the late 1990s, has grown into an impressive machine, now made available worldwide by Coulter. There will be four more sessions, highly precise yet painless, after which I will have to redefine my life, rediscover who I am, and chart a new course for the rest of my sojourn on this very curious planet.

A sense of nostalgia and doubt washes over me when I think that Flamine leaves again in ten days. Fortunately, my health continues to be surprisingly solid. This evening, I accompanied her and her artist friends, Zoila and Marla, to see *Semele*, a 1744 Handel oratorio. It played at the San Francisco Conservatory, a rectangular room with good acoustics but a soulless feeling.

Hummingbird. Saturday 11 August 2018.

Flamine and I had a pleasant lunch with Catherine in San Carlos, catching up with her life and business, then the three of us drove over to the home of my late friend and software wizard Larry Hatch on a quiet tree-lined street in Redwood City. We met with his daughter, who had invited me. A friend was helping her sort out Larry's archives.

Larry, who died ten days ago, was a skilled computer programmer with passionate interest in UFOs. Like many researchers in the 80s and 90s, notably Dr. Sturrock, I extensively used his catalog of about 12,000 carefully documented sightings as one source in my own work. He tirelessly screened and compiled the data and made the computer catalog available at a very affordable price. He had customers in every country, my colleagues in France among them.

Larry Hatch has also left a unique collection of documents and other items, recorded on floppy disks, old IBM drives, and pages of program analysis that may be of some interest to the Computer History Museum for their display of innovative software before the web. So I've put the family in contact with that institution to see if there's any value. It would lie in the software.

There are about 200 'well used' UFO books in this home, in various states of use, from which he drew his data. The best part of the collection consists of over a dozen file boxes of hard-to-find 'fanzines' and research documents in every language, produced by local investigators who fed their data to Larry. There are binders, letters, and folders of notes that could be a treasure trove for a UFO researcher looking to build a new archive. I am not in that market; on the contrary, I continue to index my own archives to preserve them. But UFOs get publicity these days, new people are becoming interested.*

The house was poorly maintained and looked decrepit. There was no

---

* The documents I was able to 'save' are now part of my donation (on the Hatch family's behalf) to Rice University's vast UFO collections.

grass in front, only dust. The gutters were disconnected and eaten by rust, so that one could see the sky through them. Inside, the smell of smoke from years of tobacco use by old Larry, grated on the throat and the brain after a while.

Hummingbird. Sunday 12 August 2018.

New questions arise around TTSA (aka To the Stars), this time because of bothersome leaks. On his *Unidentified Anomalous Phenomena - scientific research* blog, Keith Basterfield writes: "On the morning of 7 August 2018 my attention was drawn to the existence of documents and videos about the Advanced Aerospace Threat Identification Program (AATIP)... There is an image of a white padded bag with... 'To: L. Elizondo' handwritten on it in blue and 'Department/Mail Station'... 9/7/17.'" Evidently, the slides detail the AATIP program. But what are they doing splashed on the Internet, with names and addresses?"

Serious records show that $10M was appropriated and authorized in 2013, and the money was released, but two senior-level bureaucrats, with a senior VP contractor, are thought to have arranged for that 2013 money to be 'redirected' elsewhere. Like my colleagues, I'd thought that BAASS was defunded in 2011, not 2012. Was there ever a 'second tranche' in 2013? In any case, these projects were very poorly managed on the Washington end.

Hummingbird. Thursday 16 August 2018.

Yesterday was my last radiation session under the Cyberknife. There was no pain from the treatment. Today I felt elated as I drove over to Garry's house to review the new isotopic analyses of samples.

Here, Flamine has been packing the clothes she'd left behind. I helped her with the suitcases, then we suddenly needed to hold each other for a long time, in silence. If we're ever going to be together again, we must start from scratch and not under any compromise, on the edge of a chasm.

Hummingbird. Saturday 18 August 2018.

Garry and I had a late lunch at The Rotunda at Neiman Marcus this afternoon, where I'd arranged for him to meet Dr. Maxim Platzer and his wife. They're regulars of the place, so the waiters hover around them and anticipate their every need, which is fun to watch.

Max holds science and aerospace degrees from Austria in the 1950s, worked for Werner von Braun on the Saturn V rocket, and served for many years as head of the faculty at the Naval Postgraduate School in Monterey. He was referred to me by a former student, one of my DC contacts, now at the Institute for Defense Analyses.

We went over our isotope plans, which he encouraged, while he told us about an idea for energy generation. It was very pleasant, with a fine and refreshing European tone.

Still no sign of Ron Brinkley (12). I have no way to contact him, except by phone and email, to which he doesn't respond. He may have had an accident on the road, or he may have gotten into a brawl with some local rednecks, although that's not his style.

Inverness. Tomales Bay Resort. Friday 24 August 2018.

I drove to Bolinas today to work on James Fox's documentary. He'd been able to get a full copy of the 1978 UN presentations after much difficulty, hesitation, and even denials from the studios. We watched the footage of his recent interview with David Fravor, very clear and credible. Fravor wouldn't have known about the fleets of drones that were flying slowly from Catalina to Mexico at the time, and one wonders whether they played a role. The pilots might not have been briefed about a separate electronic warfare exercise, either.

I spent two nights at the Tomales Bay Resort, watching the morning fog lifting after noon over a timeless landscape of colorful brush and rolling hills. The road wanders among the redwoods, the pines, and the eucalyptus. Bolinas itself is a paradise of flowers, fresh fruit, and ancient trees framing wooden houses with widow's peaks. The whole region stands on the Pacific plate rather than the Continental plate, so it looks east at America with a suspicious eye.

James Fox has handed me the vial containing the Socorro sample Peter Sturrock had entrusted to him. It came from someone called Florence Michael on 17 June 1989.

Hummingbird. Wednesday 29 August 2018.

French statistics: There were 616,000 babies born in 1939, in my cohort, and 419,000 have survived until today, so I am within that rugged 68% of quasi-octogenarians.* The lucky ones. My recovery has been steady since the final radiation treatment, and my energy is at a high level.

The morning fog, sure sign of our coastal summer, hangs over San Francisco. In the dim light of the kitchen, as I fix breakfast, I can see hundreds of headlights of early commuters gliding from the horizon beyond Market Street, a one-way stream of bright diamonds somebody has carelessly poured down the dark channel.

One of my astute readers complains that I seem to live in an ivory tower, ignoring scientific quarrels dear to him. He's right; from this building, newly painted in pale sand color if not exactly ivory, I could foolishly pretend I'm unconcerned with passing fads as I survey the bustle below. But that's not all. These people are going somewhere; they are my peers; I will soon get into my car and join them.

Hummingbird. Sunday 2 September 2018.

After some hesitation, Garry resigned from TTSA, where we both keep valuable friends and colleagues like Hal and Jim Semivan. Now the real work begins. I drove over to his fine Peninsula home where he prepared a couple of sandwiches we ate on his deck, overlooking the San Andreas Fault while he told stories about the Marquesas Islands.

Over the next few hours, we looked at the results from a program converting the raw data from the mass spectrometer into isotope ratios for the Sierra-1 sample. This uncovered a few bugs and a better way to compute the abundances, but no firm results yet.

Mabillon. Sunday 9 September 2018.

Paris is calm, more relaxed than I've seen it for a long time. Glorious sunshine puts a glow on every tree, a luminous sheen on the ancient

---

* As of the publication of this book, seven years later, the size of that cohort has dropped near 140,000, so only about 20% of the males born the same year as me have survived to age 85. One out of five.

pavement. In front of L'Ecritoire, the café on the square that slants up to the Sorbonne, water jets add to the peace and intimacy as I work on the laptop. The trees still have leaves, but they've turned to shades of autumn, from bright yellow to sandy beige or light brown, then to rust and dark reds. They set the mood for serious conversations with my brother, deeply shaken by his wife's death, and for the trip Flamine and I are about to take to Lausanne.

Mabillon. Friday 14 September 2018.

Gabriel, my brother I hadn't seen in several years, felt well enough for a visit. We reconnected with contained emotion, and he let me do most of the talking, curious to hear how things went in California. As we discussed the trends in medicine, I lamented the fact that so much value was left on the floor of the labs in France, instead of feeding into the large-scale innovation modern medicine needs. There's some consolation, however, as I hear of the progress of the Accuray technology throughout Europe: We're saving the lives of real people.

Morges, Switzerland. Saturday 15 September 2018. Hotel de Savoie.

Flamine accompanies me to the apartment of a Swiss nurse to investigate a series of events: a bright elongated cloud with many lights, from which came a hail of material particles that hit her building. She filmed the process and recovered the stone-like debris. I lean towards a very unusual natural explanation.

Lyon. Saturday 22 September 2018.

On Wednesday my friend from the Pasteur, Pascale Altier, took me on a visit to one of the newest startup incubators. I was impressed by the energy, less by the fact that all the money came from the government. Now, back in Lyon with my grandson, visiting parts of the city I hadn't seen before (notably the remains of a Roman amphitheater) and discussing his plans for a future in technology.

Mabillon. Tuesday 25 September 2018.

The latest meeting at CNES was a closed session of our expert committee in Paris, set up to examine a dozen cases for which the investigators of GEIPAN had not reached a final conclusion.

Ominous reports continue to reach me from the US, expanding on the Utah Ranch controversy and other bizarre episodes; FedEx letters from one of the participants were intercepted, and a stolen memory stick contributes to a pattern of 'interference' in our work.

Who is spying on us? Hal Puthoff jokes: "Most likely, the usual government stooges, certainly not Aliens... They're too smart!"

18

Hummingbird. Monday 8 October 2018.

The machinery of life grinds ahead, injecting an indifference within me that feels dangerous and unwelcome. I did catch up with the flotsam of the last few months, even long-delayed letters to my kind readers like Professor John Hart, who invites me to Montana. Many others write about sightings and advice, and new projects take shape. Why complain? I'm emerging from the anguish of a cancer operation with my health intact, but the misery of loneliness lingers, and the edge of bitterness.

Tom Delonge's TTSA has just filed its first report with the SEC. It shows a huge loss, a disappointment for recent shareholders who anticipated science breakthroughs. The paper loss doesn't mean much in financial terms. It may simply have to do with the accounting of share awards promised to insiders. The filing states that 'Entertainment' is the main theme, contrary to claims of progress towards antigravity and a new UFO science. Others jump ahead: The History Channel, in alliance with Dr. Hynek's sons, launches *Project Blue Book*. They've asked me to provide case references.

Hummingbird. Friday 12 October 2018.

James Fox and his cameraman came over to IonPath in Silicon Valley this afternoon to record Garry's brilliant account of the science behind isotope analysis, and my description of the samples I've collected for 30 years from people around the world trusting me with their data.

The results so far show that the phenomenon is unlike what we've known before, but we're barely at the beginning. I left after the interview, on the way to the airport, and the 1945 crash site.

Later the same day. San Antonio, New Mexico.

Paola Harris and I met Jose Padilla at the Owl Bar Café (where I had lunch with Ron Brinkley some months ago, when he introduced me to the area). Nearby, one can still see the ruins of the compound where Dr. Oppenheimer and other scientists stayed at the time of the Trinity tests. Enrico Fermi would have known, too.

Jose, just two years older than me, remembers the first atomic blast, on 16 July 1945. His mother had looked at it through a half-open window and was partially blinded by it. Just a month later, possibly about Aug. 20, 1945, Jose then 9 years old, and his buddy Remigio Baca, 7 years old, had been sent out to check the cattle when they saw an object crashing on the property in the rain. The two kids went home to tell Jose's father about the "thing." When Señor Padilla called the authorities, he triggered the military recovery of an unknown craft.

"At the time, I already knew that thing wasn't a weather balloon," Jose told me. "We were told not to go back, but the crash was on our land, so we went back anyway. We knew how to hide. The soldiers took every opportunity to drive out to the café for meals and beer." Most debris had been taken, and the small beings they'd first seen staggering around had disappeared. But there was an intact panel attached to the inside curved wall of the craft, with a metal device that swiveled; so, the kids wrenched it out.

"I told Remigio to get a crowbar from the Army truck," Jose told me, smiling at the memory. After that, the kids kept their mouths shut, and the bracket remained a secret. Jose eventually gave the device to Paola. It was tested in Mexico on 24 Oct. 2017, at the Centro Educacional Analitico. It looks like part of an ordinary tool.

Socorro. Saturday 13 October 2018.

We went to the site today, in our rented SUV. The property lies near Carrizozo, some 13 miles away from Trinity.

"There was a rainstorm that day," Jose told us along the way. "We heard something like a sonic boom, and we saw some smoke. We got on our horses. The thing had hit a high tower, spun around and crashed. The grass was burning. The closest we came was about 300 feet. There was a hole; we saw small men running back and forth as in a panic. We must have been there two hours, and then we told our father who called the State Police."

"We went back two days later but most of the stuff had been hauled away, except for the disk that was under a tarp."

Socorro. Sunday 14 October 2018. Super-Eight motel.

Bad sleep, and worry: Why and how did Ron Brinkley disappear? He should be here with us, enjoying this investigation, adding to it with his great patience and his ability to note details in context. We'll stop again in Albuquerque on the way to the airport and visit the complex where he lives. Perhaps he's moved away, or suffered some accident? I don't want to believe he might be angry about my manuscript; he would already have told me so.

The lack of direction in my own life is problematic. I need to refocus my research interests, flee the steamy microcosm of ufology, and work on the hard problem. As a first step, I should screen the Blue Files more carefully and send them to Rice in Houston.

A note in passing: Steven Greer is said to have done a vigil for members of the French military in 2016. It took place at a château where an object plunged into a lake and disappeared. The names of "Emery Smith and Philippe Aubin de Messier" are associated with the rumor.

Later the same day.

Paola agreed to drive us to the rectangular, one-story building arrayed around a courtyard where Ron Brinkley lived. Why and how did the friend I called 'Roger Brenner'—to guard his identity while he was still alive—disappear?

We went around the complex again. A woman walked out of one of the apartments. We asked her about Ron and she said he was dead: a month ago, at 4 am, riding his bike to his job at an airport shop, he was hit by a car. His belongings were sent to his brother in Texas. The apartment where I'd left the letter asking for his whereabouts has already been rented to new people. I went to the shop where he worked as a storeroom employee. They didn't know his brother, but I got the manager's business card.

I had feared this. I had thought of a car accident on the mountain roads he drove, distributing newsletters on the reservations. I didn't know he rode his bike to the airport. I told Paola about Ron, our friendship of many years, our many expeditions together, and how sharp his mind was. As I waited for the flight, I bought turquoise cufflinks from his shop to preserve a link at least, a tangible sign of everything he meant.

When I got home, I found this letter from his brother Ken and Ken's wife Melissa, responding to my enquiry: "Your concern was apparent and true. Ron was involved in an accident on August 6th. He was on his bicycle going to work and was struck from behind by a motorist. Despite wearing a helmet, he lost his life due to extensive brain damage. It was a hit and run…"

Hummingbird. Wednesday 17 October 2018.

A new French government has been put into place by Emmanuel Macron. From afar, they seem fresh, intelligent, eager to reform ancient practices such as the dozens of complicated pension systems, an eternal source of social inequity. When I speak to friends in Paris, however, I hear a different tune: frustration and even anger at the arrogance of the new régime. As I watch them on TV, they all look the same, in black suits and black ties, white shirts, among the gilded balustrades of the Republic.

Hummingbird. Thursday 18 October 2018.

The medium, in trance before me: "Sleepy. Guardian here, blending slowly. Elder in spiritual community: Don't even think about keeping quiet. Your voice must be articulated, even for those who refuse to hear."

Then, after exhortations to learn to trust, moment by moment:

"You're not at the end of your journey. The key is altruistic service to the evolutionary fate, progression of humanity. Organize the interface: stop thoughts. What you can do, Monsieur, is to create a path, a gateway, a portal. The thin dry air and other conditions make it more difficult. Take over, Monsieur, you must cross a strong river, stepping from one stone to another. The machine resonates as if a jackhammer shook your body..."

Hummingbird. Saturday 27 October 2018.

Nine days have passed, uneventful, but something brief and terrifying happened last night, far more tangible than a nightmare. As I review it, I have to accept an experience outside my body.

All I know is that in the middle of the night I found myself propelled away into the kitchen, a tall being between me and the window. I couldn't see a face, but it was tall like me, human-like, solid. A few seconds later I woke up suddenly, back into my lying body, panting and whimpering, terrified.

It had been absolutely, shockingly *real*. No transition, none of that slow lifting of the soul people describe. I had been translated from the bedroom, standing up, totally aware, next to the entity that hovered near me without touching, yet encompassing me in a formidable presence. This is as close to a full out-of-body experience as I have ever had. It left me whimpering and scared.

Hummingbird. Wednesday 31 October 2018.

Garry and I worked through the afternoon today, reviewing all the samples this time. I didn't tell him about the tall being; I need to place it in context before I try to integrate its message. It was as tall as I was. It didn't speak or threaten me, but it was too close, too unexpected.

Garry and I worked on my recent correspondence with the original owners of the samples. We now have the luxury of three analytical machines and a specialist in mass spectrometry to manage them.

The fractured LoneStars, scattered over the Internet, continue to break up for lack of management and continued sponsorship. Las Vegas on standby: Hal increasingly withdrawn, Kit frustrated (so is Garry), and Eric and Colm facing the fragmentation of the data: too classified, connections broken. George and John rarely step in with new proposals to regain momentum. That's all right with me. Our work is done in this phase.

Hummingbird. Sunday 4 November 2018.

Wary of the touristic excesses of Halloween, the Spanish community held its own celebration of the *Dia de los Muertos* in the Mission district. With respect for Baron Samedi, I went there with Cousin Karen.

Back home, I worked again through hundreds of folders that hold case interviews, testimonies, field investigations, photographs, and maps. A lot of it could feed new physics. Who will make sense of it? New work was going to be done in Austin, where we were supposed to participate, said Kit, but it never happened: funding priorities?

I have uncovered a few keys to the various phases of the phenomenon, some small ones and some big ones, yet I can't transmit them: the Curse of the Archivist.

Hummingbird. Sunday 11 November 2018.

Dinner and shopping for food with Jack Katz. His ability to make kind, instant contact with strangers remains amazing.

Later, a warm call from Penny. We spoke of her magnificent books about the stone circles in the Hebrides (13). In a timely gift to lovers of ancient monuments, Gwendolyn Awen Jones, as she's known, researches the age-old mysteries of Callanish. From the misty shores of the Hebrides, her book poses big questions. What inspired early humanity to control its destiny? What did we miss about antiquity across the planet? What message do the stones teach us as our own civilization trembles at the edge of space?

Hummingbird. Monday 12 November 2018.

Unstoppable tornadoes of flame have reduced the California town of Paradise to smoking rubble in a few hours, trapping many people in their homes. It's a disaster that elicits nasty comments from cynics. The greed of developers (a single road through the town, no evacuation route) and the ineptitude of early warning agencies combined to turn the fire into a major crisis, but the main cause lies in the global changes of the weather patterns. Every agency now reports catastrophes on a far greater scale than previously. In Southern California, too, the winds have started blowing towards the ocean and firefighters will need several days to control rushing walls of flame. The sky of San Francisco is full of smoke, eyes cry in eternal sunset, a glow of doom. People wear white masks and dark glasses over their eyes, so it looks like the City has been taken over by zombies, once again.

Hummingbird. Tuesday 13 November 2018.

Stan Lee, the great pioneer of Marvel Comics, died today at age 95, following Steve Ditko, who died in July. This leaves Jack Katz standing alone as the last surviving master of the American graphic novel during the Great Era of the Comics. He's working at the peak of his inspiration. *Beyond the Beyond* is increasingly abstruse and etheric, a work of metaphysics rather than science-fiction. "There's never been anything like it," Jack says, proud of his own work.

There's never been anything like Dr. Garry Nolan, either. I set up a lunch today (at the Rosewood) with Federico Faggin, just back from Italy. I showed him the new binder with our records of 31 samples.

Federico's work proceeds along lines that could converge with Katz' drawings. All of physics, he says, is based on objects that emerge from the great ocean of quantum consciousness; we know nothing, in the end. As for Garry, he's busy with plans for the Atomscope, a device that will image the tiniest bit of molecules and could enable science to comprehend protein structures that have eluded medical research.

Hummingbird. Wednesday 14 November 2018.

Lunch with Dr. Sturrock, at Il Fornaio in Palo Alto. He gave me the capsule that had contained the Fe and Ti samples from the Sierra case. Along with the capsule was a paper that established the chain of evidence. I need to study the samples. Peter is still working at Stanford on the hypothesis that the Sun's original body may have captured a traveling dense cloud, which rotates in a different plane than the core. That central star, the proto-sun, is only 10% of the whole volume, he estimates.

Hummingbird. Thursday 15 November 2018.

The horror of two more fire zones, unprecedented in the history of the state, continues to spread over California as more and more people—elderly couples, families in cars blocked on the road—burn and die in conflagrations they can't escape.

Today again, the Sun rises as a red reminder of the catastrophe through ever-thicker haze that waters the eyes and stings the lungs.

Hummingbird. Friday 16 November 2018.

'Les Gilets Jaunes,' this potentially explosive popular movement is sweeping France. The people wearing yellow jackets as a sign of defiance against the government have never joined together before, least of all in politics.

They're rising against an autistic ruling class that has betrayed them once too many times. President Macron tries to reassert control, but he finds himself marginalized, and this won't be forgotten. There were one thousand meetings today over the whole territory, coordinated through social networks, while Macron inaugurated an archaeology museum in splendid contempt.

Hummingbird. Friday 23 November 2018.

Thanksgiving dinner with Catherine and Rebecca's relatives, animated and joyful. Driving home in the rain last night for the new car's first excursion along these roads of Silicon Valley I know so well, the radio played Borodin in rhythm with the windstorm racing across the glass. Garry was on the road, too. He called me on his way back from Reno, where he'd had a 5-hour dinner conversation with the head of one of the fastest-growing firms in 'New Space.'

I have felt happy for the last few weeks, simply happy, even as I fumbled with furniture, planted flowers, filed reports, made coffee, ran errands, or wrapped Christmas gifts. A dumb single guy.

Hummingbird. Sunday 2 December 2018.

Yesterday the French TV devoted its entire news report to protest and confrontation between riot police and the 'Gilets Jaunes' on the streets of Paris and at the Arc de Triomphe, which they vandalized lustily.

Today, again, stunning images of demonstrations in French cities. Fresh from a summit with Trump, Putin, and the Chinese, Macron showed up at L'Etoile, his pompous words lost over the riot clashes.

Here I continue to sort out the archives. The papers of researcher Linda Strand, and her correspondence with Sagan and others, are especially emotional, evoking sadness, deep and silent. She died of multiple sclerosis, long ago. She'd visited Janine and me and said it was one of her happy moments in a phase of despair.

We had gone with her and her companion to a Buddhist ceremony. She chanted for hours, trying to keep her mind under control, as well as her slippery sense of time. She left confidential letters with me to be opened if she ever fell into a coma, and an envelope with keys to her apartment so we could save her research papers.

I have recompiled the first 100 cases of my field work, ready to be mailed to Jeff Kripal. They carry the immediacy of our meetings with witnesses, roads travelled. There's a perfume of wet fields and apple trees of Normandy, and of the sage in bloom in Texas. Lots of photographs, too. We were young, searching. I'm still searching.

Hummingbird. Monday 3 December 2018.

Peter Levenda's *Stairway to Heaven* (2008), which I'm currently reading, is an erudite review of theories and practice in the "Art of Spiritual Transformation." He reflects deeply on the tension between science, which continues to argue on the basis of strict 'rationalism,' and the flight of the soul on a different path (see p.192 of his book).

My son, who turns 55 today, has chosen to live in France where he works very hard to build an enterprise in the difficult climate of the abusive bureaucracy that drives many citizens to despair and confrontation. Tonight, French TV news shows that three people have died in riots because of accidents, hundreds have been arrested. France staggers into a fifth week of demonstrations. Europe itself, on the verge of Brexit, sounds like a sputtering machine.

Hummingbird. Thursday 13 December 2018.

I read again what I wrote above, then a memory rises from the 1950s in Pontoise: my parents listening to the radio news, *Ici Paris* or Radio Luxembourg, and lamenting the chaotic politics of infighting factions. My father would comment that only the formation of United States of Europe could get France and her neighbors out of the mess. Years have passed, even decades, and Europe has stumbled ahead, weakly connected, with no trust in a common vision.

Now, even that primitive structure is shaken. Predictably, Great Britain broke off first, then the eastern countries veered off course in spite of their high exposure to Russian ambition. My parents, after two world wars, would look at this sadly. I too feel concerned for my own children, in spite of the education we gave them.

Hummingbird. Wednesday 26 December 2018.

Our bright sunrises are back, after a few days of strong, healthy rain. On Christmas Eve after a warm, memorable Indian dinner with Jack Katz near Berkeley, I drove back over the Bay Bridge in a storm that shook the car. At home, there are packages and books, presents from many friends, a picture puzzle from Olivier, fine Zuni cufflinks from Paola, and a thousand origami paper cranes, all bright colorful precious birds flying in symbolic groups. Yesterday, the weather was sunny enough, cheerful and warm, to allow a nice walk around the lake.

I resume work with Garry, who just returned from Paris and Stockholm with amusing memories of the Nobel celebrations. We discussed his Harvard presentation on the caudate 'antenna.' The research represents new insight into that unique structure in the brain, but the unstated effort to 'explain' the phenomena and some confusion in the selection of cases worry me: Who are these subjects in the cohort? Why were those selected out of some 200 others, and what did they see and experience? None of that has been made clear, even within our own group.

Hummingbird. Thursday 27 December 2018.

Gilda Moura came over for coffee and Brazilian stories of spirits, mediums, and abductions in the Amazon. She was with her son.

Gilda has organized investigations with Gevaert's study group in Brazil. One big issue is under debate there: Do mediums contact ETs, as well as deceased humans? Appropriately enough, my guests left just in time for me to walk to the Spiritualist Church, where a service was held.

Hummingbird. Monday 31 December 2018.

A friend and I strolled down the street to the Symphony today for their annual Gayety performance, a unique event of fun and provocation. We both fight lingering colds, so this was a welcome as well as a festive diversion, completed by a formal blessing by the Sisters of Perpetual Indulgence from the stage of the San Francisco Symphony.

Hummingbird. Tuesday 8 January 2019.

Paul Hynek called today to discuss tonight's premiere of the *Project Blue Book* TV series. They've posted the first three 'vignettes' (short snapshots of data) where Richard Dolan and I provided the references behind the fictional series.

In the meantime, the research faces a bleak assessment. The team at TTSA is broke, and even with Bob Bigelow's efforts, we are unlikely to see any progress lobbying the government. After one takes into consideration the history of BAASS, subsequent events with the publicity around the Ranch, the Tic Tacs, and any future interest by various arms of government, no significant decision is likely to happen in Washington this year.

There was a time when Senators Ted Stevens and Daniel Inouye were willing to take risks with Harry Reid, knowing the topic was real. Stevens even recalled his own experience as a fighter pilot.

Today Congressional staffers and Intelligence officials may be willing to be briefed by scientists as relief from their normal DC routines amid the Trumpian chaos, but that's as far as it goes.

Hummingbird. Friday 11 January 2019.

Two books, which were brought over by Santa Claus in anticipation of the next conference at Esalen, have re-ignited my long-term interest in synchronicity. First is the correspondence between Carl Jung and Wolfgang Pauli, which I should have studied a long time ago. Second is *Time Loops*, a recent book by Eric Wargo (14).

Time passes, and people as well. Yesterday I learned of the recent death of Larry Roberts, the man who executed so brilliantly the vision of Paul Baran for an unstoppable computer network, the Arpanet.

Hummingbird. Saturday 12 January 2019.

Mixed news from my friends. Eric Davis has started a new job search, while working on a small TTSA contract that pays through March 1st, his last day at EarthTech. (15)

I asked if there were plans for all of us to meet again as a group. It's unlikely: "The bank in Toronto refused to accept the transfer of $500,000 from a sponsor in Europe because of Canada's strict banking rules on how foreign money is to be held."

In real terms, this means that, in the current media turmoil, those who once supported the LoneStars have decided to let that phase of the research quietly die.

19

Hummingbird. Monday 14 January 2019.

A great and occasionally hilarious privilege, listening to the wide-ranging dissertations by Jack Katz calling me at all hours as he races towards the completion of his 500-page magnum opus.

With great humor, combined with a sure command of his art, Jack Katz expounds for me on the "dual infiniteverses" underlying a debate pitting Virtue against Reason, and the abrasions of the brain he observes among humans and cyborgs alike.

This scene takes place outside the cosmos, of course. So, we talk of digital consciousness, the modern media, and those blasted smartphones Silicon Valley has designed as 'ultimate pacifiers' for a declining human race incapable of trusting its own intelligence.

I simplify, of course, failing to do justice to the breath of Jack's space opera and to the circuitous notions he uses to test his theories on me, knowing I am an indulgent audience and that I love and admire him.

I am happy to be thus promoted as the guinea pig and witness of the last surviving giant of the Great American Comix Era. Not everybody has the luxury of late night, hour-long dissertations about what lies beyond the Beyond. We are rushing, of course, towards the Great Calamity because

"Our life is an artificial thing created by the State before Existence."

It's astonishing to hear this from a 93-year-old artist because it dovetails with what young scientist Rizwan Virk tells me as he develops his own theory. Both converge towards the notion that we live in an altered state that creates, nurtures, and occasionally destroys both our unreliable senses and what we think of as our own existence.

Esalen. Tuesday 22 January 2019.

Lunch on Saturday in Menlo Park with Pasha Roberts, the son of Larry. I've put him in touch with Riz Virk; they both came from MIT and have converging interests. Then on Sunday a friend and I left the City in rain and breezy fog on the way to Esalen, but it all cleared up when we passed San Jose and the sky displayed a luminous array of high clouds and the gloriously sunlit trail of the dying storm over the coast at Monterrey.

The road was magnificent, all the way to Big Sur.

Esalen greeted us with a happy mood. Recently reopened after the landslides that forced a year-long closure of the coastal highway, the retreat had the freshness of young grass under the rain. We found a torrent bouncing under the foot bridge. Michael Murphy was walking on the path, friendly and open as ever. In our lovely room at the edge of the cliff, the surf was the only sound.

Garry gave the first lecture before the dozen members of the symposium, with a stunning report on the caudate-putamen statistics he'd seen at Harvard. I followed in the afternoon with updated analyzes of synchronicity.

Now the storm is just a memory. The big Pacific birds swooped over us and played in the light as we walked along the trails of Esalen to Jeff's cabin where we were joined by the three graduate students who will work on the archives I've turned over to Rice, and the analysis files I promised to organize.

Esalen. Thursday 24 January 2019.

Anger has erupted around Diana Pasulka's book, *American Cosmic*. Her friends blame *The New York Times*, where Leslie Kean, helped by Isaac Koi, demands details and names with increasing urgency. Garry has been exposed as one of the scientists Diana quotes, and Koi found Tim Taylor on the records of Vatican visitors, so he'll soon be tracked by hordes of Twitter users. All that was to be expected.

The conference continued with vibrant presentations adding color, information, and cultural extensions to the proceedings. Through all this, the enigma has only become denser, however. Whether we talk about synchronicity, the continuing experiences of Whitley and his alter ego 'Whitley Prime,' or the potential for robots with unconditional love (an idea that our sponsor Jim pursues with the assistance of Julia Mossbridge at Noetics), I find little concrete data to take home, although some presentations were completely new to me, like one bright talk about the Nation of Islam.

Hummingbird. Sunday 27 January 2019.

As I reload the system, I find new information on the web. To counter the turmoil in the media, my colleagues have reassessed the products we once generated under BAASS. Jim Lacatski, in charge of the DIA program, based at the time in Alexandria, Virginia, where I briefed him about my work and was processed for my clearance, scanned, and created electronic files that he archived before he left. Were they randomly deleted at the program's end?

The only files submitted to the sponsor as part of the program were the 38 science monographs that came from EarthTech, plus 75 reports from the project itself. None of the medical files were submitted, but a few reports (Blue Orb, North Carolina cases, and Colares) were sanitized. Perhaps much of that will get released in time, I'm told, except for anything classified TS/SCI, such as the records of Capella.

Mabillon. Wednesday 30 January 2019.

A dusting of snow fell over Paris overnight, but it had melted by breakfast time, replaced by blue sky and pleasant sunshine. I re-read Diana's manuscript of *American Cosmic*, wondering again about the impact of the book on academics who never seriously considered the phenomena. Rumors about Tyler and James and the artefacts they dug up in New Mexico have already inflamed imaginations well ahead of publication, all the way to the ever-suspicious hallways of *The New York Times*.

France is freezing, temporarily becalmed in the expectation of another government popular consultation, which seems a bit fake, and fearful of economic troubles ahead in Europe.

Mabillon. Thursday 31 January 2019.

Naval radar officer Kevin Day writes to me about after-effects of the Tic Tac events he observed from the USS *Princeton*. Lue Elizondo has answered "no comment" on the matter, and I am not knowledgeable enough to take a stand. In the meantime, an NRO senior officer who'd resigned from Trump's staff last year as director of cryptography strategic planning, has joined the TTSA board last week. I'm told he may have kept a copy of Capella.

According to Bernie Haisch, Elizondo discovered the crash retrieval program via official channels in the 1990s. There was a claim that it was housed in major aerospace companies such as Lockheed, TRW, Raytheon, and Aerospace Corp. He was refused access because his AATIP project wasn't a SAP.

As we know, Harry Reid petitioned the DoD to elevate the status, but the request was flatly denied to the senior Senator, another indication of how seriously the subject is held.

Bayeux. Saturday 2 February 2019.

The changing skies of the Atlantic cleared up as the train reached Bayeux, and there was Annick, cheerful and eager to show me her new house. I recognized the sturdy Norman furniture, stately as ever after decades of family use, and the artwork she creates, colorful, humorous patchworks.

We drove to Port-en-Bessin for lunch on the harbor with her neighbors, a couple of retired educators, solid scholars who had known Janine. It hurts to remember that I met her in Paris sixty years ago to the day. A cold, wet wind is blowing. Trawlers at rest are crowded together, tied along the quay.

Bayeux. Sunday 3 February 2019.

Fat pastures, green with an envy of spring; hazelnut trees eager to burst, and an expectation of daisies and mimosa. The rains have prepared the land, opening the way for warmer days. The cemetery at Yvetot was gleaming under clear skies.

Annick maintains her good spirits and clear mind, but emotions overtook us at the grave because of all we'd lost. Yet it was Janine's wish for us to go on living and building upon what she had created for the people we loved.

Annick's neighbors enliven a hamlet of retired intellectuals: engineers, artists, and professors in a quiet corner of the city where they paint, sculpt, or plan long walks around the countryside. Yet every door opens up on hidden dramas: old single men forced to relocate in poor health and couples with desperate money problems or for the lost love of a precious son who became a woman in a society that grows along obsolete standards.

Bayeux preserves its souvenirs at the WW2 Memorial, complete with tanks painted pale green (!), and those long lines of trees all the way to the horizon at the German cemetery, recalling the miserable soldiers of the boastful Wehrmacht, short of time and gasoline, pulverized by Patton.

Bellecour. Saturday 9 February 2019.

After a few days in Paris spent typing parts of this manuscript, the train brought Olivier and me back to Normandy in time for a fine dinner at his restaurant. Only one tourist couple came in and sat by the fireplace, the staff nervous about the future. They served us a gastronomical meal that deserved the applause of a larger audience.

I didn't sleep well, again rehashing my mistakes, and lamenting the shallowness of modern spiritual movements. What remains of that idealism we pursued with the energy of youthful belief? Only a few teams claiming psychic superiority, and a vague promise of a peculiar brand of happiness. It's all combined with an inability to accept blame and to offer kindness free of judgment—that 'unconditional love' programmers now try to implement into robots!

Here, in the large forests on the historic *Marches* of Normandy, nothing moves except for an isolated timber man on a tractor, piling up harvested trunks by the side of the road.

Mabillon. Wednesday 13 February 2019.

Before I fly back to California, Flamine and I went to see Adam McKay's film *Vice* (centered on Dick Cheney's career), and I need to close the loop with Nathalie, the lady who told me of a detailed observation in Brittany (complete with a photograph of the object) with her two daughters. Paris is busy, exciting and almost spring-like.

Last week we saw *The Mule* by Clint Eastwood, and we attended an early music concert at the church of Saint-Germain-des-Prés. In faraway California, Garry's hard at work on the analyses, sending me encrypted videos of honeycomb microscopic details.

Hummingbird. Saturday 23 February 2019.

Last night, Jack Katz seemed to sense my dilemma. We argued about everything like two college kids, but he's the one who first spoke about the nature of love and gave me a much-needed lecture on jealousy and relationships. He also stunned me again with his exceptionally clear memory and his grasp of European fine music and painting.

In Paris, I had bought for him a large framed reproduction of Cormon's *Cain fleeing before Jehovah Curse*, which he considers the high point of the genre (the original is a giant painting at the Orsay Museum) so I grabbed a nail and a hammer to hang it on his wall, after which we went to a Thai restaurant and he gave me a psychic glimpse into my own future, pointedly reminding me of Flamine's patience, and his keen observations of us together on earlier visits.

To recover my sanity, I've been reading Diana's book with care. It is now in final form, published by Oxford University Press. She refers to my work with clarity and zest, leaving me with a deepening sense of responsibility. *American Cosmic* has gained public attention, putting the issues in a light that is both rewarding and intimidating.

Garry is now fully engaged as well. Under the fine professional Japanese microscope he's lent me, I gain a glimpse into an unknown world. Then I think of the entity that pulled me out of my body on 27 October last year, and I realize that like a bad student, I failed to note some of the details.

The unique feeling of terror has remained with me. There was no overt threat from the tall being, but the wrenching out of my sleep and of my body was brutal, as if I had been summoned to the place where we'd spent hours trying to set up 'precipitation experiments' that we thought didn't pan out. We had carefully covered the southern-facing window with thick drafting paper, and we had used various devices to invite a materialization or an apport.

The trance guide had instructed, "Medium can create an electromagnetic field, a vibration system to receive an apport. Once established, the ability can be transferred to someone else…"

We had wanted a demonstration, a communication device, to reach other entities if they existed. The cryptic answer had come, "We send what you need, not necessarily what you want. You stand at the forefront; like bliss, running alongside you…"

Hummingbird. Thursday 14 March 2019.

A shocking development. Kit writes to Garry and me that he plans to tell our closest colleagues: "I have decided that Jacques Vallee was 100% correct starting five or six years ago. UFO-ology 'research' is so contaminated, bastardized, trivialized, and used by unethical groups, as to be unavailable for serious study. He was correct when he withdrew and said: 'I will do my own work.'"

It appears that, like me, he felt under pressure to change direction, with advice from an attorney: "He essentially agrees with you (Garry) and with Jacques. I will also continue my work, if possible, on a limited basis, with a program … to support medically important data on … patient cases for which I have sequestered files. I end my associations with any other organizations."

Something traumatic must have happened on his side, as it did on mine. He adds a warning with which I emphatically agree *there is true evil afoot*, including "lies; work products that are harmful to kids; manipulation of memes…"

Hummingbird. Friday 15 March 2019.

Kit writes again to Colin, Hal, Garry, and me about medical practice changes, adding: "As a result of a number of events in the last month, the culmination of six months turmoil and angst, and direct and clear advice from my corporate attorney, I am making immediate communication and security changes."

My understanding is that the medical files, specifically including the caudate putamen records, are now frozen to prevent them being used by rogue groups with no license and no care. It's a wise move. He goes on with specific instructions about the data. None of that touches me directly, since I've only been involved as one of the patients within the cohort and had no say in methodology. It was appropriate for us, as researchers, to submit as the first test cases. But the research may have been compromised.

Albuquerque. Thursday 4 April 2019. Airport Best Western.

Strangeness hit as soon as Paola and her friend Bill Crowly greeted me at the airport, shortly after noon. We sat down for a snack at the restaurant, a few yards from the shop where Ron Brinkley used to work. Bill had just begun telling us where to rent the equipment we were going to need (magnetic detectors and a radioactive sensor, primarily) when a man walked over to our table and shook hands with Paola. He was George Waldie, a retired friend of Hynek's old associate from Boulder, Richard Sigismond (16). What was he doing there? The coincidence was never clarified, but it cast a shadow.

While renting the metal detector and piling up water bottles in the back of the car, I got to know Bill and liked him. His trajectory in life has common points with that of Ron: He spent time fishing in Alaska, loved to be alone in nature, and volunteered with his wife as a watcher in forest towers to alert helpers in case of fires or lightning strikes. He's also spent time in Japan and understands cultural subtleties. He's a member of MUFON in LA with investigative experience.

There was a recent flood in the area. It damaged the roads. Now we're driving off to meet Jose and go to the property. Perhaps the flood will have brought more pieces to the surface?

Socorro. Saturday 6 April 2019.

Once you leave the main highway, veering off towards the old Army base, the landscape unexpectedly embraces you in silence and majesty. You're driving along an immense seabed of a prehistoric ocean, with no other boundaries than the far-off mountains, hours away, dark and fuzzy in the heat.

To relieve the vague anxiety weighing on us, I turned on the radio, rewarded by a Spanish-speaking music program, so we were treated to an accordion and a trumpet that restructured the desert with hopeful melodies of love and the sorrow of separation: *Cuando podria verte?* When will I be able to see you?

Somehow, it all fit together: a simple love song, this unforgiving land of infinity, and the straight road to a far-off Army gate. Beyond it, the ever-present awareness of atom bombs.

The young soldiers at the fence checked our IDs, and we were back in the desert with miles to go before the parking lot. To the right (north) was another road leading to the McDonald ranch where the scientists had assembled the plutonium device. To the left, a wide fence perimeter, and beyond that, a black obelisk, awkward and stocky, marking the spot of the first explosion. People kept arriving in mobile homes, cars, and motorcycles. Families with kids, tourists, a little dazed like us, disoriented in the flat uniform landscape with its strange vegetation, its abnormally short mesquite bushes.

Hummingbird. Friday 12 April 2019.

Back home, sorting out samples and detailed maps, recompiling the series of readings we took at the Trinity ranch. I'm going back to France soon, on a search for the higher states of a reality that science insists on pulverizing: my search for "the unknown region." (17)

> *Darest thou now O Soul,*
> *Walk out with me toward the unknown region*
> *Where neither ground is for the feet nor any path to follow?*

Bourges. Sunday 21 April 2019. Hotel d'Angleterre.

French mood is balanced between fascination with the continuing drama of the fire at Notre-Dame, controversies about its reconstruction, and its spiritual role as a reference for the faithful. On the other side is the grim reality of economic injustice in French society. That social imbalance is felt deeply here in Bourges, with its narrow streets and wonderful countryside, as it is in Paris with scenes of clashes between riot police and the 'Gilets Jaunes' in their 23rd week of demonstration and frustration.

This is the city of Jacques Coeur, whose family home has been turned into a four-story inn leaning dangerously above the pavement (we enjoyed a simple lunch there yesterday) while his palace, now a museum, reflects the past glory of a royal treasurer who was also an erudite lover of alchemy, a master of esoteric arts.

Bourges. Easter Monday. 22 April 2019.

We walk around the ancient streets, watching the spiral staircases in the towers of Jacques Coeur's intelligent, airy castle, so far ahead of its epoch. From these rooms he directed a vast ocean empire that introduced 16th century France to the luxuries of Asia. Maps show his vessels sailing through the Mediterranean to the many counters he established in Lebanon, North Africa, and Egypt, bringing back rare perfumes, expensive silks, and spices.

Flamine and I walk hand-in-hand through those ancient streets, trying again to imagine a future together. In America, Diana's book continues to be hailed by intelligent critics with warm comments.

Mabillon. Tuesday 23 April 2019.

My first job here was to print out and study some 250 pages of new documents summarizing CNES' five new cases to be reviewed by our 'expert' group. In the throes of an early spring, Paris is relaxed and smiling again, the first green leaves shading its avenues, the menace of the 'Yellow Vests' in decline, to the relief of Macron's *Comité* formed to "study their demands," a cheap trick to drown their claims in a very expensive bathtub of tepid water and flowery discourse.

Our scientific advisory meeting at CNES was uneventful, with Rospars, Weinstein, and me wearily watching over the doubters. Jean-Pierre and I followed another interesting report from witness Nathalie, at lunch at Vegenende. She doesn't want to report anything to the CNES, "a waste of time, talking to your scientists!"

Chateau-Chinon. Saturday 27 April 2019. Au Vieux Morvan.

This old place, with faded photographs of Mitterrand and his wife Danièle on the walls, has a gorgeous view of the steep hills of Burgundy from the fine dining room. This is a world of forests and rich pastures, with a solemn reminiscence of Roman antiquity. As Mitterrand wrote, the region "isn't ready to tell its secrets. It knows how to remain silent… an old story coming up from the bottom of centuries tells the past as well as the present." But Mitterrand's brand of socialism has not done much for the town. Many houses are shuttered on every street, with fading FOR SALE signs all over.

There used to be a fortress at the highest point of the town, but it was razed like all other forts in Burgundy by orders of a king who feared his vassals. The view, and the wind, and a powerful sense of the integration of nature take over one's mind and soul.

Echoes of California reach me even here. Garry writes about the isotopic study; we're going to need a mass spectrometer with triple quad, he says, to do a proper job on our samples.

The Navy's decision to take sightings seriously again, while keeping its conclusions secret, has caught the world's attention: the *Nimitz* is making waves again. For me, time to fly back.

Hummingbird. Friday 10 May 2019.

Silicon Valley has fully embraced a model best described as 'sustainable disruption.' From the busy 'shared work' office buildings of Mission Street and the Tenderloin to the funky restaurants of Woodside, all the talk is of sky-high IPOs and ambitious startups consuming billions (millions are so *passé!*) to build the next economy…or a forgettable anecdote.

Yesterday I met with Garry at Bucks among the owner's collection of rioting clowns and 12-foot biplanes, to review our plans to meet venturer Peter Thiel in West Hollywood in ten days.

Hummingbird. Saturday 11 May 2019.

Pursuing my tests of the 1945 case, I have again measured the dimensions of the metal piece Paola left with me with a precise caliper. I found them consistent with the metric system, including one cylindrical part 1.00 centimeter high, with a hole that accepts a standard screw I just bought at the hardware store. That, of course, doesn't make sense for a sophisticated 'Alien' device.

Hummingbird. Monday 13 May 2019.

If Jose Padilla tells the truth, and the piece of metal from Ground Zero is genuine, the object that crashed on the property may be human, not Alien. But what about the short beings, disturbed and running around wildly, presumably picked up the next day by Army grunts? Paola has found original drawings by Jose, showing a typical ovoid face and those terrifying large eyes.

Today I resume my normal schedule in Silicon Valley, dropping off a book for Aunt Helen in Palo Alto, having lunch with Graham Burnette to discuss AlterG, then driving south to San Jose to review the CEO's plans for Peaxy Corporation.

Hummingbird. Tuesday 14 May 2019.

News of the death of Stanton Friedman is spreading. I am sorry to hear it. Stan had integrity and a sense of humor, a rare combination. He always stuck to his guns about the reality of the MJ-12 story and the hardware theory of UFOs, to the exclusion of multi-dimensionality and psychical effects, so we were occasionally at odds, but he never resorted to the back-stabbing comments and poisonous innuendoes so common in the field.

Stan knew aviation and mastered enough physics to understand that much of ufology was foolish. He spent time in archives and libraries, extracting good data, and went into the field for serious interviews. We shared information on cases, privately.

Hummingbird. Saturday 18 May 2019.

On Wednesday, Garry and I met with Professor Paté-Cornell at her Stanford office, the walls covered with impressive plaques. She chairs the engineering department, among other duties. She listened courteously to our presentation, but she said she knew nothing about the phenomenon. So, I explained why we had both a scientific and a private interest in the observations, and Garry described the process for scaling up our experiments.

A lot of rain, unseasonable and definitely unwelcome on this big San Francisco weekend of foot races and celebrations, comes as part of a weather system that will dump a lot of snow on Donner Pass: I can't drive up to retrieve the supposed UFO hardware someone is saving for me in a zipper bag in Reno.

Todd Pratum had a guest last night in Oakland, a student of the occult, so I joined them in his library-museum-dog-kennel apartment to admire the new books. Then we drove out in a battered Mercedes to join a dozen would-be scholars at the splendid Scottish Rite temple on the shores of Lake Merritt.

Christopher McIntosh, an authentic scholar on the Rosy Cross on the order of Joscelyn Godwin or my departed friend Serge Hutin, gave a soft-spoken lecture on Hyperborea, Ultima Thulè, and the conquests of the Vikings. When we left the lodge, a perfect yellow-orange full moon was rising over the lake.

Later the same day.

Bill Calvert just called. He had dropped by to visit our No.1 witness, finding him alone and relaxed. They spoke for half an hour. The man still had my letter, but he was tentative about a response. "That was my life before, I don't want to mix things up," he said, perhaps afraid the phenomena would start anew.

Hummingbird. Sunday 19 May 2019.

California spring was never like this: rainstorm after rainstorm and the wind howling down the elevator shaft. Images are coming back, of excursions to Nevers, Bourges, and Chateau-Chinon, delightful in their originality and in the powers of history, best appreciated in your company, Flamine; even when I was somber, and doubted.

West Hollywood. Tuesday 21 May 2019.

This town, one of the wealthiest in the United States, was never shy in its display of luxury and vanity, but some enclaves used to maintain a semblance of whatever passes for 'class' in California.

To my eyes, visiting again after a few years away from the area, even those faint traces of restraint have been eradicated. Opulence was *de rigueur* as Garry Nolan and I politely awaited our turn to present our speculative deal before venture tycoon Peter Thiel, an investor genius and

Donald Trump supporter of Paladin fame, who made his well-deserved fortune (in part) with PayPal.

Under EuroAmerica, we had tried hard to build a similar company and failed, so I recognize the business talent we're meeting here.

Indian Wells. Saturday 1 June 2019. The Renaissance Hotel.

The Contact in the Desert conference has grown up, along with the American public's fascination with ufology. It now gathers close to 5,000 people in this stunning resort, far from the real desert where we used to rough it among dust-covered cacti and unforgiving heat. Everybody is here, luminaries of web podcasts like Linda Howe and Whitley Strieber but also the cosmic energy researchers, the crystal people (the Crystal Skull has re-appeared!), and a disheveled old Yogi among a colorful display of clothes and fabrics, candles, and jars and glass wands, complicated machines that demonstrate positive energy, negative gravity, or *something*.

Somehow, it's all wonderful, far from the dubious cheap riffraff of decades ago. The gadgets, the books, and the conversations show understanding and care for humanity, if nothing else. There's a genuine, almost childish desire to enjoy the mystery. If the Aliens have not achieved anything more, we already owe them a big show of gratitude. World peace arrives later.

Alan Stivelman is here, so I introduced him to Paul Hynek. We show the movie tomorrow, after my own lecture where I'll talk about AI, and present five cases. Other things come into the conversation. Yesterday, Tom Delonge and his team showed the first episode of their TV series: thundering fighters zoom over stunned villains and silly scientists, so skeptics now use the hype to counterattack. Some podcasts have even begun assailing Elizondo, unfairly doubting his record.

Hummingbird. Wednesday 5 June 2019.

An enormous leak has taken place. Someone has stolen research documents from Eric, possibly from his old computer at NIDS in Vegas, and posted them on the Internet. The leak also includes various claims about the reality of the 'Alien Autopsy,' when people may have been tricked by their own Intelligence sources.

Two of the private investigations by Eric have also been outed, including his notes after a parking lot meeting with Admiral Thomas Wilson (18), who went on to head up DIA. The documents were found in the papers of Ed Mitchell after his recent death.

Hummingbird. Sunday 9 June 2019.

The late California spring suddenly blesses us with a deployment of foliage and colors, greens and reds; an exuberance of bushes, flowers everywhere; clouds have been expelled from the sky; people are smiling again; and my corner Italian restaurant is overtaken by noisy fans watching the basketball game.

My time has been spent down in the deep Valley, reviewing transcripts of the oral history sessions at the Computer History Museum: They've asked me to restore some chronological truth to the development of Arpanet, and later the Internet, into a medium for social networking. I spent four hours before their camera, skillfully interviewed by Marc Weber, well-informed and precise. We went over the various transformations of the later Internet and my work with Paul Baran, the various projects I initiated: Forum, Planet, and the swift implementation of Notepad. Then the Congressional Hearings, the inception of Library automation in England, and our greatest unsung software achievement, Nuclear Notepad, when we linked together 64 nuclear plants in the US, Japan, and five other countries to help manage accidents. We could not have helped at Chernobyl, but could the Japanese have used us to avoid Fukushima?

Predictably, the very trashy, more public web of today vibrates with discussions of Eric's extensive notes after his meeting with the Admiral back in 2002. Its impact on social networks amounts to an indirect Disclosure, and the document has ignited the web. The text itself, a tribute to Eric's remarkable memory, throws light on the dark recesses of Defense research where 'the hardware' hides.

Tasked with a review of secret R&D appropriations of the Pentagon, Wilson had supposedly come across a large undefined budget item, so he called the company and was faced with three executives, including the General Counsel, who reluctantly told him that the subject had to do with pieces of hardware that could not have been made by humans, but were none of his business. He was reportedly told his career would be destroyed if he persisted with his inquiries. Wilson dropped the case and went on to head up the Defense Intelligence Agency, but he never swallowed the insult.

Hummingbird. Sunday 16 June 2019.

The dust is slowly falling over the leaked memo controversy. Trump was asked about the *Nimitz* incidents and the possible existence of UFOs and calmly brushed it all off.

My friends seem increasingly bitter, shooting from the hip at random targets. Every week, some TV channel announces yet another UFO series, and the public's appetite seems to have no limit, swallowing plain fiction alongside significant fact.

Then, a happy dinner with Catherine and Becky for Father's Day; we laughed as we toured the hillsides in the setting sun. People want to forget about the trouble in the Middle East, China's shenanigans, and the Russian efforts to undermine democracy everywhere.

Bolinas. Saturday 22 June 2019.

Back to the little studio hidden in the woods. James Fox and I worked hard most of the day, editing his big documentary with Lance Mungia who previously compiled Russell Targ's movie about remote viewing at SRI, *Third Eye Spies*. This house belongs to the man who edited *Apocalypse Now*. He's in London at the moment and lets friends of James use his place, so my neighbors are the deer, a few other nocturnal animals, and two guys we just saw in the backyard next door, restoring a 1955 motorcycle.

James' product, simply entitled *The Phenomenon*, is taking shape slowly; it needs a good 'haircut' and serious reorganizing. The Socorro sighting emerges as one turning point in this historic plot, a complete case when one reviews all the personal interviews patiently assembled from original material. We recalled how Allen Hynek had kept secret the actual shape of the red insignia on the side of the unknown craft.

Bolinas. Sunday 23 June 2019.

The Sun filtering through the big redwood tree woke me up, as it gently explored the wide room with light beams of rose and gold, so I got up and drove to the little coffee counter across from Smiley's Hotel, then to the end of the street to look at the Pacific and walk on the beach, recalling what James had told me over dinner about his own sighting (he showed me the spot in Bolinas, next to the old wooden church) and Lance Mungia's extraordinary experience, when his car was lifted and deposited on a remote hilltop with no access road.

Sunday morning. In this laid-back community, only surfers and ufologists wake up this early. I got back to the car, turned on the radio, and was rewarded with a superb rendition of *Meine seele rühmt und preist* by Melchior Hoffmann as I waited for James.

Hummingbird. Wednesday 26 June 2019.

Paola is in town again, so we had another quiet dinner at Balboa Café. She has gone up to the Andes and got altitude sickness while the video team got stranded, but they were ultimately saved and she went home to Colorado, only to fly on to Sardinia to explore old monuments.

Paola has devoured *Forbidden Science* from the beginning, so she wanted to know many details, including my memories of SRI and the dark verities we learned, the cynical models through which the truth gets betrayed. "Always give 'em something for nothing!" says the carnival barker. Isn't that the way Google works? And Facebook, Twitter, the great corporations upon which our future is being erected?

Hummingbird. Thursday 27 June 2019.

The Sun blesses us with nature in bloom and the Pacific in ecstasy. My spirit bounces back on its own, rejecting the mean voices. Jack laughs at the current squabbles about artificial intelligence. "What about digital consciousness, now hyped up by Elon Musk?" I needle him.

"The ultimate pacifier," Jack replies. "Those scientists of yours are just babies who don't trust their own intelligence."

"You've got a point," I say, "but what created it?"

"There was a Great Calamity," he replies calmly. "Human life is an artificial thing created by The State before Existence. But there's something beyond happiness, an ethereal point."

Hummingbird. Friday 28 June 2019

Over lunch in Palo Alto with Russell Targ and Patty, we spoke about his trips to Italy and his vast experience teaching about remote viewing to audiences around the world. As we sat down at a quiet table at Il Fornaio, he brutally asked, "Why do those Aliens come all the way from Andromeda, just to buzz the *Nimitz*?"

The answer, of course, is that they don't come from Andromeda, so why are we still stuck in that obsolete, bankrupt model?

"You're a scientist, Russ; what made you jump to premature conclusions? They could come from a parallel world, or a hundred other layers of what science views as reality today."

We got nowhere. I admire Russell as a true pioneer (he claims to be the one who founded the SRI psychic program, not Hal, but it was a tandem, and Hal was the initiator and long-term true manager).

I brought the discussion back to the SRI psychic program and the curious death of Pat Price. It still puzzles all of us.

"Well, there may be a simple explanation," he finally said. "What do you do with a gifted psychic subject who can not only see buildings across the world but can locate classified dossiers and even read the contents... when he reports daily to a foreign power?"

"There weren't many options." I let that statement sink, and Russell went on: "Authorities have a murky role here. Why did they say *on the record* that Price was cremated as soon as his body was found, when Hal and I saw him very plainly in an open casket?"*

Later the same day.

In my recovered 'spare time,' I now turn with ravenous delight to mathematician Gian-Carlo Rota's *Indiscrete Thoughts*, disorganized as they are. He writes: "You're within your rights if you decide to take phenomenology in small doses," thus showing a kindness rarely found in a member of the Academy. Especially one who has served so long as a scientific advisor

---

* When I spoke to Dr. Green on 23 July 2024, he told me he'd been equally puzzled by what he'd seen and heard. He went to the hospital emergency ward (after Russ and Hal) where the body had been received, but it was already gone. The evening before, in a restaurant in Crystal City, Virginia, famous for its convenient outside tables and checkerboard tablecloth, a waiter had gone to the kitchen food stand to pick up some spaghetti for Price; a client who travelled with him had seen the fellow dump something in the food: they assumed it was seasoning. "That same night, Pat suffered a sharp stomachache. He doubled up in bed, I called the ambulance," the man said.

to the National Security Agency.

I will always remember that rare conversation with Gian-Carlo, in that parked bus in Austin, when he told me why UFO research would never get close to the truth, and demonstrated it, in such perfect terms. He gave me a logical key, a tough one. I plan to use it.

Jeff Kripal is now the Associate Dean of Humanities at Rice, a well-deserved promotion. What they teach has to do with ideas and *beliefs*, rarely with *facts*: Gods, spiritualities, traditions, rites. As one of Jeff's graduate students wrote an essay on my work, starting from my 1955 sighting in Pontoise, he treated it as an imaginary vision—interesting because it motivated me, but not as reality. He conveniently forgot the other two other witnesses, one of them looking through binoculars.

Hummingbird. Friday 5 July 2019.

Russell Targ's movie, *Third Eye Spies*, presents an overview of the remote viewing program at SRI with the complexities of dealing with Washington fundamentalists, liars on the payrolls of the Intelligence agencies, and narrow-minded scientists protecting their little turf.

The movie also leaves major gaps. It ignores how the program got very narrowly approved at SRI, although I've told Russell and Hal the real story of my intervention to get it approved; it never mentions Keith Harary, whose exploits were important; or Ed May (seen very briefly) who ran the program for years. It forgets that I gave Ingo Swann the key to use coordinates as an information 'address.' How convenient, to use the media to touch up history, and the beautiful painting!

Even with such flaws, the film is remarkable. Once again, we're left with the spectacle of men and some brilliant and courageous women trying to reach the realm of the spirit with calibrated instruments and the light of reason, and failing, of course, but not completely.

Hummingbird. Saturday 6 July 2019.

A book by Marvin Harris, bought off the sidewalk for three dollars (faster, cheaper, and more fun than Amazon) has captured my attention. It's entitled *Cows, Pigs, Wars, and Witches: The Riddles of Culture.*

The author writes: "Ignorance, fear, and conflict are the basic elements of everyday consciousness."

From these elements, art and politics fashion that collective dream-work whose function is to prevent people from understanding. Everyday consciousness, therefore, cannot explain itself. *"It owes its very existence to a developed capacity to deny the facts that explain its existence."*

Hummingbird. Sunday 7 July 2019.

Another death among the psi researchers, that of Elizabeth Rauscher, has triggered a stream of comments among leading parapsychologists. Their reactions inspire reflections of my own. Why would I want to survive? I thought. What about the knowledge I've missed in this present life? What about the regrets, the hard lessons I could cure if given another chance, in a higher realm, with a better mind? And what about those fleeting moments when I almost recall lives from the future, not just survival impressions dug out of old graves?

Hummingbird. Tuesday 9 July 2019.

My brother has suffered a stroke but survived. The news sent me into renewed sadness. He stays in my mind for his courage and his integrity. My excuse for contacting him again had to do with the study of isotopes I'm conducting with Garry. Gabriel was the pioneer of nuclear medicine in France in the 1950s and 1960s and led the radioisotope service he created at Necker Hospital. He could help us. He would grasp the significance of our samples.

Hummingbird. Wednesday 10 July 2019.

All of Diana Pasulka's friends are worried like me about security offices of various Intelligence agencies scrutinizing her life because of what her book implies. My own work is more subdued, but I'm at the end of a massive sorting project, going over the few remaining files. Everything will soon be safe in Texas.

Now Garry tells me he's in touch with Bruce Fenton, a blogger who tracks an old crash in Cornwall. It happened in 1984, at Llanilar, near Aberystwyth. The witness, Eurwel Evans, first thought an RAF plane had an accident. When his farm was roughly taken over by military grunts who tried to intimidate him, he realized the honeycomb he'd picked up wasn't ordinary metal. There is more, as usual.

Hummingbird. Wednesday 31 July 2019.

Eric Davis reports on recent developments in Washington impacting space projects. The Congressional Budget Office has just fixed the spending authorization for the new Space Force HQ (not the Force itself) at $720 million per year for five years, but the House DoD appropriations bill that passed this month doesn't allow adding any new personnel, so the Air Force will have to transfer existing staff into Space Force Headquarters.

Eric adds: "Trump doesn't have a space war program but DoD does; it was developed during the Bush Administration and refined twice during the Obama Administration. Mattis and Shanahan never changed it. The new Secretary of Defense (Esper) is not going to change it either because its DoD acquisition and logistics infrastructure already exists."

Hummingbird. Friday 2 August 2019.

San Francisco has become one giant building site, with cranes and bulldozers everywhere. At night it turns into a gloomy labyrinth of complicated barriers where even the buses get confused. A persistent rumor claims that much of the money for those mighty towers comes conveniently from illegal trade, but who cares? The next morning, the fog lifts over the lovely hillsides.

James Fox has asked me to help him complete and edit his new movie, so once again we'll be moving up to a borrowed house on the Bolinas peninsula for a few days, next to his own home.

In his short story "Beyond the Wall of Sleep" (1919) H.P. Lovecraft writes about dreams: "We may guess that in dreams life, matter, and vitality, as the earth knows such things, are not necessarily constant; *and that time and space do not exist as our waking selves comprehend them.* Sometimes I believe that this less material life is our truer life, and that our vain presence on the terraqueous globe is itself the secondary *or merely virtual phenomenon.*"

In this, anticipating 'virtual reality,' Lovecraft outlined an intellectual movement that reinterprets the findings and failures of science, and (in the writings of Rizwan Virk, most recently) argues like Jack Katz that our very existence on Earth may itself be the result of a simulation, a vast advanced videogame. Such theories can be seen as compatible with the 'Control System' model I proposed in *Invisible College*. Can it be challenged? We used to have these arguments in Chicago when Bill Powers and I speculated about ways of detecting the Control System. I may have a practical way to test this, if I live a few more years.

Bolinas. Saturday 10 August 2019.

Flamine and I have settled down in a magnificent house at the edge of the cliff, near James' own place. In anticipation of this trip, I read again Michael Murphy's book *The End to Ordinary History*, where the action starts between Olema and Stinson Beach, among clairvoyants and KGB spies. He speaks of joint psychic research between bright Americans and brave Russians redefining global politics and peace: Is there such a thing as 'The Earth of Hurkalya,' the mystical dual world, close to us?

We woke up in the fog this morning; it drowned the gorgeous landscape of the bay and the views of Mount Tamalpais. After breakfast, we drove over to the secret garden where James works on the movie. We sat down with Lance Mungia to resume editing.

Bolinas. Sunday 11 August 2019.

Nine in the morning, white fog again, suspended above the sleepy lagoon, pleasure boats at anchor, swerving slowly. Last night, over dinner in Olema with James and Lance, we went over the history of the field. I recounted how the psychics of the SRI remote viewing project all wanted to talk about UFOs, and the confusion that created among all the secrets with the unmentionable sponsor in DC.

I'm early at the studio, a mere shack, two rooms among the trees with a few steps in front, enhanced by multiple spider webs. Silence. The fog dissipates slowly, passing in dreamy volumes across the pine trees. A Mexican gardener gathers dead branches and reeds in a wheelbarrow he takes to a dump down the dusty road. The hedges are topped by an effervescence of nasturtium, pale pink roses, and the wildflowers that grow everywhere, profusely. Flamine has stayed at the house on the cliff, writing an essay for her psychology group.

The French media have caught the ufology bug in the wake of the

History Channel, so they launch a cheap TV series to mock the research with the gross laughter of buffoons, while the Macron government sets aside 4.5 billion euros for a new generation of satellites that carry beam weapons, in line with the Russians, the Americans, the Chinese. The polite fairy tale about non-militarization of space (and the Moon) is dead. Who was kidding, anyway? How long could our dream of a pristine sky be maintained by this awful terrestrial species? In Bolinas, the Sun has just pierced through the last streams of fog, and everything comes to life. Wrapped in an old red coat, a woman walks by with her dog and says "Hi."

Hummingbird. Tuesday 20 August 2019.

Over the weekend Flamine and I visited Karen Breschi and her gifted artist friends, and last night we had dinner with Jack Katz who gave us the final pages of *Beyond the Beyond* and the first six pages of his new work, ambitiously entitled *The Engenderment of the Void's Exegesis*, where he recollects the first great love of his life, an episode he's described to me with genuine, warm tears.

Bolinas. Saturday 24 August 2019.

Flamine called this morning from sunny Paris, where she just returned, after which I packed a light overnight bag and drove over the mountain again. At the studio, we covered another important series of cases.

The rest of the world isn't doing well. Ranchers have set afire wide swathes of the Amazon and waves of new settlers rush to rape the land. Commercial relations get stretched as China retaliates against Trump's tariffs, and Putin's Russia has launched a giant barge to float a nuclear power plant in the Arctic, capable of powering up a city of 100,000 people; things to expect as the ice melts.

In a few days I'll be in Paris again. That will be good for me, away from an artificial American scene, Silicon Valley hype, and the absurd White House, proclaiming untruths at a rising pitch of urgency. In the background, neo-Nazi marches and frustrated loners with guns shooting their neighbors dead, as one just did again in Texas.

Hummingbird. Sunday 1 September 2019.

Yesterday, driving over Mount Tamalpais, we worked for six hours straight on the story of the Rockefeller Initiative. I suggested cuts, hard edits of expensive, painstakingly accumulated high-resolution footage that no longer fits. I lost some battles, but we parted as friends.

Tonight, I had dinner with Suma Gowda at her invitation (at the Jeanne d'Arc restaurant in Hotel Cornell). She is a young Indian woman, a software professional who runs a podcast on the nature of the universe, the phenomenon, and everything around it.

I am struggling in that context. One might expect that as I approach the dangerous age of 80, I might have fixated my ideas on some basic realities, but the reverse is the case: the encounters, carefully studied, reveal a pattern that best fits the Simulation Hypothesis, which I still resist on ethical and theoretical grounds. Another example of such an experience is Lance Mungia's encounter experience: absurd, yet Lance recalls the date, 28 October 1999. He was driving on the I5 Grapevine, about 3 am on a foggy, icy patch of road. Suddenly his car was picked up and set down atop a hill, among cultivated fields.

Two highway patrolmen rescued him politely. They didn't even ask to see his papers, or to get his name. Later, at home, there were problems with his keys, an alarm that kept ringing, very weird.

20

Mabillon. Friday 6 September 2019.

Paris fortunately presented us with a most delicate face today, under a blue sky worthy of a California morning. I walked everywhere, surveyed old haunts and took my afternoon coffee at *Père Fouettard* in Les Halles. People are dynamic, slim, funny; they seem to enjoy life, a healthy rebuttal to everything heard on television. Half are vacationing tourists enjoying the end of the Summer, however, so the light-hearted impression is deceptive.

Mabillon. Saturday 7 September 2019.

We were guests today at the wedding of Flamine's cousin Valance in a small town near Fontainebleau. It had rained earlier but the Sun was out when we arrived in time to join some fifty guests at the town hall, after which we all walked down to the Seine along cobblestone streets in lovely autumn sunshine. (One house along the way had served as suburban retreat for guitar genius Django Reinhart, well hidden from fans and critics.) The couple, middle-aged and well-supplied in children and past dramas, owns half of a sizeable island in the river, where we were buffeted by barques and a formidable *péniche*. It was a blessed afternoon of chatting, playing with kids, relaxing in hammocks in the woods, and a feast among the greenery.

I came home to find a flurry of messages that made me happy to be away from the US. Eric Davis has unleashed a verbal arsenal against the lies of disinformation squads sowing discord and using dirty tricks in Washington, protecting what dark secrets? There's rumored progress to re-ignite a Pentagon study. But Chris Mellon says the Administration doesn't want to give it any money.

Caen. Friday 13 September 2019.

Normandy is in a pleasant mood, quiet and cool. Annick will pick me up here after my business meetings, and we'll drive home to Bayeux.

Chris Mellon, looking very relaxed, came over for dinner last night with his soft-spoken wife, Laurie. We hardly spoke of UFOs but Chris asked me about Brazil; he's going there soon. Leslie Kean has traded messages with me about a new set of documents that a group in Brazil wants to publish in the US; they seek an endorsement. I am sympathetic to their efforts, but I will stay away. In the meantime, Jimmy Church, popular podcaster of *Fade to Black* and *Open Minds*, teamed up with John Greenewald on 30 July to expose TTSA as a CIA front, listing the background of the many DC insiders on their roster. They even figured out the mistakes about the supposed Italian sightings promoted by Delonge on the History Channel, and the exaggerated claims about 'metamaterials' hyped on TV.

Killing time before my meeting, I'm reading Gilder's very interesting *Life After Google* about future software structures, the ravages of the blockchain in the future, and the stunning intuition displayed by Peter Thiel.

Mabillon. Monday 16 September 2019.

The weather has changed. Paris just experienced another week of unusually hot temperature, barely tolerable, tiring and sweaty. I sought asylum in a theater that showed a crude comedy with Gérard Depardieu and Michel Houellebecq that had some witty dialogue. But the warning signs are here, after three heat waves in France this summer. On June 26 a temperature of 46 degrees Celsius in the province of Hérault, the highest ever recorded there.

Last night I called Garry, who'd just had lunch with Larry Lemke and came away very impressed. It was a mutual feeling because Larry was floored by Garry's images of the honeycomb. He can bring expertise in planetary astrophysics and neutron activation, a technique that could back up and certify our mass spectrometry measurements, each isotope emitting a specific gamma ray frequency. These tests only require about 200 micrograms of matter.

Larry's father had an experience with abduction in 1918 and became part of a mysterious group that studied the issue. The idea that UFOs began in 1947 was just a silly cover.

Mabillon. Saturday 21 September 2019.

From the blue bedroom, I delight in watching the skies, their unusual transparency, and (last evening, as I waited for Flamine) their gradual washing of pink and gray turning mauve, true darkness coming over like a flimsy sheet kindly spread over a child by a caring mother as sleep arrives, with only the Sacré Coeur still glowing on the horizon.

Diana has told us of her concerns, noting that she didn't have a clearance, yet "some of us *unfortunately* have intimate knowledge of the so-called Intelligence Community, and some don't." So, she asks Dr. Kripal: "Jeff, remember when I asked you, before Garry, to come with me to the desert with Tyler? You said no. Smart move. Now I know, and Garry does too, *a lot of stuff that got us on the radar of some nasty people* (remember last year, Garry and Jacques?)"

That last sentence hit a nerve. Yes, I know the threats. So far, I've managed to escape the 'nasty people' who invaded Garry's privacy, but they're still around, planting fakes and ugly rumors, spying.

Mabillon. Monday 23 September 2019.

Met with Jean-Claude Bourret and Olivier Younes over a wonderful dinner Flamine had whipped up. She made it a joyful occasion. Jean-Claude is a wonderful storyteller, close to high authorities in Paris, and Olivier is the smartest mind in the new generation of French finance.

This afternoon I met with Leslie Kean, in town for a session with our friends at the Institut Métapsychique. She was simple and open for a nice chat under the rain and a simple French dinner in a little café near Gare de l'Est. She gave me a copy of her fine book *Surviving Death: A Journalist Investigates Evidence for an Afterlife*.

Mabillon. Tuesday 24 September 2019.

Philippe Favre and Yves Messarovitch, my favorite economists, came over last night for dinner. We spoke of Macron's misplaced expectations to develop technology unicorns. Most of the companies his bureaucrats have selected are run-of-the mill service companies.

Yves and Philippe (to whom I recommended *Life After Google*) spoke of macro-economics, the predictable debacle among holders of convertible debentures in the current interest rate environment, the looming failure of at least one central bank, and the uncertain spending attitudes of the American consumer, the only hope for recovery.

Now it rains peacefully over the Latin Quarter. Beyond the Louvre and its *mansardes*, the Sacré Coeur is just a gray, bulbous oddity. I'm waiting for my dear Maxim, on the train from Lyons.

Mabillon. Wednesday 25 September 2019.

Maxim, tall and determined, is well-organized, very kind with us, ready to complete his *licence de Science* (bachelor's degree equivalent) in Lyons and to spread his wings. We had a fun evening together, but he had to catch a train early this morning to rejoin the campus. Now I am sad, and so is Flamine, because we'll be apart again, in a cryptic knot of fate we can't untie, even as we hold each other, and I kiss her hair, and the rain falls softly over Paris, which doesn't care.

Mabillon. Thursday 26 September 2019.

Suddenly, the sound of sirens close-by, and a traffic jam under our windows. Jacques Chirac, the former French President who lived two blocks away, died this morning.

The area has been sealed off (rue de Seine, rue de Tournon) as a small crowd assembles to watch celebrities come and go. I'm told Macron is expected to pay his respects to a man who was his first political patron. France has moved on, of course. In time, historians will tell us that Chirac's apparent good humor and easy-going personality (he loved every minute of his visits to country fairs, tasting every cheese and charcuterie) was hiding ruthless manipulation.

I pack my suitcase, updating memory devices. I slept badly last night. I don't like the path research takes. Leslie told me about the plight of Lue Elizondo. He courageously relocated his family to Arizona. What was he thinking? What were we all thinking?

Hummingbird. Sunday 29 September 2019.

An early breakfast at Bucks brought Garry, Larry Lemke, and me to review the hard data and make fresh plans. I spoke of my meeting Leslie Kean and Chris Mellon in Paris and gave Larry three of the 'research contributions' books I had compiled.

Larry has novel ideas for studying the materials, so we agreed to meet on Tuesday to join Mike Angelo at the blood testing facility where one of the new mass spectrometers is installed: He promised to run all our samples in a single massive batch mode. It's now up to me to upgrade the documentation, so that the trail of evidence can be made explicit. Larry told us that the mechanical properties of honeycombs were known since the 1930s, driven by the need for materials resisting bending in aviation manufacturing.

Both aluminum and fiberglass have been used, as well as bakelite, so he doesn't think what Tim and Garry found in the field is unusual: "The fibers could be asbestos, but why are they double woven? Glenn-Martin tried those structures (in the UK?) during World War Two. Hexagons also found in radomes, aluminum skin on both sides, glued with adhesive..."

Hummingbird. Thursday 10 October 2019.

Over lunch with Dr. Sturrock today, we reviewed the extent of the samples I'd received from him, and we filled out the required due diligence forms for the benefit of our lawyers. Peter's going ahead with his model of the Sun as a star combining two protostars within one another, spinning around different axes. It's a very intriguing concept I've never encountered anywhere. After billions of years, wouldn't the plasmas have mixed?

Hummingbird. Sunday 20 October 2019.

After the required site visits to several high-tech companies, I attended a celebration of Jack Katz' completion of his grand work, *Beyond the Beyond*. We met for lunch in Emeryville (The Townhouse on Doyle Street), with Thérèse and his other friends including another celebrated graphic artist, Liam Sharp, who draws *Green Lantern*.

Hummingbird. Tuesday 22 October 2019.

Diana Pasulka was in San Francisco today, in a warm afternoon of perfect blue sky. We had lunch at my little Italian place and spoke secrets for three hours, eating grapes and watching the ballet of the hummingbirds over the flowery planters. We were able, finally, to get into the more esoteric aspects of her book, which we hadn't dared discuss over the Internet. This hints at levels beyond the usual patchwork of theological beliefs that feed people's aspiration for a higher world, yet there's a darker side to her book.

In a similar vein of absurdity, 'they' didn't want her to mention hybrids creatures, yet the idea has been in print for a long time and picked up in movies. Garry reminds us that DNA databases are complete now and don't point to any deviations in the Earth's population. So, the interdiction may be a reverse play: Use the forbidden privately to mislead publicly?

No wonder Diana remains on guard, a bit tense, although California quickly exerted a relaxing influence. She was interested in my old books and the artifacts, curious or funny or ominous.

People who work on the launching and orbital management of satellites report that the devices are often approached by "something else." In related companies supervising the programs, even top managers don't know any details.

Garry, invited to one such company, saw a large basement full of people working on complex devices.

Hummingbird. Thursday 24 October 2019.

Alan's movie is out through six distributors, and so are the interviews. It's in the hands of reviewers who've called *Witness of Another World* the best film they'd ever seen on the subject.

Indian summer in California means new fires, disaster striking Geyserville, threatening Healdsburg, prompting evacuations, in spite of the questionable decision by the Utility Company, still stuck in the obsolete Industry culture, to cut off all power up north to prevent sparks in the high winds.

I've attended the Institute for the Future's ten-year forecast, a pleasant reunion with friends, from Bob Johansen and David Pescovitz to Lawrence Wilkinson. The meeting with their clients was held in the grand old Ford plant in Richmond, one of my favorite special areas around the Bay with intense memories of Rosie the Riveter and the Pacific war.

Hummingbird. Sunday 27 October 2019.

The fires have spread again, making the air dirty and the breathing short. But it's the depressing idea of another disaster shaping up that takes hold. So soon after the devastation of Sonoma County, with the symbolic burning of the ancient Rosicrucian Red Barn, we're unable to preserve our own cultural legacy.

The end of the year is in sight, and with it the term of the second decade of the 21st century and the completion of this volume that tried to capture and preserve some of it. It's too complex a story, unresolved and frothy with evil, yet we survived. I am tired but I can still cover a lot of ground, I thought as I pushed the canoe into the roots of an old fallen tree at Turtle Island and invited some peace into my heart and mind. The wind was picking up, spinning the barque, confusing me.

Hummingbird. Tuesday 29 October 2019.

There was a discreet but momentous meeting this morning in the law offices of Wilson Sonsini, on Palo Alto's Page Mill Road. Larry Lemke brought up his new point of view, searching aerospace history for analogies to the unusual structures we've found, particularly the now-famous honeycomb recovered in New Mexico. And here we had some surprises. Larry, astute in technology history, again told us that such hexagon structures came into the aviation factories about 1938, the first being experimented by Martin in 1945. Later Hexcel, formed at UC Berkeley, offered the first viable product.

On his laptop, he showed us the weaving and knots of those mysterious 'wires' that so perplexed us.

The implications of what Larry found are obvious: What we are looking at is not extraordinary material from Alpha Centauri, only human hardware, exploded. But that's not all. It's unlikely that the honeycomb technology with the nylon thread fixing the epoxy would have been in common use before the 1950s. Who planted it at the 1947 crash site? Garry will have some questions for Tyler. The next step must be a full review of our recent runs through our entire collection.

Hummingbird. Sunday 3 November 2019.

After a day of rest, I'd recovered enough today to drive down to Scotts Valley for lunch with Valerie, whom I hadn't seen in more than two years—two years of pain and anguish with her daughter for her, and two years of health uncertainties for me. We had a fine meal of Indian food and later walked through the nearby redwood forest, a magnificent grove of massive trees where we found intriguing simulacra of bark and moss (19).

It was a welcome breath of kindness putting some distance between my mind and all the twisted games.

Hummingbird. Thursday 7 November 2019.

The fog is back, a bland whitish screen around the building. It has smothered the smoke that blew in from the north and muffled the urban sounds. Just as well. If it weren't for a pleasant visit by Judy Shadderdon who took me to lunch for a belated birthday celebration, life would be dull indeed on my lonely perch above the City, watching the news: the boastings of converts to Alien saviors, the quarrels of programmers about AI, and the repeated warnings about the climate.

In all three cases, we have the threat to our world, the exponential rise of civilization terminating among masses of drifters, and the grime of automated ghettoes, managed by intelligent connected objects.

None of that is new in principle, but it now looks like a real future.

James Fox's movie, which touches on those ideas with a warning, had its first private showing in Sonoma on Monday before an exclusive New Age audience.

Hummingbird. Saturday 9 November 2019.

Faced with a series of wildfires, and evidence that global warming was now a permanent factor in California, the electrical power company periodically cuts off power to towns in the North Bay. As a result, James Fox has been forced to relocate his editing team and his computer to a house rented by one of the staff in Berkeley. I joined him there this evening to help edit the movie with the changes suggested by Thursday's showing.

Hummingbird. Sunday 10 November 2019.

James and I met again, and he told me they had a near-disaster yesterday, a short time after I left. As they were still busy editing, he thought he smelled a curious odor, which the others didn't notice. Stepping into the kitchen he found the gas fully turned on but unlit, an oversight by one of the team members who had cooked something on the stove—90 minutes before! James turned it off, opened all windows, and they evacuated the house, ready to explode. He was still shaking as he told it, and I drove home, stunned.

Hummingbird. Sunday 24 November 2019.

Much progress over the weekend. I've produced a decent edited version of my oral history for the Computer History Museum (20). At Garry's home, where Larry joined us for a useful review, we spent the afternoon going over the material study. The plan now is to attack the isotopic ratios.

Hummingbird. Tuesday 26 November 2019.

Arriving early for a haircut in anticipation of a flight to Paris, I found little Kelly in the shop, busy refreshing the offerings in a magnificent three-level display case of rare wood, with statues of various Chinese divinities on each level. As she lined up tiny porcelain vases in front of a graceful goddess, she promised to explain it all to me some day.

San Francisco is wonderful. I walked over to Lemongrass, an Asian restaurant, where an older man in the back of the room bowed his head respectfully in my direction, so I did the same. When I had finished the meal, and folded the newspaper, he hadn't moved and we exchanged the same friendly salute, no words. This neighborhood shows tradition mixed with many changes, an invasion of techies.

Local politics stink under a thin veneer of sophistication that doesn't impress some big companies. Schwab is the latest one to give up, moving to Dallas, as Bechtel now settles in Virginia, and others flee the high costs, the unmanaged homeless, the increasing crime, and the corrupt City Hall, temporarily forgotten in the bustle of Christmas.

Mabillon. Tuesday 3 December 2019.

Paris is cold, freezing at night, yet the sun comes out of the grayish-blue haze and throws the windows and the balconies into pink relief. Flamine and I have found each other again, in tenderness so gentle I could easily spend the rest of my life with her, and I dearly hope I will.

My son is 56 today. I have some gifts for him, but he demands, rightly, to have me deliver them in person; no FedEx this time, he laughs. Tomorrow I'll take the train to Bayeux for a few days of rest in provincial calm, fleeing the insane chaos of yet another hard strike in Paris. I'm editing the oral history of the computer revolution I have loved, and hated, and loved again.

Bayeux, Monday 9 December 2019

The streets were wet when Annick and I took our daily promenade yesterday, her funny little white dog in tow. Normandy is mild, rich, seemingly content in the abundance of her gifts: every food, every fruit of the earth, markets overflowing and, to the passing observer like me, an economy running smoothly.

As I monitor the web, Paola is always ready to supply those transcripts of witness interviews that interest me. So, I draft a scaffolding of chapters to hold the 1945 events on the Padilla Ranch. It's a heady mix of mystery and American history, reaching a strategic level with the collapse of the old regime in the mushroom cloud that blinded Jose's mother and shook the world.

Bayeux. Tuesday 10 December 2019.

The strike lingers, inflated by despair and resentment, also the bitterness of a public that enjoys none of the lush retirement benefits of the striking workers. The demonstrations are building up, and traffic is impossibly glued to the pavement around Paris.

Flamine tells me she can't even get a train to go to her job, so she 'teleworks' from home, redrafting a project the staff has savagely written in bad French, testimony to the failure of schools over the last 20 years. Beyond the language, what gets torn apart is a certain way to understand the world and life itself, the very structure of thought.

I gave up on catching one of the rare trains to Paris today.

Bayeux. Wednesday 11 December 2019.

Julien Green's *Journal* entry of 27 February 1934: "There's always a France to be loved, said Renan. I'm afraid the France I love may end up disappearing. Every day seems to take a bit of it away. In the current tumult is the birth of a world where I don't find my place. General sadness. I only hear about people who are afraid and want to leave." Five years later, the War.

The cathedral: The glorious light of stained-glass splashes around me on the pillars and arches of this magnificent edifice, first erected in the 11th century under William and Mathilda, and continuously enhanced since then. The greatest window is by Thévenot (1847), a life of St. Vigor, Bishop of Bayeux in the sixth century.

Bayeux. Thursday 12 December 2019.

Blood in the French political waters. The promised announcement of an integrated retirement system for all Frenchmen (by Prime Minister Édouard Philippe, yesterday) came as a rational exposé of proposals designed to build a consensus. The most egregious injustices, recognized but never cured, like the scandalously low level of women's pensions, would be addressed head-on.

Obvious questions remain, which should have been clarified and firmly moved off the table, but they were left to fester, so the embattled unions now stand ready to drive their bulldozers through the gap. If the new system is so great, why not adopt it right away?

Other items in the government's presentation appear as *trompe-l'œil* elements of fake shrubbery on the stage of Macron's theater. Yes, social justice could be improved if 'The Rich' took a greater part of the burden. (Who's kidding? 'The Rich,' under Macron, report making over a million euros a year, not counting the stock market gains, more realistic than the stupid Hollande estimate of 48,000 euros, which hardly defined the bottom of the middle class.)

Mélenchon, in a rare flash of clarity, easily exposing the con game on television, before a befuddled starlet masquerading as a political analyst. The guys at the train depots, besieged by cameras, gave no indication of restarting the big locomotives as they paraded in the bright green sleeveless vests of the strikers.

More rains and winds come from the west, sweeping right across France. Blood in the water. Merry Christmas.

Later the same day. La Taverne des Ducs.

Lunch with Annick: *souris d'agneau* and the chef's special desert, *Sainte Eve*. That pastry is a 'mystère' with a core of hazelnut cream and a wonderful special sauce. Some good things are still firmly in place in France.

The rains have warmed up the air somewhat. Rare passersby on the street. A storm is coming, the strike is still on, and Paris is nearly paralyzed. The contrast between this quiet city, gray and wet, and the hellish stress of the capital is stunning. The TV shows a mix of scenes where stranded crowds of befuddled travelers complain meekly while scattered troops of union demonstrators fill the screen with red and yellow flags, smoke bombs, and strident complaints.

Bellecour. Saturday 14 December 2019.

"One has to experience boredom in any country, in order to know it well," writes Julien Green. That's probably why I know France so deeply, but there's pleasure and rich lessons in the smallest things.

The strikes triggered by the retirement reforms are mutating into a revolt against Macron. The situation could slip into a Christmas disaster and a political free-for-all. Paris is Hell, anger mounts in provincial capitals, train depots get physically assaulted, and I have no secure means of getting home on Monday night.

The storm came, but this is not a bad place to be stranded: simple French country dinner with my beloved son and Annick, of hot pumpkin cream, haddock *avec mousseline potatoes and mushrooms*, dessert of fresh fruits in pastry, and an excellent Chardonnay, on the level of our lunch two days ago, before retiring to a much-improved bedroom.

Bellecour. Monday 15 December 2019.

The storm swept through last night, leaving the town soaked and the pastures flooded. It also left Normandy clean and pure, all freshened up in her green dress of magic days. Valérie brought me pastries, fresh juice, and steaming coffee. I felt relieved and thankful to find my son relaxed and happy at last. Yet it will be a strange Christmas this year, hostage to the strikes in Paris.

As Annick drove me back across the plateau of Perche, one side of the sky remained a tortured mess of billowing grays and blacks, while the opposite horizon was all gentle blue brushed with wisps of white in sunshine, a sign of better things to come?

On the last train to Paris. Monday 16 December 2019.

A receding view of Normandy, Caen suburbs in the gray rainy dusk, as ugly as any disheartening area of blight. Annick left me at the station with a sisterly kiss. This is one last train to Paris, a near miracle, before the general blockade. My back-up train tomorrow, for which I also bought a ticket, has just been cancelled.

Just as the power struggle was rising this morning, the unions threatening to block the country through Christmas, Macron suddenly lost his grip on the government's arguments. At noon, his champion for the cause of reforms, an old apparatchik of the régime named Delevoye, was exposed for a dozen conflicted payments 'under the table' by various institutions and boards.

The clean image of the 'smart young government' now shattered, the leader of the CGT, open shirt and bristling mustache, went on TV to re-affirm his call for a united demonstration in the streets of Paris, a display of contempt for the bosses. The workers' movement had long failed to achieve such consensus, ever since they kicked the Juppé government into oblivion, back in 1995.

Now the train plows through the darkness and the sticky, dirty rain towards the capital, where control of the crisis moves to the streets. By identifying 'clean Macron' with the rampant corruption of the past, Martinez rises like a statesman on the barricade, defending every unhappy Frenchman. Yet both sides are playing with fire because the current anger is based on despair, not just contempt for the bling-bling images of a young President and his Court.

Mabillon. Tuesday 17 December 2019.

Returning to Paris at midnight. The last train from Normandy stopped in the dark a few miles before the city, pending instructions as the schedulers planned an orderly parking order for the matériel, idled by the spreading strike. Today, no machine will move.

Flamine was waiting for me. We fell into each other's arms at last, and the night of Paris, long expected and delayed, was sweet as a gift of refreshed expectations. There are new plans to be made, and bliss to be tasted, and cherished as our treasure.

Mabillon. Friday 20 December 2019.

Paris shines under alternating rain and spectacular sunshine, her beauty revealed in shocking displays from her bridges and avenues. I did see Olivier again when he picked up the kids' Christmas gifts on his way back to Bellecour, and then I had lunch at Flamine's apartment after a brisk walk through Les Halles. But the rush to the Holidays is muted—nearly destroyed by the strikes, to the despair of ruined merchants and scattered families.

The Computer History Museum in Mountain View has the edited text of my unpublished oral history. The process forced a wide arc of past technology into a reflective context.

Far from all that this afternoon, I enjoy mailing out handwritten New

Year cards, my effort to maintain precious links in a landscape of vacuous automata. The solstice brings early darkness. I'm left among my treasures: the bells of Saint-Sulpice in the night.

Mabillon. Tuesday 24 December 2019.

Family visits for Christmas. Flamine's bright nephews and her sister, Isabelle, trained as an attorney. I found an amusing book for her about legal issues in fairy tales in a store devoted to formal treatises.

Another day, another gathering at Mabillon with little kids running around and exquisite food, yet the streets still vibrate with the expectation of social strife, hanging over everything, ready to fester into violence. Our windows look out to the changing mood of Paris, from dizzy celebrations to somber strikes.

France is changing the hard way, as does America. I will fly back to the States alone, on the 31st. The year will quietly close on a complex decade of hope, sorrow of our own making, and the deepening quest for other minds across rivers of spacetime.

REFLECTIONS

> *In his business he should know that truth,*
> *when he finds it, is always a lie.*
> — John LeCarré, *The Russia House*

There is a place of solitude and sorrow to be reached and traversed when you embark on this research, and I did just that in these pages. French philosopher Aimé Michel, many years ago, had warned the eager explorer I was then. Allen Hynek, in his frustrated retirement from the Air Force's Project Blue Book, saw the same thing. In that sorrow I heard an echo of something else, and I followed it along the path that will soon complete this quest.

The phenomenon of UFOs, whatever its origin and nature, works in personal ways that transcend attempts to redefine it merely as a military-political secret. For eighty years, governments have ignored this, and for eighty years, as a result, they have kept us in the dark.

There are valid reasons for secrecy. We all agree the atom bomb must not be left in the open, and the horrible secrets of the submarines are best forgotten in the inscrutable depth of the oceans. Such secrets can be only understood by a few people. Not so with UFOs: Everyone can see them. Their unsettling quality emerges and astounds us. No use for fences and miradors: Unpredictably, they pop out and redefine our world in a way so vast, yet so personal that we cannot relate them to anything we've ever known. On those rare occasions when they unveil part of themselves, they do so as a gift to the individual, never revealing the mechanisms and the layers beyond.

I remember a stunning scene on TV, in the mid-sixties, where a general in uniform, overcome by anger, paces his office, yelling: *"I've been lied to, and lied to, and lied to!"* He would be astonished to witness what happens now. Official reaction to the mystery has been to erect higher and higher walls to deflect the truth, even after it acknowledged its awesome reality. Within that absurdity was a precious key, easily missed. It is the essence of this decade.

We found the transcendence and it can destroy us. It nearly destroyed me, because I felt we had betrayed the teams of investigators we had hired in Las Vegas to build the most sophisticated data warehouse of sightings ever developed, and to feed it to a twin AI system I had designed. We had a five-year project that could unveil the deep nature of the phenomenon, but it was dismissed in Washington after only two years: the investigators, the translators and the programmers were all let go with two weeks' notice, no reason given. The research was in gestation within that team. It's unlikely it was killed to save taxpayers' money. Other teams cannot simply pick up the pieces and do it all again, even better; the witnesses we were going to interview are now dead or silent. We can guess at what they saw, but no robot can fill the holes. A black shroud has been draped over decades of precious data. Discoveries were awaiting us, not just statistics. That, too, is a lesson from this book.

Science must go on, right? Aware of the pettiness of our nature, we now dream of facile AI constructs. Unfortunately, AI is not conscious, nor human; it only pretends to be. Whitley Strieber has beautifully described the awesome beauty and terrible discouragement that comes from touching the flimsy fragments of such a solution—and losing them.

Many of us had taken it for granted that, once serious people (professors, pilots, generals, as in one of Leslie Kean's well-researched books) became aware of inevitable changes to come, everything would fall into place and a new era would arise for humanity. That would constitute 'Disclosure.' I and many others tried to make it happen, hoping for new leaders. Our aborted project in Vegas was a brave example, led by competent researchers, but every effort to restart the work was either diluted or derailed. We end up ten years later with no solution in sight, not even a small public study, like the French one, where one could calmly discuss the evidence.

Today, in 2024, renewed efforts are being proposed in Washington. At least one book has been published, partially reporting on the classified computing and documentation work our project accomplished by 2012 before support was cutoff (21). Other projects are proposed. I am too old, and I've grown too skeptical of secret research to participate, but I will watch with interest if anything gets reported. More likely, as open research gets redefined as treason, even bits of material recovered by witnesses may be claimed tomorrow under Eminent Domain and hauled away. What will remain for science in the end, if Disclosure feeds Secrecy?

* * *

The real Disclosure people are hoping for could have come at three critical points.

— Ideally, it should have happened in the late 1950s. It would have been seen by America as a glowing opportunity for science, in the novel enthusiasm about the conquest of space. But the opportunity was confiscated by projects that promised breakthroughs they were incapable of delivering. Project Sign, Project Grudge, and others had simply denied there was anything there. Only Project Blue Book survived, for what it's worth, because Allen Hynek was on his own private quest for a higher-level truth, and an Invisible College arose to back him up.

— The second opportunity to 'disclose' was at the time of the Condon study in Colorado, in 1967-68. An honest report would have created a positive social impact. Yet the witnesses were betrayed again with a subtle hint to shut up (as now) and the scientists' report was a mockery of science, blessed by the National Academy of Sciences and *The New York Times*. It silenced, ridiculed, or discouraged research for a whole decade. Its effects are still felt.

— The third opportunity was around 2011, as in the pages you just read, and it failed like the previous ones. I have shown how our research died; that remains controversial. There were a few shady deals along the way, best left undisturbed. The next step is only happening now. No Disclosure yet, but a few high-clearance briefings, more false starts. A lot of time has been wasted, and a lot of valuable data, too, as witnesses died with their story. Hard evidence is lost.

* * *

One major fact remains: people have been hurt, physically and emotionally; first by the ordeal, then by the rejection and misunderstanding from families, neighbours, and the State itself. Some concrete steps can be taken to help those who are suffering. An English private group that my wife and I have joined, led by a distinguished legal expert and supported by psychologists and medical professionals, has issued initial guidelines to the necessary help.

Other projects are moving forward. My archives of 50 years of investigation, scrubbed and updated, are now in the hands of librarians at Rice University, where Professor Jeffrey Kripal directs the growing collections from a dozen others who've now joined the party. Dr. Richard Haines, Whitley Strieber, Paola Harris, Dr. John E. Mack's custodians, and others have now donated their own files. It represents a massive archival effort, with an initial ten-year horizon.

Most importantly, leading professors at three major Universities (Nolan at Stanford, Loeb at Harvard, and Knuth at Columbia) are now showing the way to serious research, relatively free from the pressures that have defeated such efforts in the past. We can expect new findings from their work, bypassing the old obstacles with novel methodologies. There is an Academic revolution in progress; witness the new courses in religious history by Diana Walsh Pasulka at the University of North Carolina, daring to raise the question of non-human intelligence.

After the demise of BAASS, I updated my private files with unpublished records, and I have been hunting for hidden clues with a form of assisted intelligence. Never-seen patterns have emerged, and I can test them in the field. More importantly perhaps, as seen throughout this book, high-level communication with the phenomenon itself—interaction of a novel kind—has made itself obvious in our lives and minds. The process is hard, but we are not helpless before it. On the contrary, it is waiting for us with a new challenge. There is still time.

Jacques F. Vallée
Paris-San Francisco
October 2024

Acknowledgments

In the publication of this book, I benefited from the expert help of several knowledgeable researchers who reviewed the various stages of the manuscript. In particular, Dr. Garry Nolan at Stanford and Dr. Christopher Green at Wayne School of Medicine guided me in the reporting of biological conditions attached to the phenomenon and generously agreed for their comments to be quoted in full. Dr. Colm Kelleher at Bigelow Aerospace and Dr. James Lacatski of DIA also reviewed key sections of the text. My publisher, Mr. Patrick Huyghe, has my trust and admiration for the handling of what became a very complex undertaking.

REFERENCES

Introduction

1. Among the technology companies for which I managed startup Investments as a Board member, four became 'unicorns,' notably Mercury Interactive (in software) and Accuray, Inc. (in AI-assisted Brain surgery).

Part Twenty-One: Crossing the Chasm

2. NRO is the National Reconnaissance Office, charged with the surveillance of the Earth's environment through sophisticated, large satellites, notably those in synchronous orbits that remain above a particular point or country, indefinitely.
3. Red Planet Capital was a venture capital fund with a projected size of 60 M$, created by Administrator Mike Griffin in September 2006, after four years of careful planning.
4. BAASS, the Bigelow Aerospace Advanced Space Studies project, was managed on a daily basis by Dr. Colm Kelleher, Ph.D.
5. AFOSI, the Air Force Office of Special Investigations, has been one of the agencies involved in the handling—and arguably, the cover-up—of UFO reports since the late 1940s.
6. Stanford Professor Leonard Herzenberg and his wife, Dr. Leonore Herzenberg, have left an indelible mark on the fundamental fields of immunology and cell biology. They are best known for developing fluorescence flow cytometry and hybridomas. Dr. Garry Nolan was one of their star students.
7. ZPE, or zero-point energy, is "the lowest possible energy that a quantum mechanical system can have. Such systems continuously fluctuate in their lowest energy state as described by the Heisenberg uncertainty principle" (*Wikipedia*)

8. MUFON, the Mutual UFO Network, was founded in May 1969 as a prime investigative group.
9. LEO, the abbreviation for Low Earth Orbit, involves objects whose altitude doesn't exceed 1,000 kilometers.
10. Colonel John Alexander (b.1937), a retired infantry officer, is a leading advocate for the development of 'non-lethal' weapons and an active UFO and psychic researcher.
11. Magonia was said to be a "land above the clouds" inhabited by magicians in the days of emperor Charlemagne in France. These magicians (or 'sorcerers') flew in 'cloud-ships.'
12. Dr. Carl E. Sagan (Nov. 1934-Dec. 1996) graduated from the University of Chicago where he studied planetary physics. A brilliant professor of astronomy at Cornell, he became one of the most popular advocates for science education in the 1970s and 1980s.
13. SERPO and Exopolitics: The SERPO hoax was an unfortunate product of US Government disinformation in the early 2000s. Invented by a well-known agent of various agencies who had infiltrated the civilian UFO research community, it built on the success of an older hoax called UMMO, which invented a story of 'real' contact with emissaries of a distant star. The impact on serious research, unfortunately supported by researchers in official or semi-official capacity, was devastating, showing that the U.S. Government was not only hiding the truth about UFOs but was ready to create false tales about it and spread them through the research community. This violation of scientific and medical ethics was a primary factor in my later decision to leave the group. See also reference 11 in Part Two.
14. Dr. Claude Poher, a French scientist with a doctorate in astronomy, originated the first official analysis of UFO reports when he founded the GEPAN (Groupe d'Etudes sur les Phénomènes Aériens Non-identifiés) in 1977.
15. Jacques Lacan, French psychoanalyst and writer (1901-1981), was a major figure in cultural history for much of the 20th century in Paris and internationally.
16. Dr. Frank Boyer Salisbury was born in 1926 in Utah. He studied plant physiology, receiving his PhD from Caltech in 1955, and went

on to serve as Head of the Utah State University Department of Plant Science. His excellent book *The Utah UFO Display—A Biologist's Report* was published in 1974.

17. Brelet, Claudine, *Jacques Bergier: une légende, un mythe*. (Paris, L'Harmattan, July 2020).

18. Herzenbergs: see reference 5. Garry Nolan's recollection of their research pointed to its worldwide impact: "Entire industries have sprung up from the Herzengers' efforts." Nolan pointed out that in the 1970s the couple astounded many researchers by making their highly specific monoclonal antibodies available to anyone who asked, even their scientific competitors: "One of the most valuable lessons they taught me was that you build better bridges between scientists by sharing rather than withholding information."

19. See also Henry R. Schoolcraft, *History of the Indian Tribes of the United States* (Philadelphia, 1953-6, v, p.683).

20. Alberta Hannum, *Spin a Silver Coin* (London: Michael Joseph, 1947, p.76).

21. Haubersak, Wulf, "Genetic Dissection of an Amygdale Microcircuit that Gates Conditioned Fear," *Nature* no. 468, 270- 276, 11, Nov. 2020.

22. There are several 'Axelrod' pseudonyms, of course, apart from the real David Axelrod, former White House official and political consultant frequently seen on TV news. Two other individuals have used the name as a pseudonym, including a former head of classified UAP research at the Pentagon. This has made the issue even more confusing than it needs to be. However, I decided not to use the real name here, although it has been leaked on the Internet.

23. Joe Firmage's research company was called ManyOne. It was accused of Securities Fraud and closed down in 2018 with "no enforcement action," having failed to demonstrate physical principles that would control gravity. The recorded loss reportedly amounted to about six million dollars.

24. David Pescovitz is one of the founders of the popular website *Boing Boing* and a former researcher at the Institute for the Future.

25. IFTF was founded in 1968 as a non-profit spin-off from RAND.

26. See the book by Dr. Eric Davis, *Frontiers in Propulsion Science,* co-authored with Marc G. Millis. AIAA, January 2009.
27. *New Scientist*: 9 July 2011, p.40.
28. The Balzar incident: An amusing coincidence over lunch in Paris with a scientist who had just retrieved, that very morning, an old copy of a research magazine on operations research that happened to include my 1968 article on advanced information retrieval, next to the paper he sought!
29. TEDxBrussels -Jacques Vallée - A Theory of Everything, https://youtu.be/S9pR0gfil_0. This TEDx lecture became a standard in the field, passing a quarter-million views and re-posted on multiple sites.

Part Twenty-Two: LoneStars

1. The Institute of Noetic Sciences was founded in 1973 by Captain Edgar Mitchell and a financier named Paul N. Temple to encourage research on human potential. Located near Petaluma, California, it offers research conferences and publishes results of its own research.
2. Mitt Romney was first elected as a US Senator from Utah in 2019.
3. SAPOC stands for Special Access Program Oversight Committee. Not to be confused with SAPCO (!) the Special Access Program Central Office, which controls the physical security implemented by Congress.
4. ITAR regulations specify conditions for sharing technology with other countries. It stands for "International Traffic in Arms Regulation."
5. US Government insider control of the UFO issue by Fundamentalists has been known for decades. It results in a dangerous data bias and interference with normal agency practices for unknown data.
6. Vyacheslav Gennedievitch Turychev was working at NASA's Jet Propulsion Laboratory. He is known for his investigation of the Pioneer anomaly and other projects related to space probes.
7. *Rubicon*, TV movie: A striking American conspiracy thriller created

by Jason Horwitch and produced by Henry Bromell, broadcast on AMC television in 2010. It was mysteriously cancelled after its first year (13 episodes) in spite of its 8.7/10 rating.

8. Dick Haines, Larry Lemke, Brian Smith, and Ruben Uriarte were all members of the NARCAP group founded by Dr. Haines to study aircraft-UFO incidents.

9. The Effigy Mounds are a National monument, preserving and honoring prehistoric sites erected by Native Americas.

10. SERPO: A bizarre story dating back to 2005, allegedly invented by a BAE Systems vice-president for global analysis, with help from U.S. disinformation agents (in this case, disinforming American citizens with impunity), and several leading ufologists. In August 2008, an investigative group, Reality Uncovered, claimed it had traced the web stories to Rick Doty's computer, which also originated the fake messages from non-existent 'insiders.' According to Ryan Dube, pseudonymous characters "Gene and Paul," started the Project Serpo Internet meme as a cock-and-bull cosmic story, inspired by the success of the old UMMO hoax, about 12 US astronauts supposedly on an exchange program with another planet. A semi-secret "Group of Five" was formed around the hoaxers when several scientists joined, lending the story greater credibility. As usual, this was supposed to come from an unnamed "highly-placed intelligence insider," and the tale impressed the volatile and credulous crowd of UFO amateurs on Facebook and Twitter for a while. The fake revelations also discredited valid research on the larger UFO-UAP problem, discouraging potential supporters of the science, and casting doubt on the whole phenomenon as a possible CIA 'dirty trick' like the dreaded Mind Control experiments under MK Ultra, an operation that never really died. See also Note 13 in Part Twenty-one.

11. TDY: Temporary Duty Assignment.

12. The late Dr. Kevin Starr, our next-door neighbor in San Francisco, had a formidable appetite for scholarship. Their apartment burst with awards, including the Order of Malta, and framed reminders of meetings with Pope John-Paul the Second.

13. "Active measures" are more than 'fake news.' They are created on the basis of credible facts that did not happen, or happened in a com-

pletely different way from what the public was told. For example, lack of food in Moscow was blamed by the State on bad management by incompetent non-Communist officials, while violent action was taken by agents of the Party in opposition, with some help from the Russian mob, to block food deliveries and discredit political opponents. This inspired other countries, including the U.S., to use the same tactics.

14. The details are in *Elisabeth de Ranfaing, l'énergumène de Nancy, fondatrice de l'Ordre du Refuge*, by Etienne Delcambre and Jean Lhermitte. Société d'Archéologie Lorraine, 1956, p.124.
15. Jacques E. Blamont (born Oct.1926) is a French astrophysicist and co- founder of CNES. He contributed to the development of Véronique, the first rocket launched by France (in 1957). He is a professor of physics at Paris University. (Subsequent note: he died in April 2020.)
16. Bennewitz, Paul: The ordeal of Mr. Bennewitz at the hand of US Intelligence has been documented in several books, the best one being *Project Beta* by Greg Bishop (Simon & Shuster, 2005). It is the story of a government-condoned disinformation campaign that "defined an era of Alien paranoia" as AFOSI and other branches of US Intelligence decided to reinforce his credulity about extraterrestrial messages sent to him, masking real classified operations in the Southwest that didn't have anything to do with UFOs.
17. Vallée, Jacques, "Physical Analyses in Ten Cases of Unexplained Aerial Objects with Material Samples," *Journal of Scientific Exploration (JSE)* Vol.12 no.3 pp.359-375 (Autumn 1998).
18. Indeed, all financial venture investments in technology with Russian participation was abandoned in 2022 as Putin's army invaded Ukraine. We would have wasted our time.

Part Twenty-Three: Field Work

1. *Rapacités*, by Jean-Louis Gergorin and Sophie Coignard (Paris: Fayard, 2007). Illegal practices of modern finance around the world.
2. The 'Vienna Hypothesis' was the term used by our group to des-

ignate the concept of an invasion of the Earth by robots initially designed for specific interstellar missions by a civilization that had perished.

3. Trans-en-Provence is a town in the South of France where a 52-year-old technical employee observed the landing of a roundish object that left traces in the soil and effects on plants on January 8, 1981.
4. *Le Dernier Mort de Mitterrand* by Raphaëlle Bacqué (Paris: Grasset-Albin Michel, 2010) is a book reporting on the investigation of the suspicious death of Mr. François de Grossouvre, a national security adviser to President Mitterrand, on April 7, 1994, initially reported as a suicide.
5. *Le monde enchanté de la Renaissance: Jérôme Cardan l'halluciné,* by Jean Lucas-Dubreton (Paris: Arthème Fayard, 1954).
6. Diathermy: "a form of physical therapy and surgery that uses high frequency electromagnetic currents to heat tissues." (*Wikipedia*)
7. Battle of the Beams: an electronic conflict between the Germans and the RAF in an early period of World War II when German bombers used increasingly accurate radio navigation systems for night bombing.
8. *Ovnis et Conscience,* by Philippe Guillemant, Fabrice Bonvin, et al. Foreword by Stéphane Allix (Paris: JMG Editions, 2021).
9. *Fatima, le Quatrième Secret*, by Daniel Robin. (Québec Livres, 2015).
10. *The Secret History of Western Sexual Mysticism, Sacred Practices and Spiritual Marriage*, by Arthur Versluis (Inner Traditions Bear & Co, also Rochester: Destiny Books, 2008).
11. Venato Tuerto is a city of 90,000 in the Province of Santa Fe, Argentina.
12. *Secret & Lies: A History of CIA Mind Control and Germ Warfare*, by Gordon Thomas (JR Books: 2008).
13. Peaxy Corporation: a private company located in Silicon Valley, financed by venture capital. It develops software using artificial intelligence for large-scale optimization of complex systems.
14. *No Return: The Gerry Irwin Story, UFO Abduction or Covert Op-*

eration? by David Booher. Foreword by Jacques Vallee (Anomalist Books: 2017).

15. *Glowing Auras and 'Black Money': The Pentagon's Mysterious U.F.O. Program.* by Helene Cooper, Ralph Blumenthal and Leslie Kean, 16 Dec. 2017.
16. MSNBC and Vanity Fair.com: Ralph Blumenthal, interviewed about "materials with amazing properties inside a Las Vegas storage facility," 22 Dec. 2017.

Part Twenty-Four: Solitudes

1. Novalem Analytics, LLC, is a small company dedicated to the analysis and study of recovered materials from UFO sites. It was incorporated in California to support our research into various crash sites.
2. Sra. Irene Granchi is a long-term investigator and writer about UFOs in Rio de Janeiro, Brazil.
3. Prague conference: This interaction with the company was followed a few weeks later by a site visit in Prague in March.
4. This company survived the crises of the early 2020s and continues to sign up brands to demonstrate a total information environment based on artificial intelligence.
5. The Holloman case involved an Air Force officer named Chartle who had moved or disappeared by the time Dr. Hynek and I tried to get real information (rather than wild Hollywood rumors) about an actual movie of a landing UFO.
6. Dr. Garry Nolan rapidly determined that the genome of the specimen corresponded to a human, probably a very sick child. It contained two radical alterations indicative of a combination of rare diseases. The individual was not Alien, contrary to continuing speculation by Dr. Stephen Greer.
7. In his presentations at the time, Dr. Greer often accused Richard Doty, Robert Bigelow, Dr. Harold Puthoff, and TTSA generally, of hiding or manipulating the truth about UAP.

8. Jim Segala, PhD, holds degrees in mechanical engineering and physics, "developing custom applications for experimental and industrial applications," according to his resume, including special-purpose control systems. He is also mentioned in connection with SFL Scientific in West Kingston, Rhode Island, and with George Hathaway's technology R&D company in Toronto, Canada.

9. *The Report from Iron Mountain* is a 1967 anti-war satire written by Leonard C. Lewin. The book purports to be a leaked report from a Study Group tasked by the Kennedy Administration to plan the transition from a wartime economy and assess the potential social impacts of a "condition of general world peace." It details the analyses of the panel, which concludes that world peace could cause the United States to collapse; war, or some alternative outside threat, is necessary for social Stability. The Group recommends the establishment of "a permanent War/Peace Research Agency" to improve "the effectiveness of [war's] major stabilizing functions" and to plan substitutes for war. (*Wikipedia*)

10. DERFUM is the informal designation for a DRFM or Digital Radio Frequency Memory device, typically carried by an aircraft to detect and reflect signals in real time. By digitizing, processing, and transmitting an adversary's radar pulses, a DRFM can deceive the radar system. This is especially critical to deceive radar-guided missiles.

11. J.D. Rhine Lecture, 4 August 2018. It covered a series of remote viewing experiments I had organized at the Institute for the Future in 1978, hiring experts Ingo Swann and Richard Bach as well as a dozen researchers in parapsychology in an unprecedented experiment using an early social conferencing system to insure there were no collusion among participants, who participated from four different States and from Europe for two weeks.

12. As I have done in a few other cases, I have referred to 'Ron Brinkley' under an assumed name while he was alive, in order to protect his identity as a prime witness of a UFO close encounter.

13. The Hebrides: Gwendolyn Awen Jones (aka Penny), a highly gifted writer, photographer and psychic investigator, is an award-winning author known for her books on healing and spirituality. She was born in England with a natural gift of spiritual sight that allows her

to see the subtle energies beyond the physical world. She has travelled widely to research and record the sacred knowledge encoded in ancient places. Gwendolyn lectures around the world, occasionally attending to physical locations to restore balance to the land. Gwendolyn currently lives in the US.

14. Eric Wargo has a PhD in anthropology and is the author of three books: *Time Loops*, *Precognitive Dreamwork and The Long Self*, and most recently *From Nowhere*.
15. Dr. Eric Davis went to work at the Aerospace Corporation in Huntsville, Alabama.
16. Sigismond, Richard: A geologist by training, he was a long-term friend and occasional co-investigator for Dr. Hynek. His knowledge of the Colorado mountains was often helpful in tracking reports.
17. "Darest Thou Now O Soul," in *Leaves of Grass*, by Walt Whitman.
18. Admiral Thomas Wilson, DIA, was very upset not to be given access to certain high-security projects that he felt should be under his jurisdiction. These projects were reportedly dealing with 'non-human' technologies and employed hundreds of people, many of them full-time.
19. Henry Cowell Redwoods State Park. 525 N. Big Trees Park Rd, Felton, California.
20. The Computer History Museum (CHM) is located in Mountain View, California, in the heart of Silicon Valley.

Reflections

21. *Inside the Government Covert UFO Program: Initial Revelations*, by James T. Lacatski, Colm Kelleher, and George Knapp (Henderson, Nevada: RTMA, Inc. 2023).

Selected recordings of presentations by Jacques Vallée (in English) on YouTube.

Contact: Learning from Outer Space
Global Competitiveness Forum.
Riyadh, Saudi Arabia. March 6, 2011.
https://tinyurl.com/mt7vrjuu
102,000 views. Duration: 10 minutes.

A Theory of Everything (else): Another Physics for 2061?
TEDxBrussels.
Belgium. November 23, 2011.
https://tinyurl.com/3btytb3r
267,000 views. Duration: 17 minutes.

The Age of the Impossible: Managing disruptive futures in the new connected world
TEDxGeneva.
Switzerland. October 10, 2013.
https://tinyurl.com/5cktxnzx
91,000 views. Duration: 15 minutes.

Remarkable Men and their Mysterious Machines: Forgotten lessons from Software History
Tech Talk. LinkedIn Engineering.
New York, New York. August 1, 2016.
https://tinyurl.com/3jbww7hu
1,144 views. Duration: 50 minutes.

The Software of Consciousness: Intriguing Lessons and Lingering Puzzles on the Far Side of StarGate
61st Annual Convention of the Parapsychological Association.
The Institute of Noetic Sciences, Petaluma, California. August 16, 2018.
https://tinyurl.com/3jbww7hu
22,200 views. Duration: 1h 13 min.

The Four Garments of Aletheia: Reality Management and the Challenge of Truth
Rice University School of the Humanities: Archives of the Impossible.
Houston, Texas. March 3, 2022.
https://tinyurl.com/2tkys4hb
163,000 views. Duration: 64 minutes.

UFO, Trinity, Aliens, Religions, Consciousness
Interview with Marwa ElDiwiny.
October 14, 2022.
https://tinyurl.com/343nep4y
220,000 views. Duration: 2h 21min

He's Seen More UFO Evidence Than Anyone Alive.
Jesse Michels
San Francisco, California. November 23, 2022.
https://tinyurl.com/52279rmn
480,000 views. Duration: 43 minutes.

INDEX OF PEOPLE

A

Abraham, 261
Abrassart, Jean-Michel, 257
Acevedo, Juan, 383
Adler, Frederic, 58, 111, 455
Agamrizian, Igor, 181
Agee, Russ, 325
Aguttes, Jean-Paul, 354, 399
Ailleris, Philippe, 224
Alexander, John, 20, 70, 79, 85, 103, 254
Alfonsin, Raul, 383
Allen, Paul, 231, 238, 240-241, 252
Allen, Woody, 329
Altier, Pascale, 16, 391, 466
Alvarez, Louis, 64
Alwin, Ken, 35, 42
Americano, Jose, 100
Amorth, Gabriele, 57
Angelo, Michael, 242, 249, 261, 270, 384, 387, 505
Angleton, James J., 269
"Ann," 403-404
Anubis, 74
Arigo, 116
Arnold, James, 23
Arohonov, 372
Artress, Lauren, 13, 19, 33, 397
Ashdown, Patty, 105, 106
Aspect, Alain, 210
Assange, Julien, 386
Aubeck, Chris, 13, 39, 194, 234-235, 280, 287, 303, 333
Auerbach, Loyd, 230
Avrahani, Yaacov, 73-75
Awen Jones, Gwendolyn ("Penny"), 472
"Axelrod, Jonathan," 67, 74

B

Baca, Remigio ("Reme"), 468
Bacqué, Raphaëlle, 39
"Baker," 244, 246
Baleron de Brauwer, Marie-Rose, 228, 245, 271
Ballhaus, Bill, 19, 30
Balzac, 141
Bancel, Peter, 12, 19
Bancelin, Peter, 98
Banich, Terence G., 298
Banks, Peter, 11, 19, 38, 145, 186, 284
Baran, Paul, 11, 14, 18, 19, 30, 33, 39, 40, 72, 126, 353, 477, 492
Barasch, Marc Ian, 110
Barbey d'Aurevilly, 14, 28, 37, 38, 242
Bard, Erik, 393, 395, 397, 456
Barnett, Barney, 422
Barney, Elly and Alex, 30
Bartholic, Barbara, 53
Bassett, Stephen, 110
Basterfield, Keith, 463
Bauquet, Alain, 194-195
Beake, Pierre, 155, 156
Beckman, Fred, 177, 236
Bearden, Thomas, 237
Bekkum, Gary, 129
Bell, Art, 143
Bennewitz, Paul, 167, 169-170, 175, 269, 347
Beren, Peter, 14, 37, 49, 182, 281, 348, 400
Berezhovsky, Boris, 181
Bergier, Jacques, 14, 37, 178
Berlanda, 378-379
Bernard, Serge, 245
Berlioz, Richard, 57
Bermudez, Arcesio, 127, 154, 343
Bescond, Pierre, 256, 354
Bhabha, Homi J., 228

Bigelow, Robert, 12-13, 16, 24, 28, 41-42, 52, 58, 66, 85, 87, 101-102, 110, 112, 122, 129, 136, 139, 146-147, 182, 190, 236-237, 270, 291, 306, 334, 360, 396, 423, 429, 447, 453
Bilodeaux, Jean, 261
Bishop, Greg, 347
Bisson, Terry, 130
Blackburn, Ronald, 334
Blair, Tony, 63
Blamont, Jacques, 148
Blanc, Yvon, 80
Blanche de Castille, 243
Blavatsky, Elena, 337
Blériot, Louis, 242
Blumenthal, Ralph, 429
Blumlein, Michael, 130
Boaistuau, 356
Boas, Antonio Vilas, 382
Boeche, Ray, 176, 271
Boëdec, Jean-François, 12, 16, 24, 40, 41, 43, 55, 168, 224, 244, 257, 331, 399
Bokias, 100, 164
Bolender, Carroll H., 202
Boles, Vincent, 12, 33
Boltzman, Ludwig, 408
Bondolfi, Théo, 183
Booher, David, 419
Borda, Luis, 343
Borloo, Jean-Louis, 161
Bormann, Martin, 165, 179
Boston, Dave, 401
Botti, Jean, 239, 278
Boudier, Alain, 154
Bouillon, Pascal & Anne, 128, 211
Bourdais, Gildas, 99
Bourret, Jean-Claude, 504
Boutenko, Vladimir, 163
Bouvet, Romain, 257
Bowie, David, 12

Bracewell, Ron, 170
Braslavsky, Catherine, 226
Brelet, Claudine, 7, 55, 207, 225, 283
Brennan, John, 456
Breschi, Karen, 387, 411, 417, 423, 472, 500
Brin, David, 105
Brinkley, Ron, 28, 42, 45, 340, 369, 398, 421, 464, 468-470
Brodsky, Ivo, 246
Brooks, Rodney, 267
Brown, Spencer, 98
Brune, François, 197
Bundy, McGeorge, 124
Bureau, Michel, 71
Burgess, Ronald , 447
Burnette, Graham, 44, 70, 71, 106, 215, 267, 489
Burrough, Brian, 181
Burroughs, William, 135
Burton, Richard, 313
Bussing, 179
Butler, Mary Kay, 166
Butler, Lisa and Tom, 197
Byatt, A.S., 260

C

Cabrol, Nathalie, 376, 389, 397
Cabu, 286
Caesar, Julius, 32
Calvert, Bill, 79, 81, 101, 115, 127, 293, 294, 308, 312, 373-5, 490
Cameron, Grant, 421
Campbell, Art, 390
Campbell, John, 193
Canegalo, Kristie, 117
Caravaca, Jose, 295-296, 383
Cardan, Jérome, 317
Cardillo, Robert, 447
"Carmelita," 321
Carpenter, John Alden, 193
Carrion, James, 13, 23, 127

Cartelier, 218
Carter, Jack, 246
Carter, Ashton, 447
Carter, Jimmy, 122, 177
Cash-Landrum, 343
Cassone, Gabriele, 68
Castelo Branco, 317
Catala, Pascale, 38
Catherine the Great, 247
Cawdor, Angelika, 58
Céline, Louis-Ferdinand, 106
Cerf, Vincent ("Vint"), 411
Cézanne, Paul, 243
Chabin, Laurent, 408
Champion, Alexis, 71, 408
Chandrasekkar, Subrahmanyan, 89
Chartle, 436
Chasseigne, Jean, 218
Chausson, 236
Chauvin, Rémy, 197
Cheney, Dick, 483
Chernouss, Sergey, 257
Chirac, Jacques, 505
Chrétien, Jean, 63
Christ, *see* Jesus
Church, Bill, 29
Church, Jimmy, 502
Cicera Maria da Silva Almeida, 116, 293-294, 321-323, 322 (Fig.15)
Clapper, James, 447
Clarke, Ardy Sixkiller, 174
Cleto, Fernando, 313
Cleto Nunes Pereira, 319
Cleveland, Marie, 134, 425
Clewell, Don, 346
Clinton, Bill, 27, 29
Clinton, Hillary, 27, 356, 363, 391
Cœur, Jacques, 487
Cohen, Elie, 248
Colby, William, 102
Coleman, Bill, 16
Colliard, Christophe, 257

Collins Elite, 70, 164, 176, 271, 359
Collins, Larry, 114
Combe, Bruno, 300
"Condor," 175
Consolmagno, Guy, 328
Conti, Nico, 257
Cooksey, Don, 64
Corbell, Jeremy, 402
Corbin, 272
Corréard, Roger, 152, 156
Corsi, Patrick, 228
Corso, Philip, 135
Cortile, Linda, 289
Costa de Beauregard, Olivier, 70, 71
Cox, Simon, 437
Creighton, Gordon, 76
Cremo, Michael, 267
Cresson, Edith, 398
Cook, Merrill, 346
Crookes, William, 267
Crosby, Douglas, 343
Crow, Trammel, 86
Crowley, Aleister, 45, 283, 460-461
Crowly, Bill, 485
Curie, Pierre and Marie, 267
Curry, Adam, 232
Cutch, Ron, 49

D

D'Estaing, Valéry Giscard, 141
Da Vinci, Leonardo, 102
Dakin, Henry, 116
Dames, Ed, 401
Dardanus, Claudius Postumus, 157
Darling, Diane, 436
Darwin, Charles, 313, 351
Dassault, Laurent, 42, 70
Davies, Paul, 88, 372
Davis, Eric, 23-25, 29, 40, 52, 66-67, 79, 88, 96, 101, 103, 117, 120, 123, 135, 152-153, 185, 210, 273, 274, 308, 360, 395, 408, 410, 478,

492, 498, 502
Davis, Andrew Jackson, 231
Davis, Erik, 140
Davis, Daniel M., 440
Davis, Ruth, 447
Day, Kevin, 457, 481
de Bonvoisin, Isabelle, 105, 515
de Bonvoisin, Jean-Claude, 43, 265, 303, 306
de Brosses, Marie-Thérèse, 390
de Brouwer, Colonel Wilfried, 85
de Gaulle, Charles, 176, 413
de Grossouvre, François, 30
de Maistre, Louis, 283
de Margerie, 153
de Montéty, Luce, 105, 131
de Ranfaing, Elizabeth, 146
de Saint-Phalle, Thérèse, 302
de Tocqueville, Alexis, 30, 42
de Vaucouleurs, Gérard, 234, 264
de Vere, Edward, 68
de Witt, François, 328
Deforge, Gérard, 43-45, 54
Delamarre, Xavier, 35, 242
Delangle, 413
Deléage, Jean, 71, 151
Delevoye, Jean-Paul, 514
Delgado, José, 341
Deliyannis, Yannis, 157, 245
Delonge, Tom, 356-358, 361-362, 364, 368, 385-386, 398, 404, 418, 421, 423, 436, 443, 453, 467, 491, 502
Deluol, Jean-François, 55
Denzler, Brenda, 83
Derr, John, 397
Dhoharsi, Ahmed, 38
Di Queiros Meneza, 314-315, 322 (Fig.16)
Diamond, Bill, 389
Dick, Philip K., 191, 403, 407
Diffie, Whitfield, 437
Dini, Luc, 442

Dirac, Paul, 273
Dittler, Thomas, 376, 436
Dittmar, Cristofer, 161-162, 222, 223 (Fig.11), 264, 316
Dolan, Richard, 477
Don Felipe Perez, 381
Don Quijote, 23
Dorefeyev, Vasilievitch, 163
Doty, Rick, 134, 140, 145, 170, 256, 261, 409, 421, 443
Drasin, Dan, 197
Dubose, Tom, 136
Dufour, Jean-Claude, 119, 152, 195, 216-217, 366, 408
Dulles, Allen, 129
Dumontheil ("Agent"), 244
Dupas, Alain, 56, 100, 128, 133, 154, 174, 176, 184, 213-214, 222, 239, 242, 255, 264, 288, 330, 399
Du Prel, Carl, 216
Durrell, Laurence, 36

E
Eberhardt, Isabelle, 8
Eco, Umberto, 347
Eddington, Arthur, 89
Edwards, Frank, 189
Ehrhart, James, 20
Einhorn, Ira, 146
Einstein, Albert, 89, 273, 406
Eisler, Robert, 52
El Naggar, Moh, 62
Elahi, Hasan, 105
Eliade, Mircea, 191, 272, 351
Elizondo, Lou, 418-9, 424-425, 429, 439, 447, 449, 451, 463, 481, 491, 505
Ellis, James O., 459
Emerald, Cesar, 343
Enders, Thomas, 278, 288
Engelbart, Douglas, 198, 369
Eremenko, Paul, 58, 399

536   Index

Esper, Mark, 498
Evans, Eurwel, 498
Evans, Hillary, 89
Evola, Julius, 337
Exon, Arthur, 27, 135-136, 373, 460
Ezekiel, 374

F

Fadiman, Jim and Dorothy, 91, 178
Faga, Martin, 11
Faggin, Federico, 23, 122, 183, 232, 235, 334-335, 344, 347, 365, 372, 391, 398, 412, 473
Faivre, Antoine, 272
Favre, Kathryn, 44, 63, 166
Favre, Philippe, 44-45, 302, 407, 504
Fawcett, Philip, 12
Fedore, Gaëlle, 256
Feinler, Jane, 369
Felter, Oranna, 49
Fenton, Bruce, 498
Fenwick, Peter, 105
Fern, August, 139, 181
Fernandez, Tania, 183, 351
Ficino, Marsilio, 272
Fields Bookstore, 197, 215
Firmage, Joe, 64, 68, 327
Fisher, Barbara, 89, 90
Flammarion, Camille, 24, 195, 321, 355
Fludd, Robert, 337
Fontaine, Franck, 244
Fontes, Olavo, 212, 313
Foote, Arthur, 201
Foote, Simon, 438

Forge, John and Michèle, 290, 347, 376
Forgeron, Gérard, 174-175, 225, 339
Fort, Charles, 26, 57, 337
Fowles, John, 138, 140, 160
Fox, James, 413, 423, 435, 455, 464, 468, 493, 498, 509
France, Martin ("Marty"), 387, 395
Frascella, Larry, 137, 291
Fravor, David, 464
Fredrickson, Rod, 151
Freidman, Stanton, 189, 202, 390, 489
Freud, Sigmund, 36
Frey family, 425
Fugal, Brandon, 326, 360, 394-395, 397, 410, 421, 456
Fullmer, Jessica, 416

G

Gabin, Jean, 30
Gaffney, Glenn, 172, 364
Gagazrin, Yuri, 183
Galacteros, Patrice, 408
Galileo, 89
Gallego, Jesus, 304
Gannon, John, 447
Garcia, Philip, 50
Gardin, Denis, 290
Garver, Lori, 19
Gauquelin, Michel, 50
Gehlen, Rheinhard, 129
Gell-Mann, Murray, 227
Geller, Uri, 74, 75, 103, 350
Gergorin, Jean-Louis, 288
Gerovitch, Slava, 216
Gevaert, 101, 116, 476
Gilder, 502
Girard, René, 239
Godwin, Joscelyn, 25, 336-367, 344, 490
Gomez, Edmund, 237
Gomory, Béatrice, 14, 22, 287, 290
Gomory, Paul, 14, 68, 169, 287, 290, 445
Gowda, Suma, 501
Gott, Richard, 88
Gottlieb, Sydney, 291, 310, 368, 401

Graffigna, 378
Graham, Robbie, 275-277
Graham, Winston, 219
Granchi, Irene, 212, 432
Gravel, Mike, 199, 200, 203
Graves, Kelsey, 198
Green, Christopher ("Kit"), 40, 45, 83-4, 86, 101, 103, 115, 117, 120, 122,129, 136, 139, 144, 150, 159, 168, 180, 190, 196, 202, 204, 209, 220, 234, 237, 250, 253, 266, 269, 278, 367n, 384, 434, 444, 449, 451, 472, 484, 495n
Green, Julian, 511, 513
Greenewald, John, 502
Greer, Steven, 85, 110, 185, 201, 388, 443, 470
Gregor, Ed, 401
Grieve-Carlson, Timothy, 438
Guénon, René, 337
Guérin, Pierre, 156, 342
Gueymard, Pierre ("Dr. X"), 408
Guez, Laurent, 185, 194
Guffey, Robert, 135
Guillemant, Philippe, 99, 152, 156, 232, 273, 328, 329, 339
Gunderson, Maurice, 46, 70, 71, 104, 105, 107, 120, 123, 129, 131, 139, 145, 158, 160-161, 181, 214, 222, 227
Gurdjieff, 283
Guriev, Sergei, 194

H

Hagemeister, Michael, 216

Haines, Richard ("Dick"), 84, 100, 127, 142 (Fig. 9), 154, 256, 257, 394
Haisch, Berhardt, 13, 64, 126-127, 142 (Fig. 9), 327, 481
Hall, Diana, *see* Hegarty
Halperin, David, 373-374
Halpern, Charles, 178
Halt, Charles I., 186
Hamilton, Nigel, 255
Hanegraff, Wouter, 272
Hannum, Alberta, 52
Hansen, George P., 249
Haraldson, Erlendur, 332
Harary, Keith, 496
Harder, James, 404
Hardy, René, 218-219
Hardy, G. H., 221
Harris, Marvin, 497
Harris, Paola, 468, 476, 485, 489, 494, 510
Harris, Thomas Lake, 231-232, 344
Harrison, 234
Hart, John, 467
Hartley, A. J., 368
Harzan, Jan, 234
Hastings, Arthur, 47, 230, 240, 267
Hatch, Larry, 31, 462
Hathaway, George, 101, 102, 122, 142 (Fig.10), 143, 178, 180, 234, 264, 293, 309, 360, 402, 408, 409, 411, 451, 456
Haubersak, Wulf, 56
Hecate, 233
Hegarty, Diane, 72, 86, 95, 239, 289, 291
Helms, Richard, 249, 401
Henderson, Oliver Wendell ("Pappy"), 135-136
Hennessey, 153
Henry, Dick, 85
Heraclitus, 191
Hess, David, 321
Herzenberg, Len and Lee, 13, 50, 186, 198, 201, 235
Hicks, Junior, 48, 96, 395-396
Hicks, Kathleen, 117
Hill, Betty, 346, 383
Hillman, Warren, 23

Hiponen, Miko, 105
Hitler, Adolph, 154, 302, 430
Hoffman, Patricio Abusleme, 382
Hole, Jane Lute, 112, 447
Holland, Jolie, 47
Hollande, François, 99, 111, 133, 207, 225, 233, 242, 263, 286, 302, 386, 512
Hollar, John, 372
Holt, Al, 64
Homer, 372, 414
Hopkins, Budd, 53, 91, 178, 182, 241, 292, 362
Horten Brothers, 388
Horowitz, Mitch, 27, 33, 50
Houck, Jack, 181
House, Charles, 397
Howe, Linda, 140, 143, 261, 402, 411-412, 446, 491
Hubbard, Barbara Marx, 58
Hubbard, L. Ron, 45
Hudson, Gary, 145
Hufford, David, 276, 325
Hussain, Jaffer, 23
Hutin, Serge, 228
Hynek, Paul, 448 (Fig.20), 477, 491, 516
Hynek, J. Allen, 25, 39, 49, 130, 168, 302, 343, 394, 411, 435, 493

I

Identity Woman, 105
Iderkou, Houria, 411
Indridason, Indridi, 332
Inman, Bobby Ray, 70
Inouye, Daniel, 477
Irish, Jim, 45

J

Jacobs, David, 160, 182, 346, 360
Jacobsen, Annie, 349, 352-353, 364, 368, 401

James, William, 267
Janet, Pierre, 321
"Janiel," 323
Jenks, Tim, 23, 51, 405, 446
Jessup, Morris K., 182, 189
Jesus, 241
Jobs, Steve, 91
Johansen, Robert, 23, 53, 87, 229, 507
Johnson, Samuel, 8
Jones, Nate, 439
Jones, Reginald, 228
Jung, Carl, 36, 106, 272, 477
Juppé, Alain, 514
Justice, Steve, 409, 425-426

K

Kahn, Norm, 409
Kaku, Michio, 61, 62
Kalinski, 33
Kamas, Nick, 100
Kan, Kaili, 38
Kardec, Allan, 313-316
Katz, Jack, 181, 182, 187, 192-3, 201, 205-7, 229, 234, 236, 267, 268, 281, 296, 311, 328, 348, 354, 384, 400, 411, 414, 417, 424, 428, 446, 453-454, 472, 478, 483, 494, 499-500, 506
Kaufman, Henri, 328
Kean, Leslie, 39, 84, 358, 418, 429, 480, 502, 504, 517
Kelleher, Colm, 12, 40, 51, 55-56, 67, 69, 85, 93-94, 102n, 104, 118, 121, 136, 140, 149, 150, 155, 159, 167, 177, 179, 190, 217, 237, 252, 258, 264, 270, 308, 409
Kennedy, John F., 124, 197
Kepler, Johannes, 89, 441
Kernbach, Serge, 394
Khadhafi, Mouammar, 98
Kinney, Jay, 50, 130, 168
Kirienko, Sergey, 158, 163
Knapp, George, 53, 260, 276, 356-

357, 402, 426, 452
Knight, Marianne Bilham, 403
Koi, Isaac, 334, 341, 480
Kozyrev, Nicolai Aleksandrovich, 346
Krasnoi, Andrew, 352
Kress, Kenneth A., 401
Kripal, Jeffrey ("Jeff"), 24, 26, 33, 37, 53, 55, 103, 107, 188, 215, 241, 306, 324, 333, 350-351, 359, 370, 375, 388, 416, 423, 436-437, 496, 503
Kripal, Julie, 37
Krishna, 198, 414
Kroot, Jennifer, 66, 90
Krupp, 179
Kubizek, August, 154
Kuchar, George, 39, 53, 66, 78, 90, 92-93
Kuchar, Mike, 39, 52, 90, 93
Kurth, Douglas, 28, 35, 40
Kyle, Robert, 23

L

Lacan, Jacques, 36
Lacatski, James, 13, 16, 40, 67, 86, 480
Lafarge, 248, 303
Lago, Silvio, 319
Lagrange, Pierre, 257, 400
Laidlaw, Marc, 130
Lambright, Christian P., 167, 170
Laursen, Christopher, 275
Laursen, Eric, 145
Lavrentiev, Mikhail Alekseevich, 346
Lawrence, Ernest, 64
Lazar, Bob, 175, 237, 276
Le Carré, John, 211
Le Gall, Jean-Yves, 233
Le Stunff, 33, 257-258
Lead, Jane, 344
Lear, John, 170, 175, 237
Lebrun-Damiens, Emmanuel, 459

Lee, Stan, 473
Lefebvre, Nadège, 404
Leir, Roger, 160, 445
Leitenberg, Milton, 139
Lemke, Larry, 58, 64, 127-128, 142 (fig. 9), 372, 394, 503, 505, 507, 509
LePen, Marine, 157, 338, 398
Letty, Denis, 39
Levenda, Peter, 357, 366, 460, 475
Levitt, Creon, 58, 327, 372, 394
Lewis, Spencer, 174
Lewis, Ralph, 175
Lianza, Ruben, 377, 382
Lima, Hollanda; 100, 101, 116, 308, 319
Lipinski, Andy, 168
Liprandi, 378
Littrell, Helen, 261
Lloyd, James, 351
Lockyer, Lisa, 23, 58
Lodge, Oliver, 267
Long, Jeffrey, 91
Lopez Rega, 179
Lorant, Gilles, 44
Louange, François, 154, 256-257, 354, 382, 400
Lovecraft, 499
Lub, Sergio, 110
Lugosi, Bela, 37
Lundahl, Arthur, 173, 445
Lurhman, Tania, 272
Lycosthenes, 356
Mac Orlan, Pierre, 225, 243

M

MacDonald, James, 182
Mack, John, 160
Macron, Emmanuel, 303, 398, 401, 470, 474, 504, 512-513
Madoff, Bernie, 88
Maeterlink, Maurice, 9

Malone, Mike, 148
Maltais, Vern, 422
Manakov, Dmitry, 104, 107, 110, 125, 128-129, 139, 141, 160-164, 192, 241, 264, 316
Manly Hall, 302
Mancusi, Bruno, 440
Marais, Marin, 151, 226
Marcopoulo, Fotini, 99
Marilan, 317
Marincovitch, Lou, 405
Marion, Jean, 63
"Mars" (Maurice Bonvoisin), 21
Marta, 118
Martin, Antonio, 318
Martinez, Philippe, 514
Martons, João, 313
Masse, Maurice, 176, 283, 332
Massinghill, Kay, 187-188
Matloff, Greg, 408
Mattis, James Norman, 410, 424, 426, 498
Maupassant, Guy de, 354
Maurer, Eva, 216
Mavrides, Paul, 130
Maxwell, Robert, 273
May, Edwin, 37, 126, 230, 268, 350, 401, 496
McCaffrey, Nadia, 47
McCarey, Mike, 409
McCasland, William N., 358, 386-387, 389, 394, 397, 409
McDivitt, James, 234
McDonald, Alonzo, 122, 164, 177
McGarity, Bill, 237
McIntosh, Christopher, 490
McKee, Mark, 23
McKenna, 306
McKernan, Brennan, 426
McLuhan, Marshall, 135, 429
McMahon, John, 401
McMullen, Clemens, 136

McNear, Tom, 403
Medvec, Gerry, 178
Méheust, Bertrand, 26, 38, 55, 70, 257, 262 (Fig.13), 329, 331, 339, 390, 408, 442
Mélenchon, 512
Mellon, Christopher, 357-358, 361, 367, 392, 395, 404, 410, 415, 418, 428, 434, 446, 447, 502
Melo, Martir, 314
Melvin, Daniel, 115
Melzner, Ralph, 178
Mengele, Josef, 352
Menzel, Donald, 227
Mergui, Gabriel, 54, 56, 70, 82
Merritt, Abraham, 376, 384
Meselson, Matthew, 207
Mesnard, Joël, 92, 119, 150, 157
Messarovitch, Yves, 30, 133, 165, 184, 504
Messerschmidt, 179
Metz, George, 44, 99
Michael, Florence, 464
Michel, Aimé, 78, 146, 160, 176, 342-343, 516
Michel, Louis, 146
Millard, Linda, 28, 53, 72, 169
Millard, Stephen, 14, 28, 38, 126, 169, 183
Miller, Henri, 36
Miller, Jerry, 102
Mitchell, Edgar, 110, 150
Mithra, 157
Mitterrand, François, 30, 211, 228, 233, 245, 399, 460, 488
Moeran, E. J., 193, 411
Mohammed, 374
Monroe, Robert, 320, 391
Moore, Bill, 169, 175, 269
Montaigne, Michel de, 333
Monteverdi, Claudio, 118, 397
Morgan, Beverly, 322 (Fig.15)
"Morgane," 88

Mossbridge, Julia, 480
Mounier-Kuhn, Pierre-Eric, 35
Mountbatten, Louis, 76
Moura, Gilda, 432, 435, 476
Mourier, Pierre-François, 57
Mulholland, John, 249
Muller, Herbert J., 350
Mullis, Kary, 373-375
Mungia, Lance, 493, 499
Munsch, Gilles, 256, 442
Muñoz, Zoila, 118
Murphy, Mike, 26, 276, 324, 388, 479, 499
Murphy, Dulce, 276
Musk, Elon, 14, 148, 494
Myers, Frederic, 24, 26, 321

N

Nagasawa, Kanaye, 232
Nahin, Paul J., 88
Nash, John, 258
"Natalie," 123, 170, 488
Nathan, Tobie, 309
Nausicaa, 414-415
Nawalf, 62, 63
Newton, Isaac, 274
Neyland, David, 58, 96
Nicolov, 162

Niemtzow, Richard, 39, 40, 68, 279, 307, 310, 326, 342-343, 361, 363
Nieto-Vallee, Rebecca, 100, 118
Nobriga, Jorge, 314
Nolan, Garry, 185-7, 190, 198-199, 201, 203, 209, 213, 219-221, 230-231, 234, 238, 240, 242, 249, 250, 252, 254, 259, 262, 268, 271, 288, 289, 293, 297, 299, 309, 324, 336-368, 349, 361, 364, 369, 372, 374, 384, 388, 398, 404, 409, 417, 419-420, 424, 428, 432, 454, 464-465, 471, 473, 476, 479, 489, 503, 506-507

Norse, Harold, 350, 353
Noyce, Robert, 23

O

O'Brian, Chris, 65
O'Reilly, Tim, 128
O'Toole, Tara, 112
Obama, Barack, 14, 28, 58, 111, 139, 207, 324, 386, 498
Oberg, James, 39
Oliveira, Agobar, 323
Ollier, Philippe, 256
Olson, Frank, 135
Oppenheimer, J. Robert, 468
Orff, Carl, 143
Ortega, Mary, 118
Ouellet, Eric, 373, 375
Ozawa, Seiji, 57

P

Pace, Frank, 251, 392
Padilla, Jose, 468, 489
Page, Thornton, 25
Pagels, Elaine, 351
Paijmans, Theo, 303
Paladino, Eusapia, 267
Palmer, Mike, 168
Pandolfi, Ronald, 449
Panetta, Leon, 112, 172, 447
Parsons, Jack, 45, 373
Pauli, Wolfgang, 106
Passavant, Francis, 23, 415, 420
Passot, Xavier, 81, 211, 256, 302, 354
Pasulka, Diana, 215, 274-276, 279, 282, 292, 373, 388, 480, 483, 487, 498, 503, 506
Paté-Cornell, Elizabeth, 14, 23, 458, 489
Paul VI, 294-295
Pauli, Wolfgang, 477
Pauline, 54
Pavlov, Ivan Petrovich, 76

Payan, Gilbert, 39
"Penny," 472
Perez de Cuellar, 178
Perez, Juan Oscar, 67, 348, 350, 370, 377-378 (Fig.17-18), 381, 415
Peron, 179
Perkins, John, 52
Persinger, Michael, 397
Pescovitz, David, 47, 50, 68, 80, 85, 140, 507
Petersen, John, 85, 395
Petit, Jean-Pierre, 433
Peyrefitte, Roger, 184
Philippe, Edouard, 512
Philippon, Claire, 98, 118, 133, 154, 175, 332
Philippon, Jacques, 390
Piccoli, Raymond, 257
Pickup, Pete, 50
Pierce, Charles Sanders, 375
Pilkington, Mark, 45, 140, 145
Pimen, 351
Pinon, Gilles, 33, 242
Planck, Max, 89
Platzer, Max, 463-464
Podesta, John, 39, 84, 356, 386, 391
Poher, Claude, 30, 124, 279, 331, 343, 433-435
Pope, Nick, 62, 84
Powers, Bill, 259, 499
Pratt, Robert, 308, 314, 324
Pratum, Todd, 336, 353, 359, 370, 490
Price, Pat, 29, 445, 495
Price-Williams, Douglas, 103, 236
Puharich, Andrija, 320, 341, 401
Puthoff, Harold ("Hal"), 23-24, 29, 35, 37, 40, 64, 67, 74, 77, 79, 83, 103, 115, 120, 136, 196, 202, 213, 233, 237, 308, 327, 361, 369, 402, 404, 418, 420, 434, 443, 467
Puthoff, Collin, 419, 449
Putin, Vladimir, 183, 207, 240, 247, 398, 500

Py, Jacques, 256, 442

R

Rabeyron, Thomas, 257
Radin, Dean, 110, 461
Radu, Paul, 128
Radzig, Jody, 50
Rainey, Carol, 362
Ramanujan, 221
Ramey, Roger M., 136
Randall, Lisa, 334
Randles, Jenny, 186, 295
Randolph, Pascal Beverly, 175, 344
Rapp, Paul, 409
Raudive, Konstantins, 197
Rauscher, Elizabeth, 497
Raytheon Corp., 449
Redfern, Nick, 45, 117, 122, 176
Reid, Harry, 68, 90, 112, 172, 447, 450, 481
Regardie, Israel, 45
Reydi-Grammond, Christophe, 255
Rheingold, Howard, 89
Rhine, J.B., 461
Richard, Philippe, 100, 132
Richmond, Garry, 314
Rickles, Dick, 99
Rivera, Jean-Luc, 331
"Robert B," 36, 233, 248
Roberts, Larry, 353, 477, 479
Roberts, Pasha, 479
Rock, Arthur, 23
Rockefeller, John D., 392
Rockefeller, Laurance, 27, 413
Rockwell, Ted, 181
Rogan, Joe, 423
Rojas, Alejandro, 346
Rommel, Ken, 167
Romney, Mitt, 111, 291, 396
Rongère, François, 23
Roosevelt, Franklin
"Roro," 99

Rospars, Jean-Pierre, 38, 211, 225, 244, 256-257, 408, 487
Ross, Gary, 138
Rota, Gian-Carlo, 495-496
Rovelli, Carlo, 99
Row, Joseph, 226
Rubtsov, Vladimir, 205
Rucker, Rudy, 105, 130
Russo, Edoardo, 256
Rutkowski, Chris, 346
Rymer, Harry, 259

S

Saël, 10, 14-16, 21-22, 111
Saffo, Paul, 14
Sagan, Carl, 25, 376, 475
Saint Anthony, 320
Saint Augustin, 388
Saley, Alain, 30
Saley, Annick Jeane Thoby, 30, 31, 78, 80, 81, 154, 225, 481-482, 511-512
Salez, Morvan, 442, 445
Salisbury, Frank, 47-50, 74, 81, 86, 97, 278, 383, 395
Sanders, Bernie, 363
Sandler, Alan, 435
Sanella, Lee, 320
Sarkozy, Nicolas, 33, 42, 43, 65, 111, 242
Sarmont, Eagle, 166
Sarney, José, 319
Satan, 172, 327
Saunière, François-Bérenger, 175
Schamber, Ellie, 47
Schlitz, Marilyn, 219
Schmidt, Greg, 168
Schmidt, Howard, 119
Schneider Corp., 303
Schneider, Andreas, 115
Scholem, Gershom, 272
Schrock, Eric, 409

Schuessler, John, 45, 94, 101, 122, 136, 203, 208, 234, 270, 293, 325, 341, 343, 355, 408, 410, 442
Schwartz, Ronlyn, 38
Schwartz, Stephan, 38, 71
Scott, Brian, 46, 59, 60, 150
Segala, James, 237, 449, 457
Semivan, James and Debbie, 250-251, 254, 259, 269, 291, 306, 364, 372, 393, 398, 404, 418, 428, 449, 461
Servais, Simonne, 146, 175, 283, 332, 413
Shanahan, 498
Shannon, Oke, 70
Sharp, David, 188
Shattuck, Harold, 307
Shaw, Greg, 272
Sherman, Terry, 24, 217
Shirley, John and Micky, 105, 130, 168
Shostak, Seth, 85
Shostakovich, 224, 230
Shotwell, Gwynne, 148
Shoup, Dick, 98, 126, 311
Sigismond, Richard, 485
Sillard, Yves, 85, 154
Simondini, Andrea Perez, 379, 381
Smith, Brian, 127, 142 (Fig.9)
Smith, Paul, 71
Smith-Cordtz, Ruthie, 335
Sobieski, Ian, 14
Spiegel, Lee, 346, 413, 455
Stalin, Joseph, 132, 216, 224, 368
Starr, Kevin and Sheila, 141, 392
Steinberg, Meyer, 387
Stevens, Ted, 477
Stewart, Helen, 51
Stivelman, Alan, 348, 377, 415, 417, 491, 507
Stockman, Alice, 345
Stout, Yolaine, 47
Strand, Erling, 257

Strand, Linda, 475
Strauss-Kahn, Dominique, 54, 80, 133
Strieber, Anne, 26, 33, 325
Strieber, Whitley, 26, 27, 33, 53, 83, 92, 160, 325, 333, 361, 373-374, 388, 459-460, 491, 517
Stringfield, Leonard, 40
Sturrock, Peter, 68, 70, 122, 124, 169, 191-192, 200, 205, 212, 227, 232, 261, 383, 462, 473, 506
Suilzer, Josée, 71
Sulak, John, 236
Sullivan, Drew, 128
Summers, Montague, 52
Surel, Dominique, 71
Svahn, Clas, 359
Swann, Ingo, 27, 170, 242, 403
Swanson, Claude, 346
Swanson, Gloria, 403
Swigart, Rob, 14, 69, 91, 118, 290, 307

T

Taicher, Janine, 402
Tambourin, Catherine, 56
Tambourin, Pierre, 56, 82, 184
Tarade, Jean, 218
Targ, Russell, 29, 47, 71, 127, 197, 401, 493, 495-496
Tart, Charles ("Charlie"), 178
Tarter, Jill, 96
Tauscher, Gary, 14
Tayler, David, 169
Taylor, Timothy, 282, 348-349, 373, 375, 388, 480
Teish, Luisa, 37, 130
Teller, Edward, 64
Teodorani, Massimo, 257
Terranova, Manuel, 389
Théon, Max, 255
Thiel, Peter, 292, 491-492
Thomas, Frédéric, 256

Thomas, Gordon, 369
Thoth, 250
Tizané, Émile, 332, 355
Töelken, Barre, 249
Tolkowsky, Dan, 50, 73, 74, 77
Toms, Michael, 177
Tompkins, Peter, 401
Tournier, Alexis, 71
Tranzistor, 247
Trickster, 250
Trouin, Federico, 23, 122, 183, 232, 235, 334-345, 344, 347
Truman, Harry, 285
Trumbull, Douglas, 52, 54, 65, 95, 96, 97, 241, 278, 346-347, 417
Trump, Donald, 388, 393, 491, 500
Tullien, Tom, 166
Turco, Michel, 287
Turrisi, Patty, 375
Turyshev, Slava, 123
Tsiolkowsky, 216
Tweedale, Charles, 267
Twining, Nathan, 135, 325
"Tyler," 458

U

Uchoa, Moacyr, 308, 312, 315
Urban VIII, 89
Uriarte, Ruben, 127
Ursuline Order, 275
Utts, Jessica, 110

V

Vaillant, Michaël, 31, 133, 256, 277, 429, 442
Valdez, Gabe, 237
Vallee-Nieto, Catherine, 24, 45, 52, 100, 118, 206, 303, 327, 452, 493
Vallee-Nieto, Rebecca, 303, 327, 452
Vallée, Gabriel (father), 475
Vallée, Gabriel (brother), 333, 466, 497

Vallee, Janine (Saley), 8, 10, 14, 18-19, 70-71, 91, 119, 133, 145, 182, 185, 199, 263, 475
Vallee, Madeleine (mother), 336
Vallee, Maxim, 30, 39, 43, 54, 134, 154, 195, 224, 263, 311, 339, 399, 504
Vallee, Olivier, 39, 81, 98, 108, 118, 125, 133, 154, 224, 300, 329, 407, 430, 475
Vallee-Philippon, Raphael, 407
Vallee-Philippon, Xavier, 175, 332
Vargolis, Mario, 38, 225, 283, 339
Vargolis, Sophie, 38
Vaubaillon, Jérémie, 257
Velasco, Jean-Jacques, 68
Venter, Craig, 58, 67
Venturini, Jean-Claude, 30, 44
Vergé, Brigitte, 442
Verlet, Raoul, 36
Verluis, Arthur, 344
Versins, Pierre, 184
Vézina, François, 328
Vian, Boris, 36
Vian, Kathy, 86
Vilas-Boas, Antonio, 313
Virchow-Robin, 326
Virk, Rizwan, 479, 499
Visse, Léon, 207
Vogel, Marcel, 171, 173, 348
Volovna, Maria, 307
Von Braun, Werner, 464
Von Lucadou, Walter, 390
Von Neumann, John, 373

W

Wade, Alan, 283
Wade, Chuck and Nancy, 398, 404, 421-422
Wagner, Karyn, 112
Waldman, Ariel, 95, 96, 97 (Fig. 8)
Waldie, George, 485
Walker, Eric, 102
Wallace, Alfred Russell, 267, 313, 351
Wallace, Smokey, 53
Walleczek, Jan, 76
Wally, Augustus, 396
Ward, Maurice, 152
Wargo, Eric, 477
Warner, Mark, 443
Weber, Marion (Rockefeller), 423
Weber, Marc, 369, 397, 492
Weil, Thierry, 407
Weiner, Amir, 137
Weinstein, Dominique, 31, 131-132, 153, 157, 165, 211, 257, 265, 282, 302, 400, 439, 487
Weiss, Rob, 387-388, 397, 409
Wellaide, 319
Wells, H. G., 248
Wendell, 117-118
West, Jolya, 310
Westrum, Ron, 257, 346
Wheeler, John Archibald, 227
Whitehead, A. E., 7
Williams, Pharis, 237
Wilhelm, Harald, 296
Wilkinson, Lawrence, 507
Wilson, Thomas, 492
Wilson, Colin, 138
Wolf, 218
Wolfe, Dan, 283
Wolinski, 286
Wolken, Christopher, 96
Wood, Nathalie, 96
Woody Allen, 329
Worden, Pete, 96, 357
Wright, Frank Lloyd, 320

Y
Yeltsin, Boris, 158
Younès, Olivier, 504
Young, Arthur, 320
Yukov ("Zhukov") Georgy, 194

Z
Zell, Oberon, 130, 232, 236, 260, 351
Zell, Morning Glory, 260, 351
Zerpa, Fabio, 348, 382
Zhukov, Sergei, 107, 123, 125, 139
Zilinkas, Raymond, 139
Zlotnicki, Jacques, 442
Zorba the Greek, 36

INDEX OF SUBJECTS

A

AATIP, 75, 434, 451, 463, 481
AAWSAP, 12, 75
abductions, 6, 15, 53, 57, 83, 91, 145, 160, 178, 204, 210, 211, 250, 270, 272, 276, 277, 292, 313, 314, 322, 323, 334, 346, 373, 378, 382, 432, 444, 476, 503
abductive reasoning, 375
advanced propulsion, 92, 153
Aerospace Corp., 11-2, 239, 481, 529
AFOSI, 13, 84, 152, 252, 269, 520, 525
Alcatel Corp., 303
alchemy, 32, 272, 441, 487
aliens, 26, 45, 53, 66, 93, 99, 116, 155, 199, 212, 213, 231, 235, 240, 254, 258, 261, 270, 271, 280, 287, 288, 289, 298, 334, 346, 351, 357, 360, 367, 420, 423, 467, 491, 495
alien tissue, 79
Alstom Corp., 303
analysis of UFO materials, 76, 87, 143, 206, 212, 222, 227, 228, 241-242, 263, 364, 371, 379, 383, 388, 432, 438, 454, 458, 468
antigravity, 6, 17, 18, 92, 143, 186, 433, 467
apparition, 267, 355
Arcelor Corp., 303
Areva Corp., 303
ARPANET, 71, 229, 353, 369, 477, 492
artificial intelligence (AI), 7, 267, 280, 494, 526, 527
assassination, 194, 245, 291
asteroid, 174, 212, 452
astronauts, 147-148, 524
atom bomb, 302, 516
autoimmune, 149, 309, 311

B

BAASS, 12-13, 16, 23, 28-29, 35, 40-41, 43, 75, 91, 93, 112, 116-117, 127-128, 136, 144, 159, 168, 170, 180, 190, 196, 203, 217, 221, 237, 243, 252, 269, 281, 307-308, 310, 319, 349, 390, 416, 429, 440, 447, 456-457, 463, 477, 480, 519
balls of light, 33, 59, 96, 244, 318
banking, 128, 192, 224, 478
beams, 10-11, 21, 74, 122, 218, 280, 314, 318, 323, 326, 373, 398, 406, 440, 526
behavior, 19, 35, 55, 83, 93, 342
beings, 22, 55, 89, 116, 119, 212, 216, 283, 294, 308, 315, 323, 354, 357, 373-374, 375, 451, 460, 468, 489
biology, 48, 50, 92, 96, 101, 102, 190, 254, 351, 520
bio detector, 29
blood samples, 95, 149, 198, 202, 250, 254, 259, 270, 271, 288, 505
blue files, 40, 349, 455, 469

C

Caltech, 48, 56, 521
cattle (mutilations), 65, 69, 150, 167, 178, 418
caudate, 309, 336-338, 349, 384, 419, 476, 479, 485
Charlie Hebdo, 302
CIA, 46, 112, 129, 135, 139, 146, 164, 170-172, 188, 195, 207, 209, 246-247, 249, 250, 266, 269, 283, 306, 310, 312, 341, 352, 364, 368-369, 382, 401, 443-444, 456, 502, 524, 526
classification (of cases), 52
classified
  - documents, 35, 40, 135, 245, 269, 308, 310, 385, 425, 480, 495
  - projects, 66, 91, 117, 177, 243, 252, 364, 447, 522

clearances (secret), 23, 69, 106, 172,188, 253, 332
close encounter (CE), 44, 147, 149, 150, 156, 326, 213, 309, 349, 361, 460, 528
Cluny Monastery, 43-45
CNES, 30, 33, 54, 80, 86, 128, 133, 154, 211, 224-225, 233, 241, 244, 256-257, 259, 260, 262 (Fig.13), 265, 273, 278-279, 302, 319, 330, 354, 377, 391, 400, 421, 439, 442, 467, 487, 525
coincidence, 77, 99, 103, 111, 122, 197, 258, 309, 310, 406
comic books, 26, 181, 189, 430
communication
   - with the Dead, 197, 321, 484
   - with Aliens, 58, 212, 295, 392, 519
computer
   - networks, 115, 229, 452, 477
   - revolution, 194, 510
   - security, 118
consciousness, 14, 23, 72, 114, 145, 150, 172, 178, 231, 232, 273, 295, 314, 334-335, 338, 341, 347, 372, 376, 391, 432, 473, 478, 494, 496-497
corruption, 128, 220, 290, 514
Cosmos Club, 19
coverup, 325, 520
craft, 76, 77, 101, 115, 120, 136, 147, 148, 171, 181, 182, 187, 245, 304, 340, 362, 364, 390, 394, 422, 426, 435, 468, 493
crashes (UFOs), 13, 35, 132, 141, 209, 228, 246, 271, 358, 367, 379, 388-390, 398, 421-422, 468, 481, 497, 508, 527
creatures, 16, 18, 48, 65, 69, 100, 115, 140, 195, 211, 260, 298, 396, 419, 446, 506
crop circles, 25, 152, 346

cryptids, 101, 136, 140, 144, 181, 185, 195, 198, 208, 211, 258, 259, 397
cults, 45, 157, 177

D

DARPA, 56, 58, 67, 69, 95-96, 229, 326, 352-353
database
   - Blue Book, 28
   - Capella, 12, 35, 121, 252
   - CNES, 400
   - government UFO documents, 310
   - Larry Hatch, 31
   - medical, 334
   - NIDS, 28
   - pilot cases, 28, 278
   - *Wonders*, 280
deception, 23, 93, 145, 165, 209, 251, 252, 292, 422, 450
Dell Computers, 96
delusion, 66, 155
demonology, 122, 137
DERFUM, 458
DIA, 99, 492, 519, 529
directed energy, 158, 167, 172, 425
disclosure, 62, 84, 114, 135, 143, 167, 182, 189, 199, 200, 202, 270, 325, 346, 408, 423, 492, 517-518
disinformation, 46, 99, 134, 139, 245, 254, 269, 291, 347, 350, 402, 408, 502, 521, 524, 525
djinns, 260, 309
dowsing, 20
dreams, 8, 36, 42, 51, 68, 130, 189, 201, 267, 284, 307, 316, 379, 454, 498
drones, 56, 77, 397, 439, 450, 457, 461, 464
drugs, 49, 134, 295, 314, 382
Dugway Proving Grounds, 66

E
EADS, 11
Edwards AFB, 10
energy
- "free" 13, 326
- directed, 158, 167, 172, 425
- "zero-point" 150, 346, 395, 430, 520
- dark, 338
ESA, 56, 82, 100, 224, 354
Esalen Institute, 26-27, 34 (Fig.3), 36, 53, 269, 272-273, 276, 282, 306, 388, 477, 479
espace-autre, 217
experiencer, 250, 361
extraterrestrial, 55, 84, 133, 187, 240, 329, 352, 443, 525

F
false flag, 388-389, 408-409, 450
fake
- documents, 214, 245
- messages, 524
- news, 167, 524
- projects, 347
- stories, 27, 132, 145, 337
- UFO cases, 302, 451
FBI, 46, 238, 309, 349, 40
financial crisis, 31, 181, 220
flashes (of light), 219, 374
folklore, 27, 89, 249, 282
French Air force, 55, 168, 244, 257
fundamentalist, 45, 112, 122, 137, 272
fusion, 18, 41, 83, 159, 186
futures research, 46

G
galaxies, 75, 210, 264-265
"Gateau" (samples), 442
GEIPAN, 68, 80, 128, 211, 224, 319, 354, 442, 467
gendarmes, 33, 258

General Dynamics, 40
genetics, 219, 235, 374
Genopole, 35, 54, 56, 82, 120, 214, 385
ghost rockets, 228
Gnosis, 50, 168
Golden Gate Spiritualist Church, 19, 20, 168, 476
Google, 47, 85, 121, 234, 349, 494, 502, 504
gravity, 12, 99, 171, 210, 328, 347, 365, 432, 434, 491, 522
gyroscopes, 143

H
hallucinations, 86, 91, 397
hardware (UFO), 29, 43, 67, 69, 74, 77, 79, 99, 102, 114, 152, 170, 171, 173, 196, 209, 251, 258, 263, 389, 391-393, 397, 419, 424, 428, 456, 460, 489, 490, 492, 508
hauntings, 113, 195, 332, 355
healing, 116, 250, 295, 321, 528
helicopters, 28, 59, 331, 399
helium, 268
Hexcel Corp., 508
hoaxes, 99, 225, 244, 521, 524
human deception, 23
humanoids, 51, 89, 132, 133, 290, 388
hybrids, 53, 101, 160, 261, 270, 430, 506
hydrogen, 339, 387
hypnosis, 178, 210, 292, 319, 378

I
immune system, 149, 155, 190, 209, 213, 234, 270
implants, 152, 160, 325, 348, 430
inertia, 96, 406
INFOMEDIA Corp., 14, 70, 327, 369, 370
infrared, 59, 102, 120, 124, 355, 439
initiation, 174, 206

injuries, 101, 115, 136, 137, 190, 238, 250, 306, 319, 326, 410, 440, 450, 451
innovation, 14, 42, 61, 70, 110, 132, 141, 161, 173, 222, 226, 288, 297, 399, 466
Institute for the Future (IFTF), 69, 86, 96
intelligence
- Air Force, 278
- agencies, 46, 496, 497
- agents, 271
- British, 228
- Brazilian, 382
- Czech, 247
- classified, 12, 245
- counter-, 197, 233
- French, 460
- machine, 217
- military, 258, 357, 429
- Naval, 124, 288, 458
- non-human, 519
initial public offering (IPO), 23, 51, 65, 220, 488
Institut Métapsychique Int'l (IMI), 70, 79, 225-226, 332, 339, 390, 407-408, 504
interplanetary, 232, 414
investigations, 44, 93, 127, 133, 203, 219, 226, 238, 254, 324, 332, 439, 455, 472, 476, 491
investor, 13, 65, 71, 95, 103, 111, 120, 129, 139, 146, 148, 162, 192, 195, 320, 326, 347, 389, 409, 433, 436, 490
invisible, 62, 67, 240, 352, 361, 385
ionization, 152
IRIS, 70, 408
isotope, 192, 213, 222, 242, 278, 379, 398, 404, 407, 408, 438, 445, 458, 464, 465, 468, 503

J K L
Kona Blue, 419n
landings (UFO), 87, 160, 288, 296, 314, 526, 427
lasers, 101, 152, 457
Lawrence Livermore, 63, 64
"Le Horla," 354
lies, 99, 145, 172, 248, 302, 347, 388, 484, 502
lithium, 41
Lockheed, 10-11, 111, 142 (Fig.9), 239, 263, 364, 397, 401, 409, 418, 459, 481
Long John Nebel (show), 189

M
magnetic, 123, 485
manipulation, 92, 116, 133, 139, 178, 227, 259, 357, 367, 459, 484, 505
material samples, 262, 525
Maxwell AFB: 35
media, 14, 26, 61, 62, 119, 132, 167, 183, 240, 264, 277, 356, 363, 387, 410, 418, 423, 443, 459, 478, 480, 496, 499
medical
- cases, 143, 234, 308, 450, 451
- charts, 310
- codes, 309
- community, 235
- data, 440
- examiner, 84
- effects, 308
- ethics, 350, 521
- files, 388, 480, 485
- firms, 148
- fund, 351
- implants, 348
- information, 451
- interferences, 424
- investments, 111
- meeting, 214

medical *cont.*
- project, 252
- records, 186, 202, 254
- research, 238, 333, 473
- sequelae, 250
- studies, 350, 361, 231
- technology, 240

medium, 263, 267, 271, 312, 315, 317, 332, 470, 484

mediumship, 20

metal
- alloy, 209, 429
- ceramics, 348, 364
- coatings, 152, 263
- detector, 422, 485
- device, 468
- molten, 387
- pieces, 422
- residue, 133
- slag, 229, 261, 278

meteorites, 154

microgravity, 92, 200

microwaves, 308, 325, 334, 343

mind control, 248, 291, 311, 324, 349, 524

Mitre Corp., 11

movements, 135, 137, 272, 344, 482

MUFON, 13, 23, 29, 52, 84, 127, 178, 234, 346, 373, 455, 521

mutilations, 65, 84, 115, 150, 167, 190, 209

mysticism, 289, 351

mythology, 241

N

NASA, 11, 19, 23, 39, 51, 58, 84, 87, 95-96, 116, 139, 142 (Fig.9), 143, 146, 148, 152, 182, 184, 190, 212, 220, 234, 312, 344, 357-358, 372, 376, 433, 523

NeoPhotonics Corp., 23, 64-66, 405, 446

neutrinos, 68, 93, 169

NIDS, 28, 41, 65, 69, 72, 94, 135, 153, 178, 180, 243, 334, 491

nightmare, 326, 471

NRO, 11-12, 18-19, 33, 358, 362, 481

NSA, 15, 125-126, 194, 200, 283, 375, 457

nuclear
- accident, 70
- carrier, 436
- centrifuges, 119
- cooperation, 228
- crisis, 70
- decay, 220
- energy fund, 103
- engineer, 234
- industry, 106
- material, 209, 214
- medicine, 497
- power, 208, 492, 500
- radiation injury, 237
- site, 35, 56
- threat, 214, 161
- war, 359
- weapons, 25

nurse, 93, 187, 318, 439, 466

O

occult, 228, 375, 430, 460, 490

orbs, 48, 52, 65, 87, 95, 103, 115, 121, 122, 144, 158, 170, 179, 212-213, 218, 234, 257, 275, 373, 392, 397, 410, 440

out-of-body experiences (OBE), 266, 365, 376, 391, 405, 471

P

pagans, 178, 232

Paladin Corp., 23, 491

paralysis, 11, 326

paralysis (sleep-), 276

paranoia, 139, 185, 314, 525

paranormal
- conferences, 62, 78
- events, 27, 87, 155, 248, 292, 361
- experiences, 13, 115, 272, 351, 400
- experiments, 247
- field, 89
- hotspot, 156
- research, 41, 81, 246-247, 390

parapsychology, 6, 38, 47, 66, 178, 181, 225, 230, 242, 311, 320, 339, 350, 528

photograph (UFO), 113, 170, 171, 195, 217, 256, 436, 483

pilot sightings, 28, 31, 39, 62, 128, 132, 208, 265, 282, 301, 302, 310, 426, 435, 436, 439, 449, 452, 457

planetary, 25, 415, 432, 503, 521
plant, 18, 29, 48, 194, 200, 521
plasma, 331, 506
police
- tribal, 48, 51
- France, 218, 286, 297, 460, 474, 486
- USA, 22, 80, 258, 300, 469

politics, 8, 63, 80, 100, 190, 199, 243, 283, 403, 421, 447, 455, 457, 474, 475, 497, 499, 510

poltergeist, 65, 87, 113, 181, 218, 230, 275, 290, 293, 375

precognition, 98

propulsion, 58, 83, 92, 101, 124, 144, 153, 173, 199, 209, 289, 292, 331, 433-434

psychiatry, 160
psychic, 6, 27, 76, 115, 116, 117, 313, 341, 348, 355, 361, 390, 411, 482, 483, 495, 499, 521, 528

putamen, 309, 324, 336-338, 349, 384, 419, 479, 485

Q

quantum
- consciousness, 473
- effects, 153, 274
- electrodynamics
- entanglement, 172, 210, 274, 345
- events, 99
- fields, 274, 335
- gravity, 328
- illusion, 428
- mechanics, 99, 126, 273
- memories, 307
- nonlocality, 210, 274
- phenomena, 153, 274
- physics, 329
- technology, 345
- theories, 75, 87, 273-274

R

radar, 85, 128, 354, 367, 449, 450, 457, 481, 503, 528

radiation, 120, 121, 132, 203, 209, 237, 270, 279-280, 308, 342, 398, 449, 453, 457, 463, 465

rationalism, 253, 475

religion, 8, 27, 116, 189, 193, 198, 241, 272, 277, 282, 292, 351

remote viewing, 16, 47, 66, 70, 129, 210, 248, 268, 309-311, 338, 369, 374, 401, 408, 459, 493, 494, 496, 499, 528

reverse engineering, 129, 152, 237, 364, 389, 423

radio frequency, 86, 325, 528

robots, 102, 181, 270, 335, 348, 480, 482, 526

rumors, 35, 59, 76, 84, 99, 112, 116, 117, 127, 167, 172, 205, 345, 364, 371, 386, 388, 389, 469, 480, 498, 503, 527

S

samples
- biological, 140, 209, 213, 226, 259
- implant, 152, 388
- material, 6, 200, 206, 212, 221, 227, 231, 241, 242, 249, 261-263, 349, 381, 383, 387, 403-405, 408, 417, 420, 422, 429, 438, 439, 445, 454, 456, 458, 460, 468, 471, 473, 486, 488, 497, 505-506, 525

satellites, 12, 141, 168, 372, 500, 506, 520
saucer, 99, 112, 141, 173, 287, 348, 352, 362
scandium, 87
schizophrenia, 137
science fiction, 61, 130, 168, 184, 232, 234, 296, 334, 363, 430, 460, 473
scientific community, 30, 183, 452
semiconductors, 231
SETI, 25, 58, 85, 95, 352, 357, 376, 389, 391, 397-398, 420
shadow people, 355
shaman, 28, 240, 249
shuttle, 87, 120, 283
skeptics, 61, 220, 313, 357, 374, 382, 435, 439, 491
skin marks, 158, 238, 239, 261, 270, 346
Skinwalker, 12, 41, 48, 99, 102, 211, 394, 396
software, 7, 41, 104, 212, 229, 235, 257, 293, 301, 373, 389, 391, 433, 440, 462, 492, 501, 502, 520, 526
solar, 31, 68, 77, 130, 172, 192, 362, 452,
space station, 48, 82, 110, 129, 181, 277, 443
spacetime, 124, 153, 210, 273-274, 515
special effects, 195
sphere, 73, 74, 128, 318

spirits, 232, 315, 317, 476
SRI, 53, 64-66, 91, 98, 114, 135, 246, 268, 311, 350, 369, 401, 493-496, 499
stained glass, 10, 26, 266, 511
Stanford Univ., 13, 50, 85, 171-172, 177, 185, 187, 189, 201, 219, 235, 239, 242, 249, 261, 278, 298, 308, 337, 379, 384, 417, 424, 429, 434, 437-438, 443, 454, 458, 473, 519
stealth, 6, 8, 17, 91, 457
suicide (suspected)
- Colonel Hollanda Lima, 100, 116
- Franck Olson, 135
- Marine Corps Intelligence officer, 84
- Monsieur de Grossouvre, 30, 526
- René Hardy, 219
survival (of death), 7, 24, 232
synchronicity, 105, 272, 477, 479, 480

T

theology, 122, 141, 239
theory
- abductive reasoning, 375
- aliens in search of genetic repair, 258
- conspiracy, 135
- control, 259
- crashed saucer, 132
- cosmic information, 254
- Dirac's, 274
- Distortion, 295-296, 383
- DNA origination, 454
- Egyptian civilization, 225
- electrodynamics, 273
- Everything Else, of, 104
- Evolution by Natural Selection, 267
- extraterrestrial, 133, 329

theory *cont.*
- Faggin's, 372
- general relativity, 153, 273
- gravity, 432
- hardware of UFOs, 489
- imagination, of the, 276
- Lorentz's
- Magonia lore, 18
- memory factors, 296
- metallurgical exceptions, 181
- Minkowski's, 273
- Morris Jessup, death of, 189
- multi-universe, 402
- nuclear decay, 220
- Pat Price, death of, 29
- physics, 37, 41, 50
- propulsion, 124
- quantum field, 273, 274
- Special Relativity
- sun rotation, 169
- thermodynamic, 87-88
- Trickster, 250
- UFO physics, 190
- unconscious, of the, 321
- universe as information, 191
- vacuum energy (ZPE), 327, 346, 395
- Virk's, 479
threat, 102, 103, 149, 203, 252, 389, 436-437, 443, 450, 483, 528
titanium, 64, 103, 212, 372
Total Oil Company, 303
triangles (flying), 21, 65, 128, 256, 375, 461
Triformix Corp., 44, 115, 161
Two-Rock Ranch, 59

U
unicorns (companies), 7, 504, 520

V
Van de Graaff, 402
venture capital, 71, 108, 113, 141, 160, 163, 181, 221, 297, 311, 320, 347, 367, 436, 520, 526
venture fund, 11, 40, 46, 69, 110, 242, 357
Veridian Corp., 40
videos
- entity, 187
- honeycomb microscopic details, 483
- marks, of, 317
- riots, 65
- UFOs, 101, 102, 201, 334, 355, 415, 426, 437, 439, 440, 445
visions, 374, 375
visitors (Aliens), 59, 160, 290, 296, 325, 453
voices, 197, 294, 371

W
warfare, 135, 369, 464
Wikileaks, 56, 386-387, 389, 423
Windsor Castle, 76
wormholes, 210
Wright Field AFB, 29, 64, 135, 207, 358, 373, 460

Z
zero-point energy (ZPE), 150, 346, 395, 430, 520

www.ingramcontent.com/pod-product-compliance
Lightning Source LLC
Chambersburg PA
CBHW070157240426
43671CB00007B/476